Situating the History of Science

SITUATING THE HISTORY OF SCIENCE

Dialogues with Joseph Needham

edited by
S. IRFAN HABIB
and
DHRUV RAINA

OXFORD
UNIVERSITY PRESS

OXFORD
UNIVERSITY PRESS

YMCA Library Building, Jai Singh Road, New Delhi 110001

Oxford University Press is a department of the University of Oxford. It furthers the University's objective of excellence in research, scholarship, and education by publishing worldwide in

Oxford New York
Athens Auckland Bangkok Bogota Buenos Aires Calcutta
Cape Town Chennai Dar es Salaam Delhi Florence Hong Kong Istanbul
Karachi Kuala Lumpur Madrid Melbourne Mexico City Mumbai
Nairobi Paris Sao Paolo Singapore Taipei Tokyo Toronto Warsaw

with associated companies in

Berlin Ibadan

ISBN 019 564 639 8

Typeset by Eleven Arts, Keshav Puram, Delhi 110 035
Printed in India at Saurabh Print-o-Pack, Noida
and published by Manzar Khan, Oxford University Press
YMCA Library Building, Jai Singh Road, New Delhi 110 001

... for abou, farru and namoo

Preface

The essays in this volume were presented at a conference organized jointly by the National Institute of Science, Technology and Development Studies (NISTADS), the Delhi Science Forum and the Maison de Sciences de l'Homme, Paris, at New Delhi in September 1996. The conference, 'Science the Refreshing River', was conceived of as a homage to Joseph Needham and, inspired by and reflecting Needham's lifelong engagement with crossing disciplinary and institutional boundaries, drew on the constituencies of academic and professional colleagues with varied intellectual and political concerns. From its inception, Romila Thapar was most encouraging and helped shape the eventual structure of the conference. Abdur Rahman provided an Indian backdrop to the Needhamian engagement with the history of sciences. From the Delhi Science Forum, D. Raghunandan and Amit Sengupta, co-organizers, ensured that Needham's legacy was not confined merely to historians of science. The significance of Needham's legacy for science activism was evident in the participation of activists from the peoples' science movements. The editors and contributors are indebted to them. We are also grateful to our colleagues from academe for transgressing disciplinary boundaries and thereby revising much of what appears in this book.

Many of the contributions to the volume bear the impress of science studies, and the concerns, as will be evident, spill over into the political and ethical dimensions of science. The editors admit responsibility for the inability to accommodate all the papers presented at the conference, the reason for which has to do with the urgency of keeping the topicality of Needham's work alive. This required not losing the momentum set at the conference. We therefore take this opportunity to thank Prajit Basu, Irfan Habib (Aligarh), D. Raghunandan, Nasir Tyabji, and Punam Zutshi for their fine papers, and regret their not being included in this volume. For Gregory Blue, Christopher Cullen, Aant Elzinga, Steve Fuller, Feza Gunnergun, Ranjit Nair, Pradosh Nath, and Shiv Visvanathan a special word of gratitude for helping us put this volume together. The contributors graciously responded to our requests and deadlines. We hope the volume does justice to their efforts. The shortcomings are to be ascribed to the editors and them alone.

Ashok Jain and V.V. Krishna were resourceful in bailing us out of many tight spots, and there were quite a few along the way. We thank all our colleagues from the technical and non-technical staff of the Delhi Science Forum and NISTADS for all they have done to facilitate the conference and the volume.

The conference and the volume might not have been possible without the financial and institutional support of the Council of Scientific and Industrial Research, Department of Science and Technology, and the Indian Council of Social Science Research, New Delhi.

And finally, we express our gratitude to Bela Malik of the Oxford University Press for egging us to complete this book.

S. Irfan Habib
New Delhi — Dhruv Raina

Contents

List of Contributors

Gregory Blue, Department of History, University of Victoria, Victoria, Canada.

Karine Chemla, REHSEIS, CNRS, Paris.

Pascal Crozet, Centre d'histoire des sciences et des philosophies arabe et medievales, CNRS, Paris.

Aant Elzinga, Department of Theory of Science and Research, University of Gothenburg, Gothenburg, Sweden.

Steve Fuller, Department of Sociology and Social Policy, University of Durham, Durham, UK.

S. Irfan Habib, National Institute of Science, Technology and Development Studies, New Delhi.

Catherine Jami, REHSEIS, CNRS, Paris.

Michel Paty, REHSEIS, CNRS, Paris.

Patrick Petitjean, REHSEIS, CNRS, Paris.

Dhruv Raina, National Institute of Science, Technology and Development Studies, New Delhi.

K. Subramaniam, Homi Bhabha Centre for Science Education, Tata Institute of Fundamental Research, Mumbai.

Romila Thapar, Emeritus Professor, Jawaharlal Nehru University, New Delhi.

Shiv Visvanathan, Centre for the Study of Developing Societies, Delhi.

1

Dhruv Raina

Introduction

Even if Martin Bernal did not exist, Everett Mendelsohn writes, Eurocentrism would still be under siege (Mendelsohn 1995: p. 59). However, the challenge to Eurocentrism dates back to the end of the First World War, when historians and philosophers of science had begun to find Eurocentric historiography discomforting. A new humanism was recognizable in history of science studies during the inter-war years (Pyenson 1993), and after the Second World War, one of the most fruitful expressions of this humanism was manifest in Needham's project on science and civilizations. His *Science and Civilization in China* is significant in as much as it occasioned a revision of perspectives on the history of sciences produced both in the West, and the non-West. A more recent undertaking, namely Martin Bernal's *Black Athena* may be considered another milestone in the historiographic evolution of a programme that commenced with Needham.[1] However, while the Needhamian rejection of Eurocentrism may be situated against the backdrop of the destruction and disillusionment that followed the two wars, Bernal's work is a product of the West that is cognizant of the diverse pressures of multiculturalism.

Following Needham's demise three years ago, a number of scientists, historians, historians of science, and members of the peoples' science movements in India felt that the time was ripe to review the manner in which studies on the history of science and civilizations have evolved

[1] Martin Bernal's book has been the subject of much criticism and controversy, and his response to his critics will be published by the Duke University Press shortly.

since Needham initiated his ecumenical programme almost half a century ago. This need was all the more urgent since, in India at least, historians and historians of science influenced by Needham, have rarely met at a forum that reflected shared concerns. The conference not only provided an opportunity to reiterate a much-needed interdisciplinarity, but also provided an occasion for reckoning with the chauvinisms that presently afflict some post-colonial societies as well. These chauvinisms may be seen as the obverse of Eurocentrism, and, later, a reaction to globalization. But that addresses only a part of the malaise. The more immediate objective is to engage with the domains of history and history of science that are highly politicized fields of contest.

The new humanism that we recognize in Sarton and Needham was founded on the idea that the unity of nature was reflected in the unity of sciences and the latter was an affirmation of the unity of mankind. This idea, while highly problematic today, continues to be endowed with a pristine charm. This is particularly so when the conception of science as embodying the age-old ideals of truth, beauty and goodness (Holton 1993) has lost currency in the wake of the crises in development, followed closely, in the academic sphere, by the sociological deconstruction of science. Those who participated in the conference, largely from India, some from France and other parts of the world, attempted to engage with the study of science and civilizations from a position that was sceptical of both scientism and judgmental relativism. This scepticism runs through all the essays in this volume. True to the Needhamian vision, the attempt here is to rise above particularism, not so much to retrieve the 'big picture historiography' (Christie 1993), but to raise questions around it. For, it may well be that if the non-West appears as a footnote in the old big picture, most versions of the new big picture are possibly founded on a metanarrative that subsumes the history of colonialism.

The essays deal not so much with the history of sciences, as with the historiography of sciences. They seek to identify changing conceptions of history and civilizations in as much as they have a bearing on the renewal of the history of Sciences (Cunningham and Williams 1993). This renewal is both a sociological and cultural reflection on the nature of science and the conditions of its production. Thereby, the object of historical inquiry, namely science, is deconstructed in a manner that challenges the transcendent accounts of science and restores science to its social context.

Situated within the framework of critical studies of science and society, this volume hopes to convey the sentiment that science cannot be thrown out with the bath water, for as Pinch and Collins appropriately put it: 'Science is not to be blamed for its mistakes, they are our mistakes' (Collins and Pinch 1993: 2). Critical historical reflection on the nature of science, embedded within societies and generating particular kinds of knowledge, could enrich theorizations of science and inform future reconceptualizations of science and history. It was this sentiment that 'Science the Refreshing River' sought to capture. While most contributors to the volume might not entertain any uncritical optimism concerning science, or the resolution of the tensions that characterize the study of science and civilizations, the domain for the majority continues to remain one that concedes little space to any kind of incommensurability.[2] Hence the discussion of universality comes up in a diversity of incarnations and formulations depending upon the angle from which it is approached. These include the history of civilizations, history of mathematics, transmission studies, philosophy of science, sociology of science and social epistemology.

How have historians responded to the Needhamian oeuvre? Romila Thapar brings in one perspective from the history of civilizations, and suggests that with the appearance of the notion of 'oikoumene' in historical literature, the notion of civilization is in need of recrafting. The examination of sources of knowledge within a culture entails the recognition that this knowledge could have been internally generated within a society or was acted upon when transmitted from coexisting cultures. The generation of knowledge consequently involves transactions within and between cultures. These transactions in historical time embody two notions of universality: vertical and horizontal universality. Prehistoric societies were characterized by a greater degree of vertical universality. As Thapar points out, the neolithic revolution occurred in different societies at different times.

[2] In a relatively recent work, Shapin cogently conveyed the sentiment, by suggesting that contemporary studies on the history of sciences tend to portray science as contingent and problematic, in as far as this account of the production of scientific knowledge by 'interested morally concerned historically situated people is likely to be read as criticism of science'. But it is not science that is in the dock, but the 'pervasive stories we tend to be told about science' (Shapin 1996, p. 165).

However, civilizations over the past few hundred years have been witness to the spread of horizontal universality: revolutions such as the industrial revolution, cut across societies and nations. The changes produced are driven both by an internal dynamic and by the processes of diffusion and transmission. As we move into the early medieval period, texts become the source of transmissions and for the historian this means deciphering the function of texts in diverse cultural contexts. Cognizance of this leads us to ask how the entities 'pre-modern' and 'modern' science are constituted. In ideological terms, the rhetoric of progress provides one frame for reading these texts, which when uncritically construed is amenable to the manipulations of nationalist historiography. From this vantage point, Thapar goes on to interrogate the history of alchemy and medicine in ancient India.

Engaging with Needham's Historiography

Several contributions hark back to Needham's writings to deliberate on aspects of his work relevant to their investigation. Though many of them appear repetitive, they have been retained since they are both germane to the discussion and suggest themes and issues where the readings converge or diverge.

Gregory Blue provides us with a comprehensive intellectual biography of Needham, wherein he highlights the evolution of Needham's thought. Needham's work may be reasonably mapped along two axes: in fact they are the two key themata that embrace his work. One is philosophical and relates to the idea of the universality of science; the second is his moral activism relevant to the social responsibility of scientists and science. In terms of his intellectual trajectory, in the pre-1930s phase Needham was essentially an internalist historian of science. The post-1930s phase was marked by a gradual move to an externalist position, but as Blue indicates, this transition is not a break, but a rapprochement between the two accounts—in fact his work anticipates this very rapprochement. While committed to a rational, anti-reductionist synthesis, which was reflected in his work from the 1920s, Needham sought to integrate different levels of human experience, including science and religion, his scientific work with progressive politics, and the cultures of Asia and the West.

His moral and political activism is reflected in his participation in the Social Responsibility of Science Movement, in his later dissatisfaction with Stalinism, which he referred to as scientific opium, and finally, during his UNESCO years, the comparative studies on science and civilizations programme, which was occasioned through his elaboration of the idea that modern science is an ecumenical enterprise. Blue goes on to carefully reconstruct Needham's philosophy of science. Interestingly, despite their divergent political trajectories, Needham and Popper were one in recognizing the possibility of reconciling rationalism and traditionalism. Furthermore, as Blue indicates, Popper's views could be summoned to provide support to the Needhamian rejection of science as a purely Western invention. This provides us a clearer picture of what Needham meant by the universality of science and the notion of the ecumenical. Out of this reconstruction emerges, what Blue calls, Needham's 'principle of epistemological egalitarianism'. The principle in a way is not merely an epistemological one, bearing upon the idea of the universality of science, but also addresses the responsibility of scientists. In those years, few scientists apart from Needham were aware of the dangers of scientism and its attempt to hegemonize other ways of knowing and acting upon the world.

The dualisms characterizing Needham's work, and that Blue elaborates on, are problematized differently by Shiv Visvanathan. This problematization probes Needham's corpus for an 'alternative ethic for science', and for how an alternative science is to be constructed.[3] Visvanathan sees in Needham's enterprise: a search for plurality struggling against universalism; an attempt to understand the levels of organization without falling into the trap of reductionism or mechanism; a defence of Marxism as theological, while, at the same time, seeking in it a pluralist sensibility; and finally the quest for cognitive justice, wherein the ideal of egalitarianism is extended from the domain of politics to that of knowledge. This notion of cognitive justice and Blue's notion of epistemological egalitarianism closely

[3] An internal historical analysis of positive science as a cognitive programme points to two contending positions. The first is that positive science was the only possible science, and that the role of historical analyses was to divulge 'the contingent circumstances of its emergence'. The second position affirms that the limited development of possible alternatives was determined by the institutionalization of positive science in terms of its cognitive and social structure (Daele 1977, p. 46). Visvanathan poses the problem of how Needham would have responded to the notion of alternative sciences, in the light of, say, the feminist critique of science.

intersect, alluding in both cases to the extension of the political notion of egalitarianism to the realm of knowledge. Visvanathan finally suggests that Needham's Marxism enabled a psychoanalysis of civilizations that exorcized the dominant myths of Western society. In this sense there is a continuity, suggested earlier, from *Science and Civilization in China* to *Black Athena,* which seeks to bring cognitive justice to bear upon Africa.

Within the domain of civilizational studies of science, the past few decades have witnessed radical shifts. Aant Elzinga maps out these shifts in the agenda of civilizational studies, their social production, and suggests corrective measures to avoid the 'pitfalls' of hyper-reflexivity that mark cultural studies of science—the elements that will help us get around this are, he suggests, contained in Needham's work. Needham's work commenced in the first stage with the Sartonian programme, but soon developed into a critique of Eurocentric internalism. For his own studies on the history of science in China this meant a gradual movement away from Eurocentric formulations of the Needham question. The radical internalist turn to the social historiography of science, prompted by Kuhn's work, generated a critique of Needham that emphasized the Western categories that underlay the latter's work, though the critics remained committed to his moral activism. This was in contradistinction to the absence of partisanship of the cold war years, disguised in the ideological garb of neutrality. Interestingly, as Elzinga argues, and as does Fuller more substantially in a forthcoming book on Kuhn, the history of science promoted this image of science in science departments during this period. Needham, the modernist, moved towards what Elzinga calls 'a hermeneutically reflexive modernity', wherein the encounter with and experience of other cultures and civilizations prepared the ground for universality. This fed into his political and moral activism, his defence of the democratic agenda of modernity and the critique of relativism.

The cultural turn relativized the universal, and its extreme formulations precluded the possibility of cross-cultural comparisons or dialogue. Elzinga suggests three modalities for straightening out the difficulties with the cultural turn. The first is by transforming an apparent universality into a situated universality, which effectively means recognizing with Needham that European science is an ethno-science produced in conjunction with modern science. The second requires a reckoning with Sandra Harding's demand that studies of post-colonial theorists of science relating to the expansion of European science be integrated with science studies. And finally, formulating a central moral prescription for linking the notions of equity, justice and solidarity with the situatedness of knowledge claims. This partially responds to the uneasiness reflected in the essay by Raina and Habib, regarding the future of civilization studies in the light of the cultural turn.

Western historical scholarship since the Age of Reason, Steve Fuller indicates, discounts time in that it rewrites its past as 'an anticipation of the present'. This time discounting amounts to a 'retrospective colonization of the past'. The Needhamian project on the other hand challenges the thesis that the past is not relevant, but more importantly challenges the contingent character of the West's ascendancy. Fuller articulates this aspect by questioning the metahistoriography underlying explanations for the ascendancy of the West within Western history of science. The overdeterminationist[4] view of history, resolutely situated within the scientist's historiography of science, attempts to identify the impediments in the path of the scientific revolution or of its non-emergence—in other words it seeks to respond to the earliest form of the Needham question. Recent decades have yielded more of the underdeterminationist version, which views the scientific revolution as an event that occurred under special conditions obtaining in the West. This view, in a manner of speaking, constitutes the 'genuine spirit of the Needham project'.

The history of Islam, Fuller maintains, poses a challenge to overdeterminationist historiographies. The response to the flip side of the Needham question, relating to the conditions for the emergence of the scientific revolution in Europe, now alters radically. The three Islamic revivalists Fuller studies attribute the inability of scientific knowledge to disseminate in the Islamic world to the close link between Church and state.

Two core misperceptions underlay received metahistoriography. The first arose from the view that the explanation for the origins of modern science amounted to an explanation of the diffusion of science.

[4] Overdeterminationist and underdeterminationist theories of history are defined in Fuller's essay in this volume, and discussed in greater detail in his book *Philosophy, Rhetoric and the End of Knowledge*, Madison: University of Wisconsin Press, 1993, cited as well in the essay by Raina and Habib.

The second emanated from the failure to ascertain how the West benefited from the relative advantage of backwardness vis-à-vis Islam—principally, gaining epistemic advantage on account of its incapacity to impose doctrinal consensus. But the more important issue is that the history of science proper gets more reflexive with every milestone that the discipline passes, in which case it is likely that post-colonial, non-Eurocentric narratives are likely to 'treat the history of science and the history of the West as only contingently related'. This could well be the logical culmination of the ecumenical programme that commenced with the framing of the Grand Question.

An interesting feature marking the shift from macrohistorical sociology to the newer world systems approaches, is that the common explanations for European exceptionalism have to do with 'cultural entelechy, the peculiar centuries-old legal-cum-political heritage of the West: its multiplicity of competing power networks, its unusually strong legal guarantees for personal property, the consequent strength of market mechanisms and commercial institutions' (Lieberman 1997). These responses are not entirely unrelated to the responses to the obverse of the Needham question. On the other hand, in the context of the Needham question, Lieberman points out, in the introduction to a special issue of *Modern Asian Studies*, that all studies within this framework agree that, with the possible exception of Japan, no Asian country was moving towards the industrial revolution at the end of the eighteenth century. And four principal explanations are offered—cultural impediments, political sins of commission (market interference), political sins of omission (failure to provide capital-friendly institutions), demographically based equilibrium traps (ibid.: 464).

Examining the status of the Needham question and the cross-cultural history of science, Patrick Petitjean's essay investigates the social relations and linkages between French and British scientists, before and after the Second World War, that went into the invention of the notion of ecumenical science. Petitjean focuses upon the historical singularity of the 1930s: the economic crisis, the rise of fascism and the place of science in the transformation of the erstwhile Soviet Union. In this context, members of Werskey's 'the visible college' at Cambridge, and their counterparts from the elite institutions of France, such as the College de France (which was totally ignored by Werskey), forged ties, formed popular fronts, and mobilized scientists to defend science, socialism and democracy. Petitjean traces the evolution of these social relationships and the manner in which scientists and social scientists from both sides of the channel were mobilized into this network. The SCHM project emerged out of this network, and the little-known role of Lucien Febvre in the deconstruction of Eurocentrism, at least in the anglophone world, interlocked with Needham's subsequent inauguration of ecumenical science, that later was the core notion founding his study on the cross-cultural history of science and civilizations.

In the 1950s, this strongly linked network split into the UNESCO camp, concerned with international science, and the WFSW, concerned with scientists' trade unions and peace movements. Petitjean gets back to a concern that surfaces in Elzinga's essay, namely the impact of the cold war ideology on the social relations of science. For one, it was no longer possible for scientists working within elite institutions to sustain such united front political coalitions. However, from the late 1940s onwards, the demands raised by the Social Relations of Science movement concerning state funding for the sciences and planning for science were continuously incorporated into mainstream science. The movements themselves appeared to have lost their sharp political edge. Nevertheless, Petitjean concludes, Needham's place was ensured since his concerns extended beyond those of his circle, namely in his opposition to Eurocentrism, and his moral activism in science.

Issues in the Contemporary Historiography of the Sciences

But how did the Needhamian project figure in historical studies of science as much in China as in other parts of the world? The essays in this volume discuss the responses both to Needham's project and to the cross-cultural history of science. For Needham, emblematic of modern science was the mathematization hypothesis of nature. Catherine Jami discusses the Chinese historiography of mathematics since Needham. Chinese contributions to science instrumental in the subsequent constitution of modern science were Needham's primary concern. Contemporary historians of mathematics in China have oriented their efforts along two axes. On the one hand they have been attempting to understand the history of mathematics in terms of Chinese thought. This has led them to 'positive characterizations' of science in China, without necessarily lapsing into relativism. Jami indicates that while Needham's work is not considered central by historians of mathematics in China, historians of science do consider his work significant. Much of this history of mathematics emerged in the late 1950s, and was inclined to Marxist contextualism, but the political backdrop did not destabilize the already current tradition of the internalist history of mathematics. Jami points out that Needham turned the question itself into an opportunity for writing the comparative history of science.

Dhruv Raina and S. Irfan Habib turn the Needham question around and ask why a Needhamian history of science in India has not been produced, and even suggest the unlikely possibility of its production owing to the changing context of the place of science in contemporary culture, among other things. They then illustrate how, for long, the concerns of historians of science in India were linked up with concerns of obtaining legitimacy for science itself. While identifying a few central narratives of the history of science after independence, they outline the interfoliation of internal and external factors responsible for the non-emergence of a Needhamian history of science. They conclude by asking whether the history of science and civilizations is possible in the present era. In the new big picture that is emerging in the history of science, will the history of colonialism serve as a metanarrative for the history of science?

A constitutive historiographic element in the history of science is the transmission of scientific knowledge, and it is a key element in the construction of the Needhamian river metaphor. Karine Chemla discusses two cases of transmission in the history of Chinese mathematics to illustrate the limitations of the metaphor. The river metaphor, she proposes, is topologically representable as an inverted tree. The two cases of transmission reveal that the inverted tree is really cut out of a more complex map. The presentist angularity of the Needhamian enterprise lacks the prospect of integrating the narratives of rivers that failed to reach the sea. This provides a further context for Jami's point as to why Chinese historians of mathematics pursue positive characterizations of Chinese mathematics. The two cases Chemla studies illustrate the point that the Needhamian metaphor at best exemplifies one of many possibilities. In the first case, the transmission of a Chinese algorithm produces a favourable mutation that is assimilable in the Needhamian picture. In the second case, there is an enriching transmission of the algorithm within the Chinese tradition, but since this did not link up with the tradition of modern mathematics, its history is impoverished through the process of forgetting. Despite a hint of despair concerning internationalism in science,[5] Chemla suggests that the Needhamian perspective could still guide us beyond the dangers of relativism.

Pascal Crozet's interrogation of historiography points out that conventional accounts of the process of modernization in nineteenth-century Egypt tend to overemphasize the antagonism between traditional and modern scientific knowledge in Egypt and propose that modern scientific knowledge was introduced into an Egypt characterized by a scientific vacuum.[6] His central problematic, quite at variance with traditional historiography, is to find an historiography or comparative approach that enables the study of how nineteenth-century Egyptian scientists opened the cultural space for the assimilation of modern science, and initiated the processes for its legitimation. Crozet uses the term 'naturalization' to refer to this first project of modernization in the nineteenth century. Essential to this historiography is the recognition of the dialectical relationship between traditional Egyptian knowledge and imported knowledge. Crozet argues that traditional sciences, geared to serve traditional demands, continued to develop in Egypt until the end of the nineteenth century. On the other hand scientists trained in modern science had to take cognizance of these developments for practical reasons as well as for symbolic ones relating to the legitimation of their own enterprise.

Referring to Egypt's unique experience in finding a natural framework for modern science, Crozet suggests that the historiography of 'science and empires' minimizes differences. While that may well be the case, there is a great deal of similarity in the process of legitimation of modern science in India, China and Egypt. For one, in all three cases the process of legitimation was initiated through a

[5] This arises from the manner in which presentist historiography has discounted the knowledge claims of non-Western cultures, and thereby validated the new intellectual property rights regimes. The issue is also touched upon in the essays of Blue and Elzinga in this volume.

[6] This point concerning the widespread extent of education in late eighteenth-century Egypt when compared with contemporary Europe, is also made by Fuller in this volume. However, Crozet suggests that the nineteenth-century vision of the history of Egypt was inherited from the nineteenth-century history of science.

dialogue between the so-called traditional sciences and modern science, and this dialogue could have been motivated by very practical imperatives, or by the more symbolic requirement of neutralizing the cultural import of the Western in modern science, or by the need to postulate historical continuity. As a matter of fact, this could well be construed as one dimension of bridge building in the Needhamian sense.

Universality of Science

Reckoning with the salient critiques of science, Michel Paty makes a defence of the universality of science, albeit in a different philosophical sense. Critiques of the idea of the universality of science were founded on the reification of science from its social context. For Paty, the origins of modern science are traceable to the transmutations in Greek cosmology, within the core notions of mythos and logos. But the question of origins is not really important, since all cultures possess language, art, techniques, religion, etc. that are inherited from the past, as well as through exchange between cultures and civilizations. Just as the neolithic revolution acquired global dimensions, we are presently living through a phase where the cultures of modern science and industry are being globalized. The history of science, as with Thapar, then becomes the history of transmissions. Thus the science of the Greeks was transmuted into modern science through the transmission of notions of techniques and experimentation foreign to Greek science. The idea of universality is grounded in the processes of transmission and translation. In fact, Paty's central contention is that at the heart of the problem of universality is the translation of meaning that follows in the wake of transmission. Universality is not to be premised on the contemporary model of globalization, but through the acceptance of difference, which Paty believes is a feature of modern science. While Paty makes the strongest claim for the universality of science, he unequivocally recognizes that the claim to universality is viable only if it is a situated one.

Two key Needhamian concerns appear in K. Subramaniam's essay. The first is the idea of the disunity of the sciences that has emerged from the microsociology of science and the studies of laboratory practices. Arguing from the laboratory studies perspective that the unity of science is not to be found in its theory, Subramaniam seeks to recover it in the practices of science, which effectively means drawing technology closer to science. Secondly, he wishes to distinguish modern science from pre-modern science, not so much in terms of the hypothesis concerning the mathematization of nature but in terms of the inscription of different laboratory practices. He introduces the notion of motivated intersections to grapple with the distinction between laboratory practices of pre-modern science and modern science. The radical break occurs sometime in the seventeenth century in the discipline of chemistry, where there is not so much a change in practice as a translation in the goals of the discipline. The evolution of alchemy instantiates his argument. In reflecting on the Needham problem, Subramaniam suggests that the emergence of modern science in Europe had to do with, among other factors, the closer attention to both language and hermeneutics. The mathematization of the nature hypothesis would thus appear to require a re-examination.

Concerns for the Future

Essential to the Needhamian corpus is the departure from a historiography founded on the premises of scientism, as is the problematic idea of universality. But any critical reflection on universality that prizes its relevance will have to engage in the Needhamian programme of bridge building. Among the many failings of scientism is its inability to provide an ethic for science, and during the cold war years it more or less provided a mallet for neutralizing the popular fronts of scientists.[7] On the other hand, within history of science studies internalism did play an important role in promoting and legitimating the performance and investment in science. However, cultural endeavours for the legitimation of science did initiate, in whatever presentist manner, a dialogue between different knowledge systems. That this dialogue was inadequate and inequitable is evident from the crisis in science studies, and the post-modern turn in cultural studies.

Needhamian ecumenism required that scientists bring back culture, society and history into their reflections on science. Today we see the

[7] In what might today appear as one of the early socio-epistemological inquiries into the historiography of science, Mendelsohn wondered why historical trends that developed during the 1930s, such as the Marxist history of science, or the sociology of Merton, were overlooked in the 1940s and 1950s, and not rediscovered till the late 1960s (Mendelsohn 1977, p. 8).

assimilation of crucial Needhamian insights into investigations of history proper. This assimilation is centred around the key proposition that the history of science is to be interpreted as a history of transmissions, lending credence to the notion of a situated universality as the terrain on which the history of transmissions can be mapped. But the history of transmissions is not without its problems, because as the experience of colonialism shows, the history of transmissions is imbricated in the politics of knowledge and the project of Western hegemony, as a result of which certain cultures tend to be privileged at the expense of others. The symmetry principle from the sociology of science might offer one important mechanism for getting around it. The Needhamian corpus, as has been argued, still offers the possibility of transcending these problems. The second difficulty with the history of transmissions, as reflected in contemporary scholarship, is that the presentist orientation of historians of science tends to overlook those histories of knowledge that do not currently feed into the mainstream history of sciences as defined by the contemporary research frontier. Consequently, there is a third limitation within transmission studies and that has to do with the excessive emphasis on the tradition–modernity dichotomy in the studies of science in the non-West—a limitation already recognized in earlier criticisms of Needham's work.

Yet, despite the renewal in the history of sciences that has deconstructed science, there are virtues in avoiding the seductions of judgmental relativism and there remains the challenging task of investigating the dialogue between civilizations, in the Needhamian vein, rather than uncritically throwing out both universality and the moral ethic Needham sought to bring to science and science studies. A methodological query derives logically from the problematizations that appear in this volume: how, in the light of this renewal, are we to pursue the comparative study of science and civilizations (Elzinga, 1997)? This volume, hopefully, emphasizes the need for this endeavour.

References

Christie, J.R.R., 1993, 'Aurora, Nemesis and Clio', *British Journal of History of Science,* 26, 391–405.

Collins, Harry and Trevor Pinch, 1993, *Golem: What Everyone should know about Science,* Cambridge.

Cunningham, Andrew and Perry Williams, 1993, 'De-centring the "big picture": The Origins of Modern Science and the Modern Origins of Science', *British Journal of History of Science,* 26, 407–32.

Elzinga, Aant, 1997, 'Science the Refreshing River', Conference Report, *Science Technology & Society,* 2, 1, 151–8.

Holton, Gerald, 1993, 'Can Science be at the Centre of Modern Culture', *Public Understanding of Science,* 2, 291–305.

Lieberman, Victor, 1997, 'Transcending East–West Dichotomies: State and Culture Formation in Six Ostensibly Disparate Areas', *Modern Asian Studies,* 31, 3, 463–546.

Mendelsohn, E., 1977, 'The Social Construction of Scientific Knowledge', in E. Mendelsohn, P. Weingart and R. Whitley (eds), *The Social Production of Scientific Knowledge, Sociology of the Sciences,* volume I, D. Reidel, 3–26.

——— 1995, 'Science and the Construction of the Idea of Europe', *VEST,* 4, 8, 59–64.

Pyenson, Lewis, 1993, 'Prerogatives of European Intellect: Historians of Science and the Promotion of Western Civilization', *History of Science,* xxxi, 298–315.

Shapin, Steven, 1996, *The Scientific Revolution,* Chicago.

Van Den Daele, Wolfgang, 1977, 'The Social Construction of Science: Institutionalization and Definition of Positive Science in the Latter Half of the Seventeenth Century', in Mendelsohn et al. (eds), 27–54.

2

Romila Thapar

History of Science and the *Oikoumene**

Historians of the history of science tend to view science from two perspectives. One is the familiar linear projection of the evolution of particular disciplines regarded as scientific, such as astronomy or medicine, a projection determined by our modern understanding of science. The emphasis is on isolating inventions and discoveries and focusing on that knowledge which is seen as strictly scientific. The other is the attempt to view science as culture and as part of the social formation of a society. This perspective is obviously more wide-angled and takes in many facets involving interactions and osmosis, which bear on a particular body of knowledge. Whereas the first is frequently the close preserve of scientists, the second approach is the one in which the historian can intervene, and more so after the debate on Thomas Kuhn's analyses of paradigm shifts, as well as Joseph Needham's concept of ecumenical science.

The notion of the oikoumene has been grafted onto historical concepts of ordering the past of the world. The historian's intervention requires the recrafting of some historical formulations since they are being recrafted in historical interpretation as well. One of these is the

*Much of this essay draws on conversations with Dhruv Raina and S. Irfan Habib both at NISTADS and their publications. The imprint of David Pingree and Debiprasad Chattopadhya should also be discernible. The examples which I quote come from the early period of Indian history since that is the most familiar to me as a historian.

notion of 'civilization' as a continuing historical unit. Used more casually in the earlier centuries to refer to the softening of manners and to artistic and literary achievements, it became a widely accepted unit of history from the nineteenth century, coinciding with colonial perceptions of history. The world was divided into discrete, geographically bounded areas each with a dominant culture, recognizably different in intellectual, aesthetic, technological and religious attainments and associated with urban centres, the use of scripts and a state of organized social order. Toynbee counted twenty-six, each rising in response to challenges and declining when the response was inadequate. More recently the count has been reduced to eight in Samuel Huntington's theory that the future of the world will revolve around the clash of civilizations. As a spokesman of the American political right wing, his theory envisages conflicting civilizations as a replacement for the cold war.

Huntington's identification of these eight is, as is often the case, a confusion of various characteristics of religion, geography, dynasty, nation, race and ranking, and is unacceptable to serious historians. But more interestingly, the reaction it drew from critics indicates new parameters in the concept of civilization. Civilizations are not seen now as static or geographically bounded. Cultures, patterns of life and belief are not immutable and do not in themselves give rise to violent conflict, nor are civilizations necessarily defined by rationality as argued earlier by Max Weber. Marxist historiography and the work of the Annales school in France have encouraged the reformulation of the notion of civilization. The more important components are cultures defined as patterns of living, deriving from economic factors, social divisions and belief systems. Variations in cultural forms within a geographical area tend to erode the monolithic character of earlier views of civilizations. Consequently, civilizations are characterized both by changes from an internal dynamic and by a constant borrowing, assimilating, interacting with others, which call for a continuous process of reordering and redefining. It is less the isolated uniqueness of each civilization and more the forms which societies take and the reasons for these, which have become the basis of comparative studies.

This in turn means that knowledge emerges from the combination of an indigenous genesis together with transmission from other coexisting cultures, a transmission which can involve some contestation and some negotiation. This process takes the form of cultural transactions, within a culture and between cultures. Knowledge, viewed either as a body of information or as theories of explanation, is part of this transaction. The question of agency and of the exploitation of knowledge becomes central. This is as relevant to scientific knowledge as to philosophy and literature and requires that science be treated as part of a culture. Where the history of science is written as technological change based on rational modes of thought outside the normal, there the historiography of science frequently misses out on its embeddedness in society.

The question of the universality of science hinges on how we define 'universality'. Is there a difference between how it was viewed in pre-modern times and now? How do we perceive change? I would like to suggest that earlier there was a greater vertical comparability with a gradual growth of horizontal universality, whereas the modern experience is that of a rapid horizontal universality. Let me try and illustrate this.

Historians conveniently use the term 'revolution' for processes of qualitative and quantitative change, but the nature of revolutions differs up to a point. Gordon Childe referred to the transition of early societies to agriculture as the neolithic revolution. This has vertical universality in as much as many societies undergo a similar transformation but this happens discretely or even in isolation, and at different points of time. Historians also refer to the European industrial revolution of the late eighteenth and nineteenth centuries which, because it occurred in the context of capitalism and colonialism, spread rapidly across the world. 'Revolution' is a convenient label for indexing societies. But the two revolutions are not identical, neither as a process nor as the universalizing of a process.

In the case of the neolithic revolution similar changes occur in unconnected societies as for example the Yang Shao in China and Mehrgarh in Baluchistan. The changes have recognizable characteristic patterns of settlement, agriculture, pottery, artefacts, icons symbolizing fertility and so on. Therefore the changes are internal to each culture and not necessarily the result of diffusion and transmission, although subsequent to the neolithic revolution there is a considerable interaction of peoples, even if the neolithic societies take variant paths to the future. Is this change predictable? Or is such change more likely in societies, therefore technology becomes a central focus? The oral transmission of technology is a major study in this kind of knowledge.

The industrial revolution on the other hand has a horizontal universality when it cuts across various societies. Changes are encouraged partly by the internal dynamic of the society but also as a result of diffusion and transmission. The possibility increases in a historical situation where transmission occurs through trade and contacts, conquest and colonialism, media and communication, the market and globalization. The change is frequently in two stages and this is significant in understanding the diffusion of scientific ideas. At first there is an imitation of the dominant culture and this presupposes an unequal relationship among and between societies. The dominant culture controls what is transmitted and thus controls knowledge through power. This is generally the phase of colonialism. Subsequent to this there is the internalization of the change with the recipient society deciding how to internalize it. This is more evident in situations which are subsequent to the colonial.

The industrial revolution as a universality is linked to the scientific revolution and these in turn are linked to the existence of capitalism and colonialism. The 'scientific revolution' cannot be clinically removed from this totality and treated as an entity on its own. To ask why the scientific revolution did not take place in China or India would require asking the prior question of why capitalism did not develop on a large scale in these areas. The notion that Asia and Europe were at par in their economic activities up to about AD 1500 after which Asia dropped behind has been questioned and some would suggest parity as late as the early nineteenth century. This inter-relates science as we understand it today and capitalism far more closely than was assumed in the previous argument. The package invoked in this new historical situation presumes a quantitative and qualitative change of a specific kind.

Change in the nature of science has also occurred in stages. Parallel systems of knowledge exist in various societies but there may be a drawing out of knowledge of a particular kind, such as the contribution of Galileo and Newton, in a specific society. This is never a sharp break, as is illustrated in the writings of Newton: he wrote in Latin, he was rooted in the alchemical and Hermetic tradition and at the same time was deeply influenced by Descartes. Such a drawing out of knowledge acts as a civilization link.

The leap, as it were, to a changed condition of science, rides on other historical changes within institutions that emerge as a result. Apart from industrialization and capitalism, with its counterpart in colonialism, there are related features of considerable consequence to the change. The nation-state was a new kind of state, different from the previous feudal rulers and absolute monarchies. It permitted new social classes to come to power with their own agendas and many of these became patrons of science and technology of a particular kind. The emergence of representation and democracy as demands with ideals of equality, rational government and secularization encourages the treatment of science as value free or at least neutral, although in effect it is often value loaded in support of the new agendas. It has been argued that 'print capitalism' creates large communities since it marginalizes dialects and localized forms of the use of language. Print culture is certainly very different from manuscript culture. The need to use a widely understood language in science extends the function of mathematics. These features facilitate a sense of nationalism and this inevitably influences notions of national identities in science, which sometimes take on the label of civilizations. The historiography of science is projected in terms of civilizational achievements.

The history of science in India is characterized by many historiographical perspectives and, as with all historiography, has been influenced by events and ideas of their time. Pre-modern science was seen as a set of separate studies defined by modern disciplines, resulting in a concern with evaluating the degree to which they approximated to European science. In the race to determine the most ancient and the most superior civilization, technology, inventions and discoveries were also assessed in terms of who got there first. Non-European sciences were regarded as having ceased at a point in time, for only Europe experienced the scientific revolution.

Yet the disciplines most discussed in early India had been partially treated as separate but were essentially viewed as facets of a total knowledge. Astronomy and mathematics were linked in the conceptualization of time, and this provided fodder for astrology. In as much as astrology governed the life of a person, it was also tied to alchemy with its obsessive search for longevity and immortality. This had a logical connection with medicine which controlled bodily well-being. At one level each is an abstract search for knowledge, at another, their linkage points to a concern with the human condition. The binary separation of the rational from the irrational, astronomy from astrology, for instance, is necessary to the analysis of scientific knowledge, but the irrational cannot be dismissed or ignored. It too has to be assessed, perhaps as an alternative, as is fashionable these days, or as peripheral, if indeed it was so, or as a system within itself confronting the rational, or, perhaps most interestingly, as a parasite riding on rational analyses.

The search for the rational and for positive sciences was and is, in part, the claim to parity with the West, even if it is Eurocentric in approach. There was a determination to disprove the singularity of the so-called Greek miracle. Taken to the point where the search for the rational appropriates the irrational, it ceases to be only the recovery of the self in post-colonial times. Instead it supports the distortion of history as we have recently witnessed in the claim that 'Vedic mathematics' is superior to other mathematics.

A scientific revolution in any case requires more than just rational thought and advances in certain categories of knowledge. It also requires patronage and finance of a kind which will motivate technological change, frequently to enhance the profits of those supporting it. Thus, repeatability and standardization in technology was basic to industrialization and was a far cry from the individualism of the craftsman and artisan. That the scientific revolution occurred in the colonizing societies is not an accidental coincidence.

Since Indian civilization was earlier defined as Hindu civilization, its science was described as Hindu science, even though the more active proponents of this science were people who were often socially degraded by brahmanical orthodoxy. Because it was viewed as Hindu science, the texts read were limited to Sanskrit and the Indian languages. Even if mention was made of interest evoked in any of these sciences in China or the Arab world, such as the interest in alchemy, there was little extended attempt to discover which aspects of these sciences appealed to other cultures and whether there was a transmission back from the scholars who were translating Sanskrit texts into Arabic and Chinese. There would therefore be much to be learnt even about Indian science through consulting texts in these languages.

Detailed information is available on Arab interest in these sciences from the eighth century onwards, especially at the court of Harun al Rashid, where Indian scholars were resident and Sanskrit texts were being translated. Arab control over the trade routes in the Indian Ocean also acted as a nexus between China, India and west Asia. The overland trade through central Asia was yet another route for the merging of ideas. Arab scholars added the knowledge from these sources to translations from the Greek and together they form a remarkable intellectual transmission eventually transmuted into early European science.

The Sung period in China saw a flow of Indian texts and resident Indian scholars who arrived either as Buddhist monks or in their company, together with Indian merchants. Some are said to have brought longevity medicines and others are believed to have changed ordinary stones into lapis lazuli, which then became so common that it failed to fetch a price in the market. Alchemists clearly had no head for business. The imprint of Buddhist interests would have determined the kind of knowledge which was more readily transmitted. But much of this points to an essential feature of civilization, namely its ability to encourage a free flow of knowledge, people and artefacts.

Because science was extracted and isolated from the context of changes wrought by modernization, little attention was given to locating it as an idiom of culture, even for earlier times. The absence of a scientific revolution in colonial societies fueled the disparity in viewing the colonial as hegemonic and the colonized as marginal. It is maintained that for the colonized, modern science has come to be seen as neither neutral nor universal but largely a mechanism of control. Its practice is said to induce alienation. Non-European science in contrast is visualized as holistic. However, this rejection does not explain how scientific knowledge, even in pre-modern societies, relates to its wider culture.

What then do we mean by science as culture, or, as Needham saw it, science as integral to the economic, social and intellectual history of the culture from which it is generated? This is not a determinist view but one which also investigates the concerns and sensibilities of a society. Let me touch briefly on this with reference to medicine and alchemy in early India.

The earliest evidence of medicine are references to those who heal through herbs, incantations and chants—the *bhishaja*. It is said that herbs gather round such a person like chiefs gathering round a king. These were not doctors and were more likely shamans, who have functioned as healers in most societies and claim to destroy the demons of disease. But the healer had an ambiguous status in a socially hierarchical society. He was excluded from the ritual of sacrifice—a major expression of religious belief and practice—because he was regarded as impure.

At the turn of the Christian era, medical knowledge and practice changed with the writing of medical treatises such as those of Charaka and Sushruta. The fact that they were written in Sanskrit and therefore by scholars, gave a superior status to the information which they contained. The medical man was now more frequently called the *chikitsaka*. Still later he was called the *vaidya*—the learned man. It is stated that experience and knowledge based on skilful examination, together with rational explanation, would explain the cause of an ailment or the effect of medication. Chants and incantations were subordinated to diagnoses, at least at royal courts and in urban centres. The new knowledge circulated among those who could understand it and afford it. The bhishaja remained a permanent feature among the less well-established strata. With the formulation of knowledge as part of a discourse, the more random practitioners were relegated to the ranks of heretics—the disciples of the *shudras* with no knowledge of the *shastras,* as the *Maitri Upanishad* describes them.

The change was from experience to experiment, and analyses were based on precedent and also on anatomical and other knowledge derived from autopsy. While experience drew from customary and subjective knowledge, experiment was linked to theoretical knowledge. The theory drew in part on precedent and this required recording and availability, which in those days meant the writing of a text. This change has been described as a shift from therapy dependent on the supernatural, to therapy dependent on rational application. The shift may not have been quite so marked but is nevertheless apparent.

The medical texts were written in Sanskrit, the language of high culture, and have become essential to the Ayurvedic tradition of medicine. At the same time, the brahmanical norms of ordering society placed physicians in the lowest social category, regarding them as polluting, presumably because they handled diseased bodies and corpses. Nevertheless, this medical knowledge not only survived, but encouraged a variety of other medical interests. Buddhist monks, for instance, were less squeamish about disease, and monasteries had to ensure the health of their monks. Royal courts patronized the veterinary sciences, pertaining particularly to the health of horses and elephants, so essential to the army. Because of its utility, medical knowledge became, in effect, independent of the dominant orthodoxy. Investigating the history of medicine therefore requires more than an analysis of the available texts. Even the dialogue between the healer and the physician, or the intervention of the physician, which may frequently have to be deduced from various texts, becomes important information. The intention was not to control nature, but the intervention became an attempt to assert control.

It would be worth investigating why medical knowledge and practice moved from the healer to the professional physician, even though the former remains a part of Indian life. The period from about 400 BC to the early Christian era saw the classification of various fields of knowledge in the form of texts. These included grammar, geometry, ritual—both domestic and public—codes of ordering society, political economy and medicine. Some clearly were meant to enhance the dominance of those who formed the ruling groups and to provide them with a structured model. But such classifications also occur when there is the availability of new information. In this case there appears to have been at the time the opening up of existing Indian cultures to the proximity of other cultures, particularly of the west Asian and Hellenistic world. On the Greek side, Theophrastus's work on plants suggests information on new sources and species coming from Asia. New information needs systematizing and leads to a redefining of existing knowledge. Such an exercise is also carried out when there are contending theories, some of which may eventually be subverted.

Analysing the nature and function of texts, apart from the information they contain, becomes central to investigation. How do texts organize information and formulate arguments? Does their language relate to the segments of society which they are addressing? How do they observe the world and represent what they see? How do they reflect the relations between cultures and classes, or provide the characteristics of what we like to recognize as Indian. When an alchemical text on mercury takes the form of a dialogue between the deities Shiva and Paravati, it is surely telling us something other than providing information on mercury. These are questions which might illumine science as culture and which would require the interaction of history and science.

The pattern of alchemy is a little different. The body is again one of the arenas of experimentation, with attempts at longevity and immortality. The other arena was investigating minerals where the changing of base metal into gold brought wealth. Alchemy therefore became secret knowledge, and alchemists were part of the underground, as they were everywhere else. Alchemy of course provides one of the best examples of a truly ecumenical science both in terms of similarities across Eurasia of applications of knowledge and of the attitudes of those among whom the alchemists live.

In India, it was extensively interwoven with the more esoteric Tantric practices and beliefs. Again, at a certain point, in the late first millennium AD, it ceased to be only practical experimentation and came to be textualized knowledge as well. Initially there is a distancing by the established religious elite. Association with Tantricism was linked to heterodoxy; the search for immortality contradicted the belief in transmigration and rebirth which was by now a fundamental ideology for asserting dominance; and those who carried out the experiments were often of the lower castes. Yet the texts were in the language of the learned and were part of the cultural exports to west Asia eliciting intense interest on the part of Arab scholars. The textualizing of this knowledge appears to have coincided, even in India, with royal and courtly interest in the possible results. The text then plays a dual role: it is both a body of knowledge and a statement of claims.

One of the more intriguing transmissions involved astronomers and mathematicians and the authors of the early *Puranas*—texts incorporating legends, myths and some history and composed in the early Christian era. The dialogue concerned concepts of time. Astronomers used enormous figures for their calculations, figures which are in many instances, identical with those in the cosmology of the *Puranas*. Did the astronomers borrow these figures or did the Pauranika authors take their cue from the astronomers? Not only are the figures for the cycles of time the same but the concept of the four ages, although simplistic, is mathematically ordered. However, whereas the astronomers later came to argue about the authenticity of the measurements of time, the Pauranikas held to the original construction. Astronomy and mathematics were not only privileged in Indian society at the time but there were also dialogues with counterparts in the Hellenistic world and later with China. Possibly the authors of the *Puranas* in constructing their cosmology were peppering it with items from what were regarded as superior branches of knowledge, intending thereby to associate their construction with the authenticity attributed to astronomy and mathematics.

That explanations are of the essence in determining rationality may again be illustrated by some activities of the earlier past. Avoidance of disasters appears to have been one of the aims of pre-modern science, whether they were crises in the life of an individual or were catastrophes of a larger kind. This sometimes resulted in a particular study being deviated towards irrationality. One has only to remember the numerous records of astronomers calculating and predicting eclipses, and of astrologers encouraging royalty to indulge in expensive gift giving to fend off the evil effects of an eclipse. Needless to say, the astrologers were generally the recipients of the gifts. The credibility of scientific knowledge lies not merely in invoking the rational but in ensuring that the rational is not converted into the irrational.

Transmission of knowledge is a cultural transaction and raises questions about who does the transmitting and in what form. A theory may be expounded, texts may be translated and experiments may be demonstrated. Or, the transmission may be through a process of osmosis in a juxtaposition of cultures. Each of these undergoes its own evolution and change and what is transmitted may be at some variance with the original. We are familiar with the vicissitudes of translation in terms of what actually gets transmitted and whether the concepts have the same meaning in different cultures and different languages. The act of transmission would move the transmitted knowledge to a central role, even if earlier it had been at the periphery. There is also the question of why all societies do not treat transmitted knowledge in an identical way. Plane geometry for example, is used differently by the Greeks and the Indians. When the culture of a society is imprinted on a technology, then the analysis of the imprint is as significant as the actual technology in assessing the historical role of the technology.

The question has been posed as to whether such knowledge from pre-modern societies should be called science or would it be better to differentiate it as protoscience. Are we privileging it by giving it that label and do we then expect to find that it was also said to be based on rationality and therefore approximating truth, as has been attributed to modern science? The locating of those who controlled such knowledge becomes a crucial historical investigation and has a bearing on the nature of what is regarded as true. This has sometimes been tied to issues of nationalism and transnationalism, in the argument that protoscience is again a term which separates the scientific achievements of Europe from the efforts of other civilizations which tried but could not make it to science.

In constructing a national identity a tradition is invented through choosing from the past that which is most effective in creating the required identity. Historical reconstruction is central and there is a drawing on images of the past. This requires collating varieties of knowledge and because science is seen as the crucial variable, it gets to be privileged. Science is perceived as a historical category. If the history of science is tied to nationalist ideology, and it may be nationalism of any kind, it cannot be treated as universal. Categories such as Hindu science and Islamic science disallow a universality to science. This is underlined through periodizing Indian history into the Hindu and Muslim periods. The privileging of science permits it to become the measure of progress and progress is important to nationalism. It also comes to be used by those in authority to assert dominance, an echo of the colonial pattern which can become resonant if it is not curbed. What we are now experiencing is something which claims to be beyond nationalism and calls itself transnationalism—a synonym for globalization, enticing us all to the siren of the market. But the strongest feature of this is the insulation which follows globalization, when the imagined past is desperately invoked.

Universalism could draw on cross-cultural studies, particularly as societies and cultures become less bounded. The historical evolution of societies need be neither predictable nor linear. The units for comparison are not discrete civilizations, often romanticized out of recognition, but socio-economic cultural forms. With this, the transmission of knowledge in a historical context would be a major study. The points when transmissions take place have to be identified together with the ensuing results. A neolithic society cannot transmit much to a feudal society, and the reverse will involve the disappearance of the neolithic. The definitions of dominance in history therefore become important and have to be recognized. Universality may make cognitive knowledge available to all, but does not imply the equality of all. The politics and mechanisms of transmission demand attention.

The intersection of what I have called the vertical and horizontal universalities will remain a contentious issue. Civilization is not static and its constituents constantly change. It has therefore a kinetic form whose dominant modes require recognition as also their relationship to the form. This is perhaps why one argues that the link between civilization and science would be illumined by a greater input from history.

Alberuni visiting India in the eleventh century is generally admiring, but he does comment sarcastically that Indian astronomical and mathematical literature was a mixture of pearls and dung. He adds that both are equal in the eyes of Indians since they cannot raise themselves to the methods of a strictly scientific deduction. Alberuni was not a representative of the European scientific revolution, yet some recent debates would echo this statement. But the assessment of knowledge requires an understanding of the genesis of that knowledge, its function in society, the nature of its transmission, especially in the context of variant worldviews and of course historical change which both generates and is generated by such knowledge. This brings us to a complex historical process, involving more than just historians and scientists, but a process that is inevitable and unavoidable. One is reminded of what Needham said, almost half a century ago, when he first set out on his quest for science in China:

> For better or for worse, the die is now cast, the world is one. The citizen of the world has to live with his fellow citizens at the ever narrowing range of the aero-foil and the radio-wave. He can only give them the understanding and the appreciation which they deserve if he knows the achievements of the sages and precursors of their culture as well as his own.

3

Gregory Blue

Science(s), Civilization(s), Historie(s): A Continuing Dialogue with Joseph Needham

After opening with a sketch of Western thinking about Chinese science prior to the Second World War, this essay examines several key aspects of Needham's early career. It then discusses some general features of his work on Chinese science and considers various critiques of that work. The final sections focus on his ideas about the universality of science and the social responsibility of scientists which remained crucial preoccupations through both halves of his career.

Chinese Science before Needham

Joseph Needham is often credited with having initiated the study of pre-modern Chinese science, technology and medicine. The claim is not exactly accurate. He himself was the first to recognize that his work draws on a long line of Chinese scholars who over many centuries contributed to an extensive body of writings on natural phenomena and useful devices, and who often displayed a desire to treat those subjects historically. In the last phase of the Qing dynasty (*c.* 1840–1912), Confucian reformers showed renewed interest in such subjects as they tried to come to terms with the new wave of technical innovations coming from abroad and shaping the empire's fate.

Likewise, a few late nineteenth- and early twentieth- century Western authors, including both missionary scholars like Alexander Wylie and W.A.P. Martin and academic sinologists like Adolf Pfizmaier, made important studies of pre-modern Chinese scientific and technical achievements. Wylie drew attention to a particularly impressive body of indigenous Chinese mathematical knowledge in his *Chinese Researches* (1897), a pioneering survey of Chinese innovations and discoveries in a wide range of technical and scientific domains. Similarly, W.A.P. Martin, President of the Chinese Imperial University (Dongwenyuan), opened his *Lore of Cathay* (1901) by observing Chinese historical precedence in the development of gunpowder, printing, and the use of the magnetic compass. Elsewhere in the same work he argued that modern chemistry had roots in Chinese alchemy. Such studies were often written with the intention of showing that Chinese culture had a rational dimension, to which Western 'civilizing' endeavours, religious and otherwise, could reasonably make appeal.

Reports on indigenous Chinese technologies and nature-lore had indeed been a staple of European travel literature between the sixteenth and the eighteenth centuries, when there was a desire to borrow practically from Chinese inventions and techniques. Such an interest is clearly in evidence in the major missionary writings of that period, from Mendoza's *Historia ... del Gran Reyno de la China* (1585) to Du Halde's *Description ... de l'Empire de la Chine* (1735) and the *Lettres edifiantes et curieuses* (1703–76). It is also to be found in the writings of such early modern scientists as Leibniz, his contemporaries in the French Académie des Sciences, and the founders of the Royal Society, with whom Needham identified all his life.[1]

Yet, despite this early literature, the idea of the superiority of European civilization over all others dominated Western consciousness from the late eighteenth century until the years after the First World War. Western superiority was typically interpreted as natural and innate, rather than as the result of historical contingency or as the projection of a particular system of values. From Turgot, Hume and Condorcet onward, many Western writers saw the capacity for progress and improvement as the hallmark of Western superiority,

[1] Characteristic of his writings on China was his frequent use of early modern sources for the evidence they gave about Chinese and European practices and ideas. His fondness for such sources was often evident in his retention of their archaic spellings, to which this essay's title makes an oblique bow.

and progress in science and technology as its most visible manifestation. Though Michael Adas has perhaps underestimated the importance of religion and racial ideology in shaping the arrogance of modern Western identity, his valuable *Machines as the Measure of Man* does document how highlighting differences between European and other civilizations in terms of scientific and technical achievement was an essential mechanism in enhancing the Western sense of superiority. One genre of scholarship that Adas does not consider much is the history and philosophy of science, where the Western sense of superiority was distinctly articulated, most often through omission of non-Western materials. The monumental studies of Auguste Comte and William Whewell, which so influenced thinking through the nineteenth and early twentieth centuries, and the voluminous work of Lynn Thorndike a century later, gave little attention to non-Western civilizations. From the era of the Roman empire onward these treatments confined themselves almost exclusively to European achievements.[2] The opinion expressed in 1925 by Whitehead, from whose philosophical outlook Needham otherwise took so much, had thus come to be quite typical by this time. As Whitehead put it: 'There have been great civilizations in which the peculiar balance of mind required for science has only fitfully appeared and has produced the feeblest result.' Thus,

> ... Chinese science is practically negligible. There is no reason to believe that China left to itself would ever have produced any progress in science. The same is true of India.[3]

In a similar vein, Whitehead reproduced more conventional wisdom in 1933 when as a doyen of British science he wrote:

> Whatever of scientific or religious value has passed into modern life finally reached us through the mediation of these three countries, Egypt, Greece, Palestine.[4]

[2] A. Comte, *Cours de philosophie positive*, 6 vols, Paris, 1830–42; W. Whewell, *History of the Inductive Sciences*, 3 vols, Parker, London, 1837; L. Thorndike, *A History of Magic and Experimental Science*, 6 vols, Columbia University Press, New York, 1923–41.

[3] Whitehead (1926), p. 8.

[4] *Adventures of Ideas,* Mentor, New York, 1955, p. 107.

The generally accepted corollary was that the history of science since antiquity had essentially been a European drama, perhaps with a minor supporting role for the Islamic world.

Revisions of these assumptions began to emerge among historians of science in the inter-war period, and among the first to challenge them was George Sarton. Before the First World War, Sarton had set out to formulate a 'new humanism' which would take as its centrepiece the scientific and technological history of humanity.[5] Later, summarizing the crucial lesson that he thought the history of science held for humanity, he argued that 'science is one', because it bases itself on the postulate that 'nature is one', and 'this confirms the view that in spite of many disparities and animosities, "mankind is one"'.[6] After his move to the United States from Belgium in 1915, Sarton became the defining force in the early development of the history of science as an academic discipline in the curriculum there. His three-volume *Introduction to the History of Science* (1927–47) argued explicitly against Eurocentrism and included discussions of developments in various civilizations outside Europe, most notably those of Asia and north Africa. Sarton particularly stressed the scientific contributions of Islamic societies and argued that their creativity had been essential to the scientific revolution in Europe. Constrained by the lack of requisite languages from delving far into other traditions, he nevertheless gathered what relevant materials he could find and modestly affirmed the need for their further study.

[5] See, e.g., Sarton (1962). It can be noted that Auguste Comte's 'Religion of Humanity' also gave pride of place to science and its history. Within that framework, Comte's successor as Director of Positivism, Pierre Laffitte, had an appreciation of Chinese civilization unusual for the nineteenth century, though he still thought of China as devoid of any properly scientific achievement (Laffitte 1887). On Sarton's relations to neo-positivist circles in the early twentieth century, see Holton (1993), pp. 15, 31.

[6] Sarton (1927–47), vol.1, pp. 30–1. See also Sarton (1962), e.g. p. 30:

> ... because of the cumulative and progressive nature of knowledge, the history of science and the history of civilization focused upon it give us the impression that we are dealing not with a chaotic mass of people, but with a single man growing continually in wisdom and experience. That feeling is strengthened by the contemplation of two correlative ideas: the unity of science and the unity of mankind. Science is a mirror of nature; nature being congruent, we must expect science to be.

Joseph Needham: The First 'Half Life'

Sarton's younger contemporary, Needham was of course only one of a group of young left-wing British scholars in the inter-war period who came to recognize the history of science as a subject of acute intellectual and political significance. As with his like-minded colleagues J.B.S. Haldane, Lancelot Hogben, Hyman Levy and J.D. Bernal, Needham brought to the subject the outlook and training of a first-rate natural scientist. In 1931 he published his first *magnum opus*, the three-volume *Chemical Embryology*, a pioneering work that extended the frontiers of the new discipline of biochemistry. From his school years, when he was first influenced by H.G. Wells, Needham showed a distinct interest in historical subjects, and the first volume of his embryological *tour de force* was in fact a history of the subject. Needham and his wife, Dorothy, another distinguished biochemist, received their early tutelage in the history of science during the late 1920s from their older colleagues and friends, Charles and Dorothea (née Whaley Cohen) Singer.

In 1931 Needham became involved in organizing the Second International Conference of the History of Science, over which Charles Singer presided. There the young biochemist was strongly influenced by the Marxist approach to the history and philosophy of science espoused by the Soviet delegation, some members of which supported the critique that Needham and Hogben, from their own perspectives, offered against biological neovitalism.[7] That delegation, unexpectedly high-powered, was headed by the recently rehabilitated Nikolai Bukharin, and it included, among other prominent members, the British-trained plant-geneticist N.I. Vavilov, still celebrated today for his work on the origins and spread of plants, and the historian of physics Boris Hessen, who undertook to explain Newton's work in terms of the economic, social and political conditions of the late seventeenth century. Both disappeared in the Stalinist purges, as did Bukharin, but their work lived after them. The identification of the 'Vavilov-zones' laid the groundwork for later work in plant evolution and agronomy, while Hessen's paper served as the starting point for

[7] Werskey (1988), pp. 140–1. The Needhams had actually been awakened to the importance of the Marxist classics in the mid-1920s through their friendship with the French biologist Louis Rapkine, whose family had fled the anti-Jewish pogroms in Lithuania early in the century (Needham 1973, p. 8).

what came to be known as the internalist–externalist debate in the history of science.

That controversy, which occupied scholars in various parts of the world from the 1930s until at least the 1970s, turned around the question of whether the basic causes of scientific and technical change are to be accounted for in terms of developments within scientific thought and practice, or result from broader transformations in society. By the late 1970s the long-term outcome of that debate was a broad consensus on the idea that satisfactory explanations of scientific–technological change (or indeed lack of change) require a nuanced assessment and integration of both 'internal'/cognitive and 'external'/ social factors, as well as an analysis of the linkages between the two.[8] This outcome is one which might be seen as having been anticipated in Needham's own work. Despite the 'internalist' slant of his early writing on the history of embryology, he became firmly associated in the 1930s with the externalist side of that debate. Yet his championing of the externalist cause was due to his conviction that understanding the dynamics of scientific and technological development required both detailed examination of concrete discoveries and inventions, and analysis of broad changes in society which affected scientific interests, ideas and practices. Needham's scholarly writings on the history of science and technology combined close readings of the relevant technical sources, bibliographic virtuosity, and sensitivity to cultural differences and evolving socio-economic contexts. They thus had the character of syntheses that elaborated a Marxist perspective of a rather special sort, intended to clarify the fundamental social determinants of knowledge, while simultaneously addressing many of the 'internalist' concerns of non-Marxist writers, and drawing on evidence and arguments from a wide array of specialist literature.

This marriage of orientations was typical of Needham. It was not unrelated to the fact that his work was cast in the mould of the Cambridge college environment, and his mature outlook always retained that stamp. Assigned in the mid-1930s to fill what he saw as

[8] Historians and philosophers of both liberal and Marxist leanings were making such arguments by this time. See, for example, Geymonat (1977) and Newton-Smith (1981), both of which drew conclusions from the Kuhn–Popper–Lakatos debate and argued against the imposition of rigid explanatory schemas on the historical record of scientific development. Mikulinsky (1977) described orthodox Soviet historians of science as accepting a similar position.

a glaring lacuna in the Cambridge University curriculum, he became the moving spirit in organizing a special lecture series in the history of science. This brought together some of the leading scholars interested in the field, from across a wide spectrum of disciplines and political sympathies.[9] Yet, since the university was at least as likely to foster disciplinary tunnel-vision as breadth, his academic environment is not sufficient to explain his own remarkable intellectual breadth. He himself frequently accounted for the diversity of his interests in terms of his parents' characters and his somewhat peculiar upbringing. However that may be, it is clear that one of his deepest and most abiding concerns from the 1920s was the rational integration of distinct 'levels' or moulds of human experience, which he conceived of after the manner of Whitehead and Collingwood.[10] This commitment to rational and anti-reductionist synthesis underlay his characteristic (and to many, paradoxical) attempts at reconciliation, or 'bridge-building'— between science and religion, between scientific work and progressive politics, between socialism and Christianity, between science and history, and eventually between the cultures of Asia, especially China, and those of the West. Essential to his attitude throughout his intellectual career was an affirmation of the respective claims of each type of experience, as he defined them—religion, art, history, philosophy and science[11]—and resistance to denying the validity of

[9] The political spectrum on the organizing committee ran from the conservative Herbert Butterfield to the communist Desmond Bernal. Among the distinguished speakers in the inaugural series were Rutherford, Eddington, William Dampier and J.B.S. Haldane. See Needham and Pagel (1940).

[10] For his appreciation of Whitehead, see, e.g., Needham (1986), pp. 178–206 or (1969a), p. 124; for Collingwood, Needham (1986), p. 9 and (1976), p. 269. The notion of distinct human faculties is of course an ancient one, which was taken over by early modern thinkers including Francis Bacon. It seems to have been the French philosopher and economist A. A. Cournot, however, who in his *Essai sur les fondements de nos connaissances* (1851) devised the model of five basic 'faculties' and 'forms of development' (religion, art, history, philosophy and science) that Collingwood and then Needham were later to adopt. Cf. his 1975 *Oeuvres*, vol. 2, pp. 381, 399.

In speaking of Needham's commitment to a rational integration of the various moulds of experience, I should specify that what he affirmed was a philosophical position that allowed for their coexistence and recognized the validity of each. He opposed 'synthesis' of the various types of experience, however, and instead favoured keeping them distinct, and shifting as appropriate from one mould to another. See his 1930 statement in Needham (1976a), p. 121.

[11] Needham (1935), p. 429.

any one of them in its proper sphere. This approach was evident in his commitment to the 'Social Responsibility in Science' movement in the 1930s and 1940s.[12] It contributed important principles to his denunciation of Nazism, which he condemned for the political persecution of scientists and scientific ideas. It also underlay his quieter critique of Stalinism, in which he warned against 'scientific opium', that is the use of scientific language to justify the ruthless overriding of broader human values.[13]

Throughout the Depression years, Needham as a leading left-internationalist—a Christian-socialist, to be precise—was a prominent spokesman for the anti-fascist movement in Britain. From the 1937 Japanese invasion of China, he became increasingly intrigued by China, spurred largely through his and his wife's associations with recently arrived Chinese research students. One of them, Lu Gwei-Djen, soon started him on the language; and before long, at her suggestion, he committed himself to the idea of writing a modest work on the history of Chinese science, to fill the gap in the literature. By the time war broke out in Europe he was meeting regularly with Cambridge University's professor of Chinese, Gustav Haloun, with whom he made careful readings of ancient Chinese texts while paying particular attention to possible evidence of technical and scientific concerns.

After the attack on Pearl Harbor and the entry of the Western allies into the war in Asia, Needham was appointed director of the Sino-British Science Cooperation Office in the Chinese wartime capital of Chongqing. During the four years he spent based there from 1942 to 1946, he travelled widely in unoccupied China. These travels allowed him to gain an intimate knowledge of the Chinese scientific community, its conditions of work and its concerns; they also gave him the opportunity to engage in numerous discussions about Chinese technical and scientific traditions and to gather materials relevant to them. By

[12] On this movement, see the studies of McGucken (1984) and Werskey (1988).

[13] Many of the essays in *History is on Our Side* and *Time the Refreshing River* develop the critique of Nazism, which Needham also sharply rejected for its distortions of biology. Needham discussed 'scientific opium' as a part of Soviet Marxism in his essay 'Science, Religion and Socialism' which appeared first in the United Front-ish *Christianity and the Social Revolution* (1935) edited by John Lewis, Karl Polanyi and D.K. Kitchin; pp. 436–8. Clearer criticism of Stalinist politics, particularly of the 1936–9 purge trials, is implied in the paragraphs added to the revised version published in *Time the Refreshing River* (1943); see, e.g., pp. 69–70.

late 1942, as his correspondence to Dorothy indicates, he was planning a major work of scholarship on the subject.[14]

During the final years of the war, Needham's 'agitation' from China for the formation of a permanent body responsible for international scientific liaison on a world scale was instrumental in the establishment of the Science Division of UNESCO when that organization was set up in 1945. From 1946 to 1948 he served as the first director of that division. One of his conceptual contributions there was the notion that modern science is an 'oecumenical' enterprise in which people from all nations and cultural backgrounds could, and ought to, be involved.[15] This was a notion that he was soon using not only in policy proposals, but also in his comparative research into the history of science and technology.

Characteristics of Needham's Approach to 'Chinese Science'

Needham returned to Cambridge in 1948 to resume his duties as William Dunn Reader in Biochemistry.[16] From that time he embarked on the *Science and Civilization in China* project (hereafter *SCC*), together with Wang Ling, the research associate he brought back from China for this purpose. From its early days the *SCC* project was planned as a comparative venture requiring collaboration between Chinese and Western scholars, each with distinct viewpoints and skills. The first volume was published in 1954, and by Needham's death in 1995 a total of sixteen volumes had appeared or were in press. In 1958 Wang Ling left Cambridge to take up a chair in Australia, but Lu Gwei-Djen returned and continued to work as Needham's chief collaborator until her death in 1991. Approximately seven more

[14] Unpublished letter, 6 December 1942.

[15] Discussion of the intellectual and political origins of this notion is found in Elzinga (1994), pp. 15–17, and in Patrick Petitjean's contribution to the present volume.

[16] This was the university position he held until 1966 when he was elected Master of Gonville and Caius College. Originally admitted to Caius as an incoming student in 1918, he thereafter remained a member of the college till his death. It was in his rooms there that he and his collaborators worked on the *SCC* project until he and Lu Gwei-Djen established their own institute, the East Asian History of Science Library, when Needham retired as Master in 1976.

volumes of the series are still in progress, in the hands of the team of collaborators he left behind.

Needham brought to the study of Chinese science, technology and medicine the approach of a professional historian of science, as well as the authority and the appreciation for technical detail of an experienced experimental scientist. His work was the first attempt in this century at making a systematic survey of Chinese materials relevant to the history of science and of integrating them into the Western academic discipline. His ability to carry out that integration successfully was closely linked to his persistently comparative method. His writings give recognition to the mass of Chinese historical evidence, both documentary and archaeological, relevant to the development of technology and to the understanding of natural processes. Furthermore, his historical work is imbued with a vision of the contemporary relevance of his subject that was forged on practical experience in cross-cultural scientific liaison and in the development and administration of science policy.

In publicly announcing his intention to pursue a major scholarly project in 1946, he posited a complex of problems that continued to inform his writing over the following decades.[17] That initial statement started from the premise that science and democracy were inextricably bound together. This was a view shaped by the issues and experiences of the 1930s and the war against fascism. It was also one that reflected the historical analyses of Marxists, Weberians and others, according to whom the rise of capitalism and the seventeenth-century scientific revolution were paired phenomena. Needham's central problem as he already defined it then was the question of the 'origin of modern science in Semitic-Occidental (Euro-American) civilization' and 'its failure to arise in China'. Because of the 'intricacy and scope' of Chinese history and culture, he considered this question to be 'one of the greatest problems in the history of civilization'. Subsequently, Kenneth Boulding called it 'the Needham problem', and that identification has stuck.[18] Yet it should be noted that the question was in fact far from new. The Western literature comparing civilizations was rife with

[17] This discussion came in the last section of his essay 'Science and Social Change', first published in *Science and Society* (1946, no. 10) and reprinted in Needham (1969a). See pp. 145–53 of the reprint for the relevant passage.

[18] Boulding (1976), p. 9.

a variety of answers to it—biological, cultural, religious, and so on—almost all affirming with unhesitating conviction that the West was uncontestedly best and had been so since the ancient Greeks.

What was new in Needham's approach was the way he re-problematized the question. In doing so, he challenged two premises that underlay most of the recent writing, namely that China had neither democracy nor science. After briefly affirming that China did indeed have democratic traditions of its own, he challenged the weighty view of Feng Yulan, to which conventional Western writers, for example Whitehead, subscribed, that China had no indigenous scientific tradition. Needham countered this position by making a distinction between pre-modern and modern science that remained central to his later thought. The point, as he stated it then, was that China 'had an abundance of ancient philosophy, and of medieval technology; only *modern* science (until the coastwise impact of the West) was lacking'.[19] In later works he developed the same line of argument further. Against the cliché that the Chinese had empirical and technical talents, and hence technology, but that they lacked the capacity for abstract thought, and thus had no science, Needham had little trouble showing that pre-modern Chinese views of nature were conceptually complex and sophisticated. Although in his view, the orientation of pre-modern China was predominantly practical, 'traditional' Chinese science, like its counterparts in other civilizations, had theories of the medieval type.[20] What it never indigenously developed, he thought, were two things: characteristically 'modern' theories incorporating mathematized hypotheses that were experimentally testable, and the specifically modern relentlessness in experimentation.

This stance involved Needham in issues related to what logical positivists and Popperians call the demarcation problem, that is, the

[19] 'On Science and Social Change', *Science and Society* (Summer, 1946), p. 247; Needham (1969a), pp. 147–8 (italics mine).

[20] The difficulties Needham faced in trying to get across the characteristics of the Chinese tradition are evident from the 1961 discussion of his paper 'Poverties and triumphs of the Chinese scientific tradition'. While he was trying to draw attention to the theoretical component of the traditional Chinese sciences, in order to counter the orientalist cliché that the Chinese were only practical, Willy Hartner raised the inductivist objection that depicting Chinese culture as predominantly practical was excessive, since practical orientation was characteristic of modern science. See Crombie (ed.)(1963), pp. 157–8.

task of delineating science from other types of human activity and culture. This constitutes one of the thorniest problems in the philosophy of science, which with reference to non-Western civilizations were vested with ideological significance. In Baconian and positivist fashion, Needham thought of science as a broad continuum of experientially derived natural knowledge that included exact, experimental and classificatory sciences as well as the full range of technologies.[21] He accepted a distinction between this broadly defined realm of science and pseudo-scientific types of thought (such as geomancy), which in his eyes were characterized by ideas that were not experientially derived. More broadly, he followed nineteenth-century precedents and distinguished science from other spheres of human culture, such as art, religion and philosophy; but, as a follower of Whitehead and Collingwood, he saw these spheres as organically linked and as exerting mutual influences. His distinction between 'traditional' and 'modern' forms of science was based on the premises that science evolved through time towards greater complexity and objectivity, and additionally that the modern structuring of natural knowledge in mathematical terms makes modern science cross-culturally accessible. The traditional–modern distinction can be seen as a factor delimiting the time-frame of the *SSC* project, which generally tends to close its treatment with the emergence of modern science and the employment of the Jesuit missionaries by the imperial court at Beijing.

East–West Transmissions and the Evolution of Global Science

Following on from Sarton, and working contemporaneously with Lynn White, Jr., who was set on a similar course, Needham elaborated what amounted to a global approach to the history of science. This was no doubt facilitated by his personal experience, his relational philosophical outlook, his ecumenical religious vision and his political

[21] Summing up this view, he wrote: 'One cannot separate science from technology, pure science from applied science—the two intertwine inextricably' (see Needham [1973b], p. 3). Though he consistently held to this position over the following decades in carrying out his research, it can be noted that it did not keep him from considering scientific ideas or technologies in their own right.

internationalism. However, from the outset the force of the concrete historical evidence was for him overwhelming. As a self-professed Baconian, he found the claim that China had no science before the West arrived untenable. Concluding that that claim was a sign of Western ignorance and arrogance, he took as emblematic of his position the assertion of Francis Bacon himself that paper and printing, gunpowder and the magnetic compass were innovations which showed that the modern world was superior to the ancient. Needham could marshal solid evidence to show that these were first devised by the Chinese.

Summarizing his case in the 1964 essay, 'Science and China's Influence on the World', he noted that gunpowder had arisen from the 'systematic if obscure' investigations of Daoist alchemists, and that there had been discussions of the declination of the compass for centuries in China before Europeans showed awareness even of magnetic polarity.[22] In the same essay he went on to indicate a series of other discoveries and inventions of social and historical importance in which Chinese precedence could be documented. These included astronomical apparatuses incorporating the system of equatorial coordinates and, derived from this, mechanical clockwork using a linkwork escapement mechanism; devices for increasing the efficiency of animal power like the stirrup, and the equine breast-strap and collar harnesses; mechanical devices such as the crank, the piston rod and double-acting piston bellows all developed in metallurgical processing; the driving belt (perhaps developed first in India) and the chain-drive; the mastery of cast iron a millennium and a half before it was known in the West, and its use not only for directly consumed manufactures, but also in such innovative engineering works as chain suspension bridges, the segmental-arch bridge and deep-drill machinery; and, finally, nautical inventions such as the sternpost rudder, the paddle-wheel boat, and watertight compartments in ships. If this list is impressive, it is only the tip of the iceberg of what he suggested in the volumes of the *Science and Civilization in China* series.

The evidence for Chinese priority of discovery and invention was linked from the outset in Needham's work to several other issues. In the first place, the evidence of non-Western priority required examination of the effect of social, economic, and ideological

[22] Needham (1946), pp. 68, 73.

institutions on the development of science and technology. One of Needham's fonder notions about Chinese science (one he developed in the context of the reawakening of interest in Daoism in French sinological circles after the Second World War) was that of the various Chinese schools of philosophy and religion, the Daoist tradition had been the most conducive to the investigation of natural processes.[23] On the other hand, the centralization of the imperial Chinese state and the concentration of economic demand in an extensive, politically unified market sustained iron foundries large enough to make cast iron production technically possible and economically feasible. Similarly, the Confucian state's requirements for revenue and political order furnished material and ideological stimulus to new techniques in agricultural and textile production. Dissemination of such techniques was carried out by successive dynasties, for instance, through distribution of agricultural treatises like the *Qi Min Yao Shu*, which contributed to what Francesca Bray has called the Song Dynasty 'green revolution' in south China.[24]

Secondly, the question of the nature of the social relations of science in traditional China was for Needham never far from the comparative problem of why modern science developed only in Europe. By 1964, having found so many cases of Chinese priority, he had formulated a new comparative question: why was it that China, which for fifteen hundred years had been more efficient at applying knowledge about nature to human benefit, failed to develop modern science?[25] The implicit simplifying assumption that under pre-modern conditions a capacity for technological innovation implied potential for breakthroughs in theoretical science is admittedly problematic and in need of further discussion. Suffice it here to say that Needham's fundamental answer to the question of why China did not develop modern science seems to have always remained similar to the one he adopted in the 1930s from Chi Ch'ao-Ting (Ji Zhaoding), Wu Ta-K'un (Wu Dakun) and K.A. Wittfogel.[26] That is, the social and

[23] See *SCC*, vol. 2, sections 10 and 16.

[24] *SCC*, vol. 6, pt. 2, pp. 597ff.

[25] See A.C. Graham in Nakayama and Sivin (1973), p. 46.

[26] Wittfogel is best known for his bitterly anti-Marxist *Oriental Despotism* written in the US during the McCarthy period. However, in Weimar Germany he had been the German Communist Party's authority on China, and it was his writings from

economic conditions shaped by China's specific geography and hydrological circumstances were not conducive to it.[27] In the 1960s he nevertheless came to feel that this argument needed further elaboration or rethinking, for which he engaged collaborators to explore the issues further. Yet in my experience he always sought to retain his original basic position in some form. He thus saw traditional China as essentially dominated by bureaucratic feudalism which could promote certain types of technological development and certain traditional sciences, but which was a hindrance to the emergence of modern science. The latter seemed to go hand in hand with the growth of capitalism and the political rise of the bourgeoisie, with its tendency to analyse the world in terms of a single mathematized standard of value.[28] As he summed it up,

> Apparently a mercantile culture alone was able to do what agrarian bureaucratic civilisation could not—bring to fusion point the formerly separated disciplines of mathematics and nature knowledge.[29]

The extent to which that interpretation needs revision, in the light of recent, more conjunctural analyses of the trans-Eurasian nature of the shift to capitalist hegemony, is a subject on which discussion continues and has intensified.[30] Aspects of it are explored by Immanuel Wallerstein, Timothy Brook and myself in a forthcoming volume that brings together our contributions originally written for *SCC*.

Thirdly, for Needham the establishment of scientific and technical priority always raised the further question of possible transmission of a given invention or discovery to other societies. This was one of his most prominent thematic interests,[31] one which served to link his

---

that period, including his *Wirtschaft und Gesellschaft Chinas* (1931), that influenced Needham. In those works he gave primacy to Marx's notion of an Asiatic mode of production, just as that notion was being officially condemned in the Soviet Union.

[27] As stated already in 1944: see Needham (1946), p. 150.

[28] Probably the best statement of this comes in the conclusion to the mathematics section of *SCC*, vol. 3, pp. 150–68.

[29] *SCC*, vol. 3, p. 168; similarly, p. 166.

[30] See, for example, Abu Lughod's 1989 study of the late medieval Eurasian trading oecumene, the recent article by Flynn and Giráldez (1995) on China and the early modern silver trade, and A.G. Frank's new book (1998) arguing that Asia was the centre of the early modern world economy.

[31] On importance of thematic commitments in research programmes, see Holton (1973), especially chs 1–4.

empirical research with his philosophical vision of the fundamental universality and continuity of all science. As he put it in a 1970 essay reflecting on his own approach, in words reminiscent of Sarton:

> My collaborators and I have all along assumed that there is only one unitary science of Nature, approached more or less closely, built up more or less successfully and continuously, by various groups of mankind from time to time. This means that one can expect to trace an absolute continuity between the first beginnings of astronomy and medicine in ancient Babylonia, through the advancing natural knowledge of medieval China, India and the classical world, to the break-through of late Renaissance Europe when, as has been said, the most effective method of discovery was itself discovered.[32]

Already in the first volume of *SCC* he affirmed confidently that in the fifteen hundred years prior to the Renaissance in Europe the overwhelming direction of technology transfer was from East to West.[33] Of special interest to him were the diffusions which may have contributed to the modern scientific revolution and later the industrial revolution in Europe. Here again the inventions mentioned by Francis Bacon—gunpowder, paper and printing, and the compass—were emblems for him, but in addition to these he argued that plausible grounds existed for identifying or at least hypothesizing many other transmissions from China to the West. These included several innovations such as equatorial astronomy, the stirrup, efficient harnesses, the crossbow, the gunpowder-ignited bombard and the gun piston-bellows, cast iron, and the sternpost rudder. He also posited various instances of stimulus-transmissions: the decimal place-value system, silk-reeling machinery, the draw-loom, canal-lock gates, smallpox immunization, and, of course, porcelain.

To lend conceptual rigour to his discussion of transmissions, he drew on anthropology, particularly the works of V. Gordon Childe and A.L. Kroeber. Following them, he laid out a theoretical scheme for calculating the likelihood of diffusion in a given case according to a number of variables, including the degree of complexity of an innovation and the time-lag in its appearance in various societies. By this means he envisaged a variety of scenarios: (a) direct borrowing of a technology by one society from the material practice of another (as

[32] Needham (1976a), p. 282; similarly *SCC*, vol. 5, pt 4, p. xxxv.

[33] *SSC*, vol. 1, pp. 240–4.

in the case of Persian sericulture); (b) indirect transmission via another society (as probable with European reception of gunpowder); (c) 'stimulus diffusion', a notion developed by Kroeber for the situation in which word of an invention spurs a similar development in another society with no concrete experience of the original innovation; and (d) independent invention.[34]

Assessments and Critiques

Since the 1960s, Needham's project has been widely hailed as a monument of twentieth-century historical scholarship. Perry Anderson, a judicious and critical reader, seems to express the view of many when he refers to *SCC* as 'a work whose scope is without any precedent in modern historiography'.[35] The series remains a monument to the comparative historiography of science, providing the fullest documentary account of the Chinese record for those without mastery of Chinese original sources. Citation figures indicate that it is a valued and frequently used tool for scholars working in disciplines throughout the humanities, sciences, and social sciences. The general opinion in scholarly circles, if I am not mistaken, is that Needham's discussions can be relied upon to be erudite, intelligent and stimulating—in short, essential points of reference—even when they are in need of revision.

The need for revision is hardly surprising, since the first sections of *SCC* were composed in the late 1940s and 1950s. Some sections of the series—mathematics, for instance—have been superseded by later work, much of it stimulated by Needham's contribution. Other sections, such as those on engineering subjects, still stand up well, though this may be due in part to the unfortunate fact that fewer scholars have devoted themselves to working on those subjects. In the following sections of this essay I consider some aspects of Needham's work which have been criticized or which one is advised to treat with caution. It is a sign of his seminal role in encouraging and organizing scholarship that many criticisms of his work and reservations about it have been formulated by historians of Chinese science, technology, and medicine who have worked with, and been influenced by, him.

[34] *SCC*, vol. 1, pp. 225ff.

[35] Anderson (1974), p. 541.

One important reservation strongly expressed by Nathan Sivin, and taken over among others by Francesca Bray and Georges Métailié, concerns Needham's use of modern scientific disciplinary categories to classify and analyse traditional Chinese thinking about nature.[36] This conceptual framework is built structurally into the *SCC* project, which is organized according to an essentially positivist ordering of the sciences, starting with mathematics and astronomy, and then proceeding through 'physical', 'chemical' and 'biological' subjects. Technological fields like engineering, metallurgy and textile machineries are subsumed as applied science under physics and chemistry, while agriculture and medicine are subsumed under biology. This ordering of disciplines reflects Needham's view that the valid insights of the traditional sciences are or can be synthesized in modern science and reach a completion in it, if only a provisional one. As a mode of presenting an encyclopaedic work, this format had the advantage of being accessible to readers not specialized in traditional Chinese culture. However, as an analytical tool it is open to the serious complaint that it tends to obscure and/or seriously distort the ways in which Chinese thinkers in the past actually understood and valued the world, and likewise the ways in which various types of thinking related to one another in different eras.[37] Moreover, the difficulties with this format were aggravated by Needham's related tendency to identify traditional Chinese insights as 'anticipations' of modern scientific ideas,[38] a tendency which followed from the positivist conception which sees science as solving problems posed uniformly by nature. In my experience Needham was never able to admit the strength of these objections, for his primary reaction to them was always dominated by anxiety that accepting the notion of different constellations of knowledge would lead to the Spenglerian notion of closed and irreconcilable civilizational units, and hence to historical

[36] A recent statement of Bray's criticisms is found in the introduction to Bray (1997).

[37] An early formulation of this criticism is found in Sivin's preface to Nakayama and Sivin (eds) (1973), pp. xiii–xix and xxvii–viii; see also Sivin (ed.) (1977), p. xiv. A related reservation concerns the term 'Chinese science', since pre-modern Chinese had no unifying concept comparable to the Latin *scientia* which embraced the various approaches to understanding the world.

[38] Ibid.

pessimism.[39] This was unfortunate, for it forestalled the exploration of ways to improve comparative research and cross-cultural understanding.

A related, and now broadly accepted criticism, also put forward in the 1970s by A.C. Graham and Sivin, concerns 'the Needham problem'. Scholars who make this criticism point out that this approach contains the implicit assumption that the European experience in social development and in science was normative for global development. According to this view, Needham's focus on what Chinese science or society lacked is a negative feature of his approach that diverts attention from reconstructing the actual ideas and practices that did occupy people in pre-modern China.[40] From the early 1970s, Sivin urged historians of Chinese science, technology, and medicine to turn towards a number of different tasks, namely identifying the indigenous elements in Chinese science; identifying the sources of new ideas, techniques, and problems within Chinese society; clarifying the demarcation, within China, between science and other forms of culture; and identifying the forces that shape and constrain new developments in scientific and technical fields.[41] It can be argued that these were actually tasks to which Needham himself had given considerable attention in his own writings. The point at issue would thus seem largely to be a matter of emphasis and priorities—apart, that is, from the issue of how to treat the questions of why capitalism and modern science did not develop in China, and what obstructed their emergence. Needham continued to regard these as essential questions until the end of his life, whereas others found them pointless or misleading.[42] In the early 1980s, Sivin took the argument forward by showing that a modern-style mathematization of astronomy was carried out by seventeenth- and eighteenth-century Chinese

[39] See, for example, *SCC*, vol. 5, pt 4, pp. xxxvi–xxxvii; and Needham (1976a), pp. 283–4.

[40] Graham in Nakayama and Sivin (eds)(1973), especially pp. 52–3; Sivin (1982), p. 66.

[41] Sivin (ed.) (1977), p. xv. Sivin (1988) surveys current research projects on the history of Chinese science and medicine, and considers how they elucidate these issues.

[42] Nathan Sivin's position that it is a waste of time to discuss why the equivalent of the Western scientific revolution never developed in China is discussed by Needham in *SCC*, vol. 5, pt 4, p. xxxvii.

astronomers, though this did not have the revolutionary cultural significance of the Western scientific revolution.[43]

However, it can be noted in Needham's defence that the question of impediments, particularly to the emergence of capitalism, continues to occupy scholars. Those who consider this relevant vary in ideological inclination, but tend to have interests in social and economic history.[44] It is perhaps a sign of Needham's fecundity of scholarship that the issues he raised resonate across a variety of fields.

Reservations about focusing on the 'Needham problem' do however tie in with the sentiment, increasingly shared by some (though not all) historians of China, that the conventional depiction of Confucianism as antithetical to the investigation of nature is untenable as a generalization. Needham's position is vulnerable to this objection because he singled out Daoism as the primary source of traditional Chinese interest in nature, but more especially because, like Chinese nationalists, he tended to portray late imperial institutional Confucianism as an important factor that inhibited social and technical change.[45] At the same time, however, it should be recognized that Needham repeatedly drew attention to the fact that the Confucian state long promoted the official sciences, particularly astronomy and to a lesser extent medicine. He also appreciated Song dynasty neo-Confucian thinkers for their synthetic holism which he saw as anticipating, or perhaps even influencing, the emergence of what he took to be the typically organicist outlook of modern science.[46]

[43] Sivin (1982). The significance of the Chinese developments brought to light by Sivin is dismissed by Huff (1993), but rather too hastily.

[44] Elvin's *Pattern of the Chinese Past* (1973) was written with Needham in mind. Recent discussions dealing with the same question by historians of China unconnected with the Needham group include: Rozman (ed.) (1981) and Shue (1995), ch. 3. Among treatments by comparativists, see Nelson (1981), Jones (1987), Blaut (1991), Huff (1993) and Goody (1996).

[45] Nakayama (1973), p. 38; Spence (1992), p. 147.

[46] *SCC*, vol. 2, especially section 16, which suggests as a hypothesis that the organicist/dialectical orientation might have been introduced into the Western tradition from China via the Jesuits and Leibniz. Follow-up work on this suggestion has indicated that Leibniz formulated his key ideas with a view to new developments in dynamics, and before entering into correspondence with the Jesuits. In any case one can argue that traditions of holism and dialectics existed in the West from much earlier periods.

Criticizing Needham from a different direction, several authors have recently stressed what they see as deep-seated incompatabilities between traditional Chinese institutions and philosophies, or even Chinese culture as a whole, and the modern scientific outlook. For Needham, as for Max Weber, the proper formulation of the problem meant determining comparatively whether various traditional institutions and forms of culture in China constituted greater obstacles to the emergence of modern rationality than did their equivalents in Europe. By Needham's reckoning, it was China's social, economic, and political characteristics (i.e. 'external factors') which determined that China did not develop modern science, but he deemed the cultural and ideological obstacles it faced no greater than those found in Europe. An opposing position is inherent in Jacques Gernet's recent writings,[47] but is perhaps most clearly stated in Derk Bodde's *Chinese Thought, Society, and Science*, a work originally written to serve as the penultimate section of the *SCC* series. Bodde parted company with Needham on two major points: first, he argued that neither Confucianism nor Daoism were especially conducive to the scientific investigation of nature.[48] His criticism of Needham's appreciation of Daoism may well be telling, and seems to be fairly widely shared, though Needham himself remained unrepentant. Secondly, Bodde maintained that written Chinese itself was imbued with Confucian values and therefore constituted a formidable obstacle to the emergence of modern science. On this point, Bodde himself is vulnerable to serious criticism, because he based his conclusion exclusively on official and literary texts and failed to analyse traditional Chinese writings devoted to naturalistic and technical subjects.[49]

Finally, serious reservations have been voiced about Needham's tendency to look for technological diffusion and to embrace diffusionary hypotheses. In this regard, one should distinguish between critics who object in principle to diffusionism and those who fault Needham's suggestions about particular transmissions. In the first category, Perry Anderson argues that the mutual ignorance of one

[47] See, e.g., his contribution to Jami and Delahaye (eds) (1993), pp. 231–41.

[48] Bodde (1991), ch. 7. Furth (1994) similarly criticizes the depiction of Daoism as favouring sexual equality, which Needham borrowed from Van Gulik.

[49] Chemla (1992); Kuhn (1992) develops this point by illustrating what he terms a 'pro-technological' school of thought before the Qing dynasty.

another's social institutions suggests that frequent technical communication between pre-modern Europe and China was unlikely.[50] Though at first arresting, this objection would seem to be answered adequately by the evidence marshalled by Needham and others showing that material artefacts were historically more subject to transfer than were abstract or organizational ideas.[51] In certain cases there is good evidence that more conceptual strategies for investigating nature could travel as well, as in the case of the elixir of Chinese alchemists,[52] or Arabic mathematical astronomy. A second type of theoretical objection has been posed by the Swedish historian Per Sorböm, who identifies a number of rhetorical inflations by Needham and Lynn White regarding the westward transmission of the compass. He argues against diffusionist explanations on the grounds that anthropology has abandoned them and that they underestimate human creativity.[53] A recent influential formulation of this objection is advanced by the geographer J.M. Blaut, who dismisses arguments for cross-cultural diffusion as colonialist constructs.[54] For the historian, this position seems to be far too dismissive of the evidence supporting the view that societies do successfully borrow from (or impose on) one another. Disseminations of such things as grain agriculture, the alphabet, Indian numerals and railroads constitute boundary cases that indicate the continued need for taking diffusion seriously.

It can, however, be readily conceded that Needham did at times go beyond his evidence in claiming transmissions. One example is David Landes' much-cited criticism of the treatment by Needham, Wang Ling and Derek de Solla Price of the monumental mechanical water-clock devised by Su Song at the Southern Song capital of Kaifeng.[55] Though Landes praises their reconstruction as one that

[50] Anderson (1974), p. 548.

[51] A point first made in *SCC*, vol. 1, pp. 238–40, and revisited regularly in later volumes.

[52] On the elixir, see *SCC*, vol. 5, pt 4.

[53] Sorböm (1984).

[54] Blaut (1993), ch. 1. In fact Blaut is enthusiastic about Needham's contributions to weakening Eurocentrism and ignores his diffusionism.

[55] Needham, Wang and de Solla Price (1986), pp. 187ff.

ranks 'among the great detective coups of historical scholarship', he follows Cipolla in rejecting their speculations about a possible stimulus diffusion to the West as unfounded and far-fetched.[56] This seems prudent. Nevertheless, from the point of view of promoting cross-cultural research, one wonders whether Needham's general propensity for thinking about possible forms and routes of diffusion has not had a stimulating role even when his hypotheses have involved speculation. Admittedly, there is a serious problem, and objections are certainly in order, when suggestive hypotheses or theoretical constructs are misrepresented as simple facts.

When such distortions are avoided, however, it seems entirely reasonable to treat diffusion and independent invention on the merits of the evidence. Dieter Kuhn's discussion of the likelihood of stimulus diffusion of the silk-reeling frame from China to Europe is, in this regard, an interesting reminder. Like the author who treated silk machinery in Singer's *History of Technology*, Kuhn judges independent invention unlikely because the essential features of the Italian reeler were the same as those of the Chinese. However, where the earlier author thought the Italians showed adaptive genius in coming up with the smoke-pipe and treadle as improvements, Kuhn argues that both of these features were known in Song China.[57] Thus, the problems of such current large-scale technology transfer not-withstanding, in some cases at least, a fairly systematic diffusion of existing technology does seem to be quite solidly attested to historically.

The Threads of Universality

In its vision of science's universal vocation, as well as in its complexity and nuance, Needham's conception of science is reminiscent of Leibniz, whose optimism and subtlety Voltaire caricatured in the figure of Dr Pangloss in *Candide*.[58] One of the difficulties in appreciating

[56] Landes (1983), pp. 21–5. The citation is from p. 21.

[57] *SCC*, vol.5, pt 9, p. 422.

[58] Following the seminal studies of Couturat and Bertrand Russell, the first half of this century saw a revival of historical and scientific interest in Leibniz, who was also the most prominent early Enlightenment sinophile. In the first years of the *SCC* project, Needham drew on the recent Leibniz literature in analysing the history of Sino-European scientific communications.

Needham's conception of science lies in his tendency to knot together ideas about various features of science with the somewhat rosy-sounding epithets of 'universality' and 'oecumenism'. At times this synthesizing manoeuvre seems to obscure what he was actually referring to. I will try to unravel his argument in order to assess the strength of its individual strands and the integrity of the fabric as a whole.

Sometimes Needham used the terms 'universal' and 'oecumenical' to refer to what he saw as the progressive continuity of scientific endeavour. A strong statement of this view is given in the long quotation about the 'unitary nature of science' cited above.[59] Let us label as proposition (1) Needham's notion that science in general is both continuous in nature and cumulative over the long term. At other times he employed the same terms to designate a variety of distinct qualities which he attributed specifically to modern science. Let us attempt to state the qualities he ascribed to modern science in the following propositions of type (2):

(2a) Modern science is formulated using the 'universal' language of mathematics. To use Galileo's words, 'the book [of the universe] is written in the language of mathematics',[60] and modern science comprehends this.

(2b) Modern scientific knowledge possesses a uniform validity because it reflects nature, which is itself fundamentally uniform. Because of this uniform validity, modern scientific knowledge is in principle applicable anywhere in the universe, irrespective of its place of initial discovery.[61]

(2c) Modern scientific method is the model for arriving at scientific knowledge through the framing of mathematized hypotheses and through testing them by means of observation and relentless experimentation.[62]

(2d) Once formulated in the seventeenth century, modern scientific method took on a definitive character in the sense that it continues to retain its fundamental validity thereafter.[63]

[59] Needham (1976a), p. 282; and above p. 44.

[60] Cited in *SCC*, vol. 3, p. 158.

[61] *SCC*, vol. 3, p. 157.

[62] See, e.g., *SCC*, vol. 3, section 19(k), pp. 150–68; Needham (1969a), p. 117.

[63] Cf. *SCC*, vol. 3, p. 156: '... the experimental method, which appeared in almost perfect form in Galileo, and which led to all the developments of modern science and technology ...'. Also Needham (1969b), p. 82.

(2e) Modern science possesses an inherent cross-cultural comprehensibility, which is attributable to its being formulated in the 'universal' language of mathematics, rather than in culturally specific categories (like *yin-yang*, *karma*, and the Galenic 'four humours').[64]

(2f) Especially at its origin, but also to some extent later on, modern science involves rational synthesis or fusion of various bodies of knowledge developed by different, or indeed all, peoples and cultures, as represented in the image of rivers flowing to the sea.[65]

(2g) Scientific knowledge has grown, and will continue to do so, in tandem with advances in the scope and complexity of human society, which is becoming ever more highly integrated and tending towards the formation of a world community.

Of these propositions, I take (2a) and (2b) to be assertions of objectivity, that is, of the conformity of modern scientific knowledge with the mathematical structure of reality. Proposition (2a) may be taken as a fundamental axiom of Needham's conception, comparable to Kelvin's dictum that 'mathematics is the only true metaphysics'.[66] Propositions (2c) and (2d) are claims that the 'modern' *method* is qualitatively more efficient for revealing the mathematical structure of reality than were or are other techniques, such as those used for investigating nature before the seventeenth century. Proposition (2e) is a principle of epistemological egalitarianism, which asserts that knowledge is in theory communicable across cultural borders and that persons of any cultural background are in principle capable of utilizing it. This in turn implies that scientific activity itself tends to become a form of global social integration (cf. 2g). Proposition (2f) represents a genealogical claim, to the effect that the growth of modern science has involved the integration of insights arrived at by a multiplicity of 'sources'. This implies a moral–political claim on behalf of all peoples for access to scientific–technical resources and for inclusion in the activities of the modern scientific community.[67]

[64] Needham (1969b), pp. 12–13, 82.

[65] Needham (1966), p. 397; Needham (1969b), p. 16; (1973a), p. 20; and Needham (1973b), p. 3.

[66] Kelvin cited in d'Abro (1950), p. 343.

[67] 'Until it had been universalized by its fusion with mathematics, natural science could not be the common property of mankind' (Needham [1969a], p. 15).

Let us consider some of the connections among these statements and some of the issues they raise. Mathematics and its integration into the investigation of nature is central to propositions (2a), (2b), (2c) and (2d) which represent attempts to designate the specificity of modern science. Proposition (1) implies that a degree of objectivity is attributable not only to modern science, but also to ancient and medieval science.[68] What differentiates modern science is that it tends to a higher degree of objectivity largely because of its mathematical language and its incorporation of experimentation, as stated in propositions (2a) and (2c). Since much of the effectiveness of experimentation resides in the mathematization of hypotheses, Needham's distinction between 'traditional' and 'modern' science hinges on mathematization of the latter.

The depiction of a qualitative difference between 'traditional' and 'modern' forms of science raises the question of whether Needham's conception is whiggish. His response would no doubt have been that science did indeed undergo a major transformation in the early modern period,[69] and explaining that change cannot easily be dismissed as a false problem. Though this appears plausible, his view (2d) that the scientific method was definitively established with the formulation of Galileo's hypothetico-deductive model seems overly rigid. On my reading his position here was a primarily defensive stance adopted to counter dismissals of science. But can that end not be secured by other means? Though Needham's position is difficult to refute due to its qualitative and abstract nature, one wonders if it does not run the risk of obscuring important changes in the structure of scientific thought and practice that have occurred since the seventeenth century. As Forbes and Dijksterhuis remarked: 'The relationship between mathematics and natural science [was] not always the same to Newton as it is to us.'[70]

Needham's thesis that mathematization is the key to distinguishing between modern and pre-modern sciences leads moreover to serious problems of historical interpretation on at least two counts. In the first place, it can be demonstrated that mathematics was integrated

[68] Needham (1966), p. 418. See also the citation at the opening of the next section.

[69] Or, as he sometimes put it, from the late medieval period.

[70] Forbes and Dijksterhuis (1963), vol. 1, p. 259.

into the analysis of certain types of natural knowledge in several civilizations before the seventeenth century. Sivin indicates that this was the case in imperial China for harmonics and astronomy.[71] The objection is similarly illustrated by Chinese cartographic traditions going back to the Tang, by the physics of Robert Grosseteste in medieval Oxford, and by Grosseteste's predecessors such as Al-Hazan in the Arabic world. Conversely, it can be argued that Western disciplines such as geology, botany and zoology remained predominantly qualitative long after the moment when Needham was disposed to consider them 'modern'.[72] Secondly, because Needham's tradition–modern dichotomy was tied to the premise that modern science emerges first with Western capitalism, he may have overestimated the uniqueness of modern Western scientific development and overlooked similar scientific transformations which took place elsewhere along what have been considered until recently to be 'classical' seventeenth-century lines. Roshdi Rashed and Georges Saliba have thus argued strongly that the optics of Al-Hazan in the eleventh century already marked an integration of mathematics and physics, which served as the basis for a new and rigorous notion of experimental proof, and thereafter for precise physical modelling as a means of indirect testing.[73] If these points are accepted, the case Needham made requires revision. Several options present themselves. The fundamental linkage of mathematicized science and early modern capitalism might be called into question and that linkage might be attenuated so that capitalism would play a facilitating but not a generative role in the growth of the mathematized science. Alternatively, a degree of capitalist development can be

[71] See Sivin's preface to Nakayama and Sivin (1973), pp. xxiv–xxvi.

[72] Needham (1966) held that distinct disciplines were 'modernized' at successive historical moments, with time-lags calculable as a function of the 'level of complexity' of respective subject matters. Thus, modern astronomy and physics were elaborated in the seventeenth century, whereas botany was elevated to that level only in the eighteenth, and biology in the nineteenth, while in his opinion the 'fusion' of traditional and modern medicine is still underway. However, the depiction in that article, though a brilliant endeavour at conceptualizing the process, seems to underestimate the period required for 'fusion' of traditional and modern forms of knowledge. For instance, ancient astronomical records are still being studied by archaeo-astronomers. That process of interaction, thus, cannot be considered to have come to an end in the early modern period.

[73] For example, Rashed (1984), especially pp. 314–16.

recognized as having facilitated the new form of science in culture-regions other than the European, particularly in the Islamic countries. This latter option would appear compatible with a Braudelian notion of capitalism conceived not as social totality or a discrete stage of history, but as a type of relationship or set of institutions present in different eras. Such revisions, it can be remarked, might still remain compatible with Needham's view of science as accessible in principle across status and cultural divides.

Needham's and Popper's Critiques of Verificationism

It might be observed that Needham's affirmations of the objectivity and uniformity of modern science as well as his notion of scientific method (2a, 2b and 2c) betray a positivist or verificationist theory of science. His theoretical pronouncements are pretty consistent in this respect, though even in these he sometimes registered qualifications.[74] In any case, his historical practice was often more flexible and nuanced than his programmatic statements. Thus, despite his proudly proclaimed Baconianism and his fascination with scientific precedents and transmissions, it is difficult to label him an inductivist because of his consistent interest in thematic linkages between religious/philosophical culture and scientific/technical activities. Nor did he fail either to realize or to caution his readers that a scientific theories are only ever approximations to truth, and must always be considered open to fundamental challenge.[75] Similarly, as a practising scientist, he was quite at home in describing the aim of scientific experimentation as either the verification or the invalidation of hypotheses.[76]

This said, a high degree of verificationism seems integral to his overall vision. It was also very much in tune with the literature on the philosophy of science in the period when he was formulating that vision. By my reckoning, most of the elements of his conception of science were clearly in place before 1940. Building on those earlier ideas, he further elaborated certain views about the 'oecumenical'

[74] As implied in his self-characterization as 'more than half a Baconian' (Crombie [ed.] [1963], p. 173).

[75] *SCC*, vol. 3, p. 165.

[76] See, e.g., *SCC*, vol. 3, p. 165.

nature of science during the war, at UNESCO, and then in the 1950s and early 1960s. My impression, however, is that, after the 1956 publication of *SCC* volume 2, which dealt with the history of philosophy, he rarely felt there was much of concrete value for his historical work on pre-modern science, technology and medicine to be gained from new positions being advanced in the philosophy of science.[77]

The philosophical objections to viewing science as a process of verification were not taken squarely onto the English-speaking world's intellectual agenda until Popper published *The Poverty of Historicism* and the English translation of his *Logic of Scientific Discovery*, in 1957 and 1959 respectively.[78] That Needham as a historian had not conceived his project in terms of Popper's alternative to the then standard verificationism is thus hardly a surprise, given the timing. It is even less surprising, their differences over Marxism aside, when one considers that among the main philosophical positions attacked by Popper were the holism and evolutionary view of history which Needham's circle championed, Needham's biochemical collaborator Conrad Waddington being one of the particular butts of criticism in *The Poverty of Historicism.*[79] Thus, Popper's attacks on holism and social evolutionism together with his rejection of the notion of a

[77] Needham's orientation was at least partly due to his empirical attitude towards the study of history—what he called the 'brass tacks' approach. During the 1961 symposium on the history of science held at Oxford, at which Georges Canguilhem and Thomas Kuhn both presented papers (this was one of Kuhn's first presentation of his model of paradigms), Needham declared himself somewhat 'puzzled' by the terms 'paradigm' and 'conceptual model', which he seemed to consider too vague and non-concrete (Crombie [ed.][1963], p. 611).

[78] Though the essays that made up *The Poverty of Historicism* were first published in article form in 1944–5, Popper's falsificationism came to be widely discussed only after their publication in book form. See Passmore (1968), p. 406; Carr (1986) p. 91.

[79] See Popper (1957), e.g. pp. 72, 106–7, 114–15 where Waddington and Bernal are (reasonably enough) associated with the ideas of J.S. Mill, Engels, Herbert Spencer, Whitehead and Smuts, all of whom were among Needham's avowed intellectual forebears.

Admittedly, Needham might have been disposed to reject Popper's views because of Popper's critique of Marxism. But to assume this would run the risk of underestimating Needham's demonstrated ability to adopt interesting ideas from authors whose general approach differed from his.

specifically modern scientific method, his resistance to recognizing the constructive importance of traditions in the history of science, and his hostility to Marxist thought indicate clearly that in a variety of ways Needham and he were on different wavelengths.

It is nevertheless of interest that there were also significant convergences.[80] Most obviously, they were both concerned with differentiating science from other forms of consciousness, and they both saw progress as central to scientific endeavour. While the political contrast between them was significant, Popper was willing to accept certain of Marx's ideas, though he opposed what he considered 'Vulgar Marxism'.[81] In addition, when Popper attacked the evolutionist view of history, he specified that he was attacking it as scientistic.[82] Likewise, his discussion of holism distinguished between a type that claimed to be able to attain total knowledge, which he rejected, and another type which sought to distinguish levels of organization, which he considered acceptable.[83] Needham's conception of coexistent moulds of experience, his interpretation of holism as a theory of 'levels of organization', and his own explicit rejection of scientism, meant that the Popperian criticisms could be applied to him only with some difficulty.

Curiously, although Popper denied that there was a specifically 'modern' scientific method, his identification of 'trial-and-error' as a broader scientific methodology seems consistent with Needham's view that science can be found outside of the West and/or the modern period (see proposition 1). Popper's view might thus be enlisted to provide support for Needham against those who have claimed that science is a purely Western phenomenon.[84] Similarly, Needham's aim of countering attacks on the epistemological status of science could have been furthered by Popper's broader notion of testability as the criterion of all scientific validity, since Popper's approach underwrote

[80] Needham himself seemed to be affirming this when he once remarked in the late 1980s that he wished he had time to study Popper's works in depth.

[81] Popper (1969), p. 342.

[82] Popper (1957), pp. 105ff.

[83] Ibid., pp. 76ff.

[84] Carl Sagan (1995) uses an essentially Popperian definition of science to counter Eurocentrism, but mistakenly opposes it to Needham's position (p. 312) because he ignores Needham's distinction between modern and traditional science.

the integrity of science which Needham was so committed to upholding. Finally, and again despite political differences, there is a curious convergence between Popper's call for a reconciliation of rationalism and traditionalism and Needham's views concerning, on the one hand, distinct and coexistent moulds of experience, and, on the other, his conceptualization of the relations between traditional 'sciences' and modern science[85].

Needham's appreciation of science as a process of positive acquisition tied in with his views that science includes both pure and applied domains, and that there are important continuities between 'science' and 'technology', which are inextricably though not necessarily simplistically linked in their historical development.[86] This approach not only lent the notion of Chinese 'science' a degree of manifest concreteness, but also provided Needham with intellectual space for making historical connections between science and technology. On the other hand, while Popper's notion of science was elegant and theoretically powerful, it singled out falsification as the essence of science, a position which, in Gellner's phrase, bears 'only a rather distant relationship to the concrete activity actually recognised as science in our society'.[87] While this does not diminish the strength of his analysis, it can be taken to imply that something more is needed if one is to comprehend the actual historical development of science and technology. Needham's less austere definition of science allowed him to develop a broader and more concrete appreciation of the material as well as the intellectual culture of science, and it enabled him to focus on the complex forms of interaction between the legacy of previous students of nature and the cutting edge of scientific and technical progress.[88] While his verificationism might be considered philosophically messy, it had advantages for guiding historical research.

[85] Popper (1969), p. 376.

[86] Agassi (1987) observes that Popper too subscribed personally to positivistic and realist views, though he declared these to be private, since he considered realism an irrefutable doctrine (pp. 300–2).

[87] Gellner (1992), p. 109.

[88] Agassi (1987), p. 335, notes that in the 1970s Popper came to revise his radically discontinuous vision of science and to accept the significance of historical continuities and trends.

Universality of Science and the Responsibility of Scientists

As stated in proposition (2e) above, one of the reasons Needham called science universal was its capacity to be communicated across cultural divides. This is an affirmation in principle of the epistemological equality of persons from all cultural backgrounds, and Needham posed it explicitly in opposition to claims of innate Western mental superiority. In this sense, I think, it remains significant, valid, and relevant to current politics. If taken in isolation, however, it is liable to be construed as excessively abstract, perhaps as naive, scientistic, and controversial in ways Needham might have found rather perplexing. To conform to his original vision, this affirmation requires: (a) analysis of the obstacles from within particular cultures and societies to the dissemination of modern scientific culture; and (b) recognition of the role which the international political system and the global division of labour plays both in sustaining institutional barriers to the dissemination of scientific and technical knowledge, and in perpetuating conditions of life which make it virtually impossible for enormous numbers of people to obtain either a decent standard of living or a basic scientific education.

Important aspects of Needham's understanding of the egalitarian and cross-cultural nature of science (propositions 2e, 2f and 2g above) were effectively shown in the closing passage of his article 'The Roles of Europe and China in the Evolution of Oecumenical Science' (1966). There he wrote:

> The standpoint here adopted assumes that in the investigation of natural phenomena all men are potentially equal, that the oecumenism of modern science embodies a universal language that they can all comprehensibly speak, that the ancient and medieval sciences (though bearing an obvious ethnic stamp) were concerned with the same natural world and could therefore be subsumed into the same oecumenical natural philosophy, and that this has grown, and will continue to grow among men, *pari passu* with the vast growth of organisation and integration in human society, until the coming of the world co-operative commonwealth which will include all peoples as the waters cover the sea.[89]

[89] Needham (1966), pp. 417–18. The last clause is reminiscent of a passage by the seventeenth-century English revolutionary Gerrard Winstanley, of whom

Again this might remind one of Leibniz, whom Russell once described as having two systems of philosophy: one optimistic, fantastic and shallow; the other profound, coherent and logical.[90] Similarly contrasting judgments are sometimes heard of Needham's vision of modern science as 'oecumenical'. Though there probably is some justice to the charge of naivety regarding the beneficence of science, he has left on record clear statements of the dangers of scientism and the oppressive uses of scientific knowledge. His voice was long prominent among those calling attention to the ethical choices facing scientific workers, and he raised it consistently from a standpoint of critical social engagement with the major issues of his day.

A scientific optimist Needham certainly was, but his optimism about science was a component of his fundamental and considered philosophical optimism about humanity and social evolution. This position might be contested as a philosophic predilection, and it is now certainly out of fashion, but one should recognize that for him it constituted a call to committed action, rather than a deterministic justification for abdicating social responsibility. In 1941, he put his view as follows to a meeting of the Social and International Relations Division of the British Association, in the panel on 'Science and the World Mind' chaired by H.G. Wells:

> Summing up ..., we have seen that although physical order is everywhere decreasing, biological and social organisation is everywhere increasing. The human social organisation into which we were born has yet far to go towards its ultimate triumphs. The whole enterprise of science is a manifestation of this social organisation, itself the product of evolution and the guarantee of

---

Needham was particularly fond, and whose outlook he frequently sought to apply to modern circumstances. Winstanley spoke of restrictions on science as follows (cf. [1941], p. 564, cited in Needham [1986], p. 101):

> If the earth were set free from kingly bondage, so that everyone might be sure of a free livelihood, and if this liberty were granted, then many secrets of God and his works in Nature would be made public, which men do nowadays keep secret to get a living by; so that this kingly bondage is the cause of the spreading of ignorance in the earth. But when the Commonwealth's freedom is established, then will knowledge cover the earth as the waters cover the seas, and not until then.

[90] Russell (1984), p. 563. According to Russell, Leibniz's official face was visible in his published writings, while his inner thoughts were revealed by nineteenth- and twentieth-century scholars from his correspondence and unpublished works.

further evolution. Our business is to find out how we may best make ourselves efficacious instruments of the cosmic process as it manifests itself in evolving human society, not caring if we perish in this cause, a holy and living sacrifice ...[91]

If this sentiment now sounds overly dramatic, one might keep in mind that it was expressed in the midst of what was a life-and-death struggle against Nazism, fascism, and militarism, the outcome of which was by no means certain. Identifying the essential thrust of science as the creation of a 'unified world-picture',[92] Needham posited a fundamental contradiction between science and political authoritarianism, and he argued that there were fundamental historical, philosophical and psychological correlations between science and democracy.

In his view, those correlations were hardly such as to absolve scientific workers from taking a social and political stand. The position on the social responsibility of scientists which he articulated in 1935 when criticizing the British scientific establishment was one he adhered to throughout later decades.[93] There he argued against replacing the 'Kingdom of God' concept typical of quietist religion with a scientific 'kingdom-concept of mathematics, equally sterile with respect to human welfare, equally satisfactory to the powers of the world'. He continued,

Science does not exist in a vacuum; scientific discoveries are not made by an inexplicable succession of demiurges sent to us by Heaven; science is, *de facto*, involved with 'particular social and economic theories,' since it exists and has grown up in a particular social and economic structure.[94]

And further,

There are two ways open to the scientific worker at the present time. All the backward pull of tradition urges him to throw in his lot with the existing capitalist order, with its corollaries of nationalism, imperialism, militarism, and, ultimately, fascism. On the other hand, he can adopt the ideals of social justice and of the classless State ...[95]

[91] Needham (1946), p. 219; reprinted in 1976, p. 206.

[92] In this his vision was similar to Einstein's, as discussed in Holton (1996), pp. 160ff.

[93] As expressed by the biologist A.V. Hill, a stalwart of the Royal Society, and J.M. Keynes' brother-in-law.

[94] Needham (1935), p. 420.

[95] Needham (1935), pp. 223–4.

In short, in Needham's view a choice was there for the making, and scientific credentials by themselves were no guarantee that the right course would be chosen.[96] Later, looking back on the struggle of those years and its culmination in the war, he endorsed with approval Sarton's view that the Nazi scientists had allowed themselves to be technically seduced into finding solutions for Hitler's problems.[97]

Although in Needham's view the modern era had separated scientific research from ethical considerations and released it from their constraints—and in that sense he affirmed the 'neutrality' of science—nevertheless the notion of 'value-free' science typical of the modern West did not negate for him the idea that scientists had social and ethical responsibilities. Nor did it erase the reality that modern science and particularly modern technologies posed tremendous dangers for humanity.[98] In his 1976 essay 'History and Human Values',[99] he argued that the progress of science and technology meant that greater and greater problems, and 'all but intolerable ethical choices', were emerging as time went on. Recognizing that 'the control of applied science is probably the greatest single problem for humanity today',[100] he proceeded to enumerate threats inherent in the development and use of nuclear energy, artificial intelligence, genetic engineering, etc. His conclusion was that such dangers confronted humanity with unprecedented ethical problems.[101] This analysis, characteristic of his assessment of the world in the post-war decades, was thus a far cry from any simple scientism. While his outlook was infused with substantial scientific optimism, his expectation of the benefits of science was explicitly conditional on the realization of broader social and political values.

Needham's solutions for the ethical problems raised by scientific and technological research were typically imbued with the values of

[96] See Holton (1993), e.g. p. 181, for a similar recent view.

[97] Needham (1986, typescript), pp. 16–17.

[98] Ibid., p. 17, crediting the point to Nakayama Shigeru, who had written of the notion of 'value-free' science as a form of escapism from responsibility. Similarly, in (1983), p. 21.

[99] Later delivered in revised form in Colombo as the inaugural lecture of the Martin Wickramasinghe Trust.

[100] Needham (1976b), p. 18.

[101] Ibid. pp. 18–21.

the socialist movement and of the religiously oriented ethical systems with which he felt the most personal affinity, particularly Judaeo-Christianity and the non-dualistic traditions of China. Whereas his descriptions of current problems tended often to be concrete and realistic, the long-term solutions he proposed were framed in terms of broad normative principles, often sweepingly visionary in character, as witnessed above. Lack of organizational experience cannot be an adequate explanation for this, since he had such experience in some abundance. I believe that it stemmed instead from his sense of the scale and complexity of the human drama, which in his view was not amenable to precise prediction.

In the last years of his life, despite important victories such as the overthrow of apartheid, Needham saw the values he espoused again coming under severe assault. Though the new attacks have taken on somewhat different forms from those he had dealt with earlier in his career, he had little trouble recognizing in the new tide of authoritarianism, neoconservatism, deregulated multinational capitalism and religious fundamentalism of the 1980s and 1990s the same tendencies towards power-mongering, greed, intolerance and bigotry that he had confronted during the inter-war period and the cold war. In the new conjuncture, as in the earlier ones, he continued to centre his solutions consistently on the promotion of human solidarity, social justice, scientific progress, intellectual tolerance and creativity and peace.

One of his primary desires from the 1930s until the end of his life was the strengthening of the scientific resources and the culture of science in the Third World. This he pursued as a quest for justice and solidarity as well as an intellectual duty.

The aim of strengthening Third World science seems to have been one of the primary forces behind his early conceptualization of 'oecumenical' science. In the early periods of his career he saw Western scientific superiority used both to maintain and to justify Western domination. From that time his social and his scientific commitments together led him to join his energies to those of the Third World, especially Asian, scientists in a dual process of post-colonial reconstruction—on the one hand, bringing to light the technological and scientific histories of their respective civilizations, on the other hand, building Third World research and educational institutions of world-class calibre, which would contribute to enhancing the well-being of the citizenry of each nation.

By the 1960s, a number of Third World countries including India, China and Brazil had succeeded in developing impressive scientific institutions along these lines, a fact which personally thrilled him. Yet there were limits to the extent to which scientific institutions and a cadre of trained scientific and technical personnel could contribute to raising popular living standards, for 'science' was as always embedded in social relations. Scarcity of resources was one problem, though probably not the greatest. Ideological narrow-mindedness or hostility, such as that predominant during the Cultural Revolution in China, or the religious literalism found elsewhere, was another major constraint. Yet another was the subsumption of scientific and technological research institutions within the global power structure. In the real world of the cold war, this entailed the domination of world science and R&D by military priorities.[102] In much of the Third World it meant that development conceived in terms of 'big technology' was provided with scientistic justifications. Citing Abdur Rahman, Needham recognized that in such situations the deployment of technology was limited to the needs of capitalist development and the role of science as an instrument of criticism and enlightenment was being effectively negated.[103]

Aptly assessing the international state of science and technology in the early 1980s, P.N. Agarwala noted:

> The process of technological innovation (differentiated from invention and understood as the creation of new knowledge about production) requires organization and financial resources to make these inventions commercially feasible.' Over 90 per cent of the world expenditure in these fields is concentrated in the developed countries. The industrialization process of the developing countries, which was based on import substitution, established a technical and engineering basis linked to the light consumption industries; and, in these, sophisticated techniques and equipment are to be imported. The process of innovation in developing countries has created an imbalance within the whole scientific and technological system. Innovation is determined by the policies of transnationals that transfer and apply different

[102] Williams (1980), p. x, noted already before the Reagan armaments programmes that in the United States, the country with the highest levels of scientific and R&D resources, 51 per cent of all scientists and engineers were employed in military-related work.

[103] Needham (1986), pp. 23–6.

grades of technology to meet their own needs, and not necessarily those of the developing countries.[104]

This was a situation which Needham was well aware of during his last decades. Unfortunately it did not improve in his final years. An illustrative case seems to be that of the new regime of Intellectual Property Rights which has been enshrined in the 1994 GATT accords.[105] One should be aware that protocols of this complexion, allowing exclusive rights within a global market, were clearly not what Needham had in mind when he affirmed the philosophical proposition that 'science is something which can only be shared in fullest freedom among all the world's peoples'.[106] It is therefore scarcely reasonable to dismiss his vision and scholarship on the grounds that the terminology of the 'universality' of science is open to abuse. A better course would appear to be for people with scientific and technical expertise to join with broader social forces, as Needham and his generation of progressive scientists tried to do, in working for changes that will help fulfil basic human needs through promoting local, national and international solidarity and cooperation.

Concluding Remarks

Needham's work is too rich and variegated for me to presume to give an overall assessment of it. It seems likely that it will continue to evoke divergent responses from different quarters. Some might consider it compromised because of its underlying vision of social evolution; some will be inspired by Needham's commitment to making known the achievements of civilizations other than one's own; some will continue cautiously to use his volumes as artfully crafted but fallible tools of scholarship. These are not the only options, nor are they necessarily mutually exclusive.

In this essay I have discussed Needham's historical context, some aspects of his overall worldview, and a few of the facets of his historical scholarship.[107] What I believe this excursion shows is that he has left a

[104] Agarwala (1983), p. 34.

[105] See the thorough and measured assessment by Nachane (1995) which calls these protocols 'a definite victory for the industrialised countries' viewpoints'.

[106] Needham (1969b), p. 15.

[107] I have not analysed all his many political commitments, which would require another essay.

valuable rationalist legacy. Curiously, the strengths and weaknesses of his approach to the history of science seem to go hand in hand in certain key respects. His fundamental contribution is generally deemed to have been that he pioneered the integration of non-Western traditions and achievements into the history of science and technology, and that he based his analyses on evidence from the indigenous sources. Yet in order to make that integration he adopted (and was arguably forced by circumstances to adopt) modern scientific and philosophical concepts as organizing categories, a choice which tended to distort the thought-worlds and practices of other times and places. Similarly, his vision of a global history of science in which cross-cultural communications played an essential role carried with it a weakness for positing 'anticipations' and transmissions that have in various cases proved problematic. Finally, his notion of the universality of modern science, which he thought of as empowering non-Western societies, has the potential of being invoked deterministically to justify a blanket 'modernization' that ignores many of the social values Needham ascribed to.

These points illustrate that Needham's works stand in need of revision in a number of respects, and, like all contributions to scholarship, need to be read critically. But an attentive and critical reading of them still gives us much to think about and much to learn.

Bibliography

Abu Lughod, Janet L. (1989), *Before European Hegemony: The World-System, AD 1250–1350*, OUP, London/New York.

Adas, Michael (1989), *Machines as the Measure of Man: Science, Technology and Ideologies of Western Dominance,* Cornell University Press, Ithaca, New York.

Agarwala, P.N. (1983), *The New International Economic Order: An Overview*, Pergammon, New York/Oxford.

Agassi, Joseph (1988), *The Gentle Art of Philosophical Polemics,* Open Court, LaSalle, Illinois.

Anderson, Perry (1974), *Lineages of the Absolutist State,* New Left Books, London.

Blaut, J.M. (1993), *The Colonizer's Model of the World: Geographic Diffusionism and Eurocentric History*, Guilford Press, London/New York.

Bodde, Derk (1991), *Chinese Thought, Society, and Science: The Intellectual and Social Background of Science and Technology in Pre-modern China*, University of Hawaii Press, Honolulu.

Boulding, Kenneth (1976), 'Great Laws of Change', in A.M. Tang, F.M. Westfield and J.S. Worley (eds), *Evolution, Welfare and Time in Economics*, Lexington Books, Lexington Mass.

Bray, Francesca (1997), *Technology and Gender. Fabrics of Power in Late Imperial China*, University of California Press, Berkeley/Los Angeles.

Brook, Timothy and Gregory Blue (eds) (in press), *China and Historical Capitalism: Genealogies of Sinological Knowledge*, Cambridge University Press, Cambridge; Maison des Sciences de l'Homme, Paris.

Carr, E.H. (1986), *What is History?* 2nd ed. (Introduction by R.W. Davies), Penguin, London.

Chemla, Karine (1992), '[Review of] Bodde, Derk, *Chinese Thought, Society and Science ...*', in *Revue bibliographique de sinologie*, vol. 10, pp. 319–20.

Comte, Auguste (1830–42), *Cours de philosophie positive*, 6 vols, Paris.

Cournot, Antoine-Augustin (1975), J.C. Patiente (ed.), *Oeuvres complètes*, Vrin, Paris.

Crombie, A.C. (ed.)(1963), *Scientific Change: Historical Studies in the Intellectual, Social and Technical Conditions for Scientific Discovery and Technical Invention, from Antiquity to the Present*, Heinemann, London.

D'Abro, A. (1950), *The Evolution of Scientific Thought from Newton to Einstein*, Dover, New York (originally published 1927).

Du Halde, Jean-Baptiste (1735), *Description géographique, historique, chronologique, politique de l'Empire de la Chine*, Paris.

Elvin, Mark (1973), *The Pattern of the Chinese Past*, Methuen, London.

Elzinga, Aant (1994), 'UNESCO and the Politics of Scientific Internationalism' (Contribution to the ORSTROM/UNESCO Conference on 'Twentieth-Century Science: Beyond the Metropolis' [Paris, September 1994]).

Flynn, Dennis O. and Arturo Giráldez (1995), 'Born with a "Silver Spoon": The Origin of World Trade in 1571', in *Journal of World History*, vol. 6, no. 2, pp. 201–21.

Forbes, R.J. and E.J. Dijksterhuis (eds)(1963), *A History of Science and Technology*, Penguin, Harmondsworth.

Frank, Andre Gunder (1998), *Reorient: Global Economy in an Asian Age*, University of California Press, Berkeley/Los Angeles/London.

Furth, Charlotte (1994), 'Rethinking Van Gulik', in C. Gilmartin *et al.* (eds),

Engendering China: Women, Culture and the State, Harvard University Press, Cambridge Mass., pp. 125–46.

Gellner, Ernest (1992), *Reason and Culture*, Blackwell, Oxford.

Gernet, Jacques (1982), *Chine et Christianisme: action et reaction,* Paris.

——— (1993), 'Espace-temps, Science et religion dans la recontre de la Chine avec l'Europe', in Jami and Delahaye (1993), pp. 205–16.

Geymonat, Ludovico (1977), *Scienza e realismo*, Feltrinelli, Milano.

Goody, Jack (1996), *The East in the West*, Cambridge University Press, Cambridge.

Graham, A.C. (1973), 'China, Europe and the Origins of Modern Science: Needham's *The Grand Titration*', in Nakayama and Sivin (eds) (1973), pp. 45–71.

Holton, Gerald (1973), *Thematic Origins of Scientific Thought: Kepler to Einstein*, Harvard University Press, Cambridge Mass.

——— (1993), *Science and Anti-Science*, Harvard University Press, Cambridge Mass.

——— (1996), *Einstein, History and Other Passions: The Rebellion Against Science at the End of the Twentieth Century*, Addison-Wesley, Reading Mass.

Jami, Catherine and Hubert Delahaye (1993), *L'Europe en Chine: Interactions scientifiques, religieuses et culturelles aux XVIIe et XVIIIe siècles*, Collège de France, Paris.

Jones, E.L. (1987), *The European Miracle*, Cambridge University Press, Cambridge.

Kuhn, Dieter (1992), '[Review of] Bodde, Derk, *Chinese Thought, Society and Science ...*', in *Orientalistische Literaturzeitung*.

Laffitte, Pierre (1887), *A General View of Chinese Civilization and of the Relations of the West with China*, Trübner, London (French original 1861).

Landes, David (1983), *Revolution in Time: Clocks and the Making of the Modern World*, Harvard University Press, Cambridge Mass.

Lettres edifiantes et curieuses (1703–76), (Ch. Le Gobien, J-B. du Halde *et al.*, eds), Paris.

Martin, W.A.P. (1901), *The Lore of Cathay, or the Mind of China*, Edinburgh and London.

McGucken, William (1984), *Scientists, Society, and State: The Social Relations of Science Movement in Great Britain, 1931–1947*, Ohio State University Press, Columbus.

Mendoza, Juan Gonzalez de (1585), *Historia de las Cosas mas notables, Ritos y Costumbres del Gran Reyno de la China*, Rome.

Mikulinsky, S. (1977), 'History of Natural Science as a Science: Present State and Theoretical Problems', in *Soviet Studies in the History of Science*, USSR Academy of Sciences, Moscow.

Nachane, D.M. (1995), 'Intellectual Property Rights in the Uruguay Round: An Indian Perspective', in *Economic and Political Weekly*, 4 Feb., pp. 257–68.

Nakayama, Shigeru (1973), 'Joseph Needham, Organic Philosopher', in Nakayama and Sivin (eds)(1973), pp. 23–44.

——— (1984), *Academic and Scientific Traditons in China, Japan, and the West*, University of Tokyo Press, Tokyo.

Nakayama, Shigeru and Nathan Sivin (1973), *Chinese Science: Explorations of an Ancient Tradition*, MIT, Cambridge Mass.

Needham, Joseph (1931), *Chemical Embryology*, 3 vols, Cambridge University Press, Cambridge.

——— (1935), 'Science, Religion, and Socialism', in J. Lewis, K. Polanyi and D.K. Kitchin (eds), *Christianity and the Social Revolution*, Gollanz, London.

——— (1946), *History is on Our Side*, Allen & Unwin, London.

——— *et al.* (1954–), *Science and Civilization in China*, 7 vols (in progress; 17 parts to date), Cambridge University Press, Cambridge. (abbreviated as *SCC*).

——— (1966), 'The Roles of Europe and China in the Evolution of Oecumenical Science', reprinted in Needham (1970), pp. 396–418.

——— (1969a), *The Grand Titration: Science and Society East and West*, Allen & Unwin, London.

——— (1969b), *Within the Four Seas: The Dialogue of East and West*, Allen & Unwin, London.

——— (1970), *Clerks and Craftsmen in China and the West*, Cambridge University Press, Cambridge.

——— (1973a), (Henry Holorenshaw, pseudonym) 'The Making of an Honorary Taoist', in Teich and Young (eds), pp. 1–20.

——— (1973b), 'The Historian of Science as Oecumenical Man', in Nakayama and Sivin (eds)(1973), pp. 1–8.

——— (1976a), *Moulds of Understanding: A Pattern of Natural Philosophy*, in G. Werskey (ed.), Allen & Unwin, London.

——— (1976b), 'History and Human Values: A Chinese Perspective for Science and Technology', in *Centennial Review*, vol. 20, pp. 1–35.

——— (1983), 'Science, Technology, Progress and the Breakthrough: China as a Case Study in Human History', in T. Ganelius (ed.), *Progress in Science and Its Social Conditions*, published for the Nobel Foundation by Pergammon, Oxford/New York.

——— (1986), 'The Cosmic Setting of Human Life', typescript of a paper presented at the Tenri Foundation, Tenri, Japan, December 1986.

——— (1986), *Time the Refreshing River*, reprinted with a new introduction by the author, Spokesman, Nottingham (original: Allen & Unwin, London, 1943).

Needham, Joseph and Walter Pagel (eds)(1940), *Background to Modern Science*, Cambridge University Press, Cambridge.

Needham, Joseph, Wang Ling and Derek de Solla Price (1986), *Heavenly Clockwork: The Great Astronomical Clocks of Medieval China*, Cambridge University Press, Cambridge (Original: Cambridge, 1960).

Newton-Smith, William (1981), *The Rationality of Science*, Routledge & Kegan Paul (RKP), London.

Passmore, John (1968), *A Hundred Years of Philosophy*, Penguin, Harmondsworth.

Patterson, R. (1954), 'Spinning and Weaving', in Singer and Holmyard (eds) (1954–8).

Popper, Karl (1957), *The Poverty of Historicism*, RKP, London.

——— (1969), *Conjectures and Refutations: The Growth of Scientific Knowledge*, RKP, London.

Rozman, Gilbert (1981), *The Modernization of China*, Free Press, New York.

Rashed, Roshdi (1984), 'Appendice: La notion de science occidentale', in *Entre Arithmétique et algébra*, Les Belles Lettres, Paris, pp. 301–18.

Russell, Bertrand (1984), *A History of Western Philosophy*, Unwin Paperbacks, London.

Sagan, Carl (1995), *The Demon-Haunted World: Science as a Candle in the Dark*, Random House, New York.

Sarton, George (1927–47), *Introduction to the History of Science*, 3 vols, (Carnegie Institution Publication, no. 376) Williams & Wilkins, Baltimore.

——— (1962), *The History of Science and the New Humanism*, Indiana University Press, Bloomington (originally published 1931).

Shue, Vivienne (1988), *The Reach of the State: Sketches of the Chinese Body Politic*, Stanford University Press, Stanford.

Singer, Charles and E.J. Holmyard (eds) (1954–8), *A History of Technology*, 5 vols, Oxford.

Sivin, Nathan (ed.) (1977), *Science and Technology in East Asia*, Science History Publications, New York.

——— (1982), 'Why the Scientific Revolution Did Not Take Place in China—or Didn't it?', in *Chinese Science*, no. 5, pp. 45–66.

——— (1988), 'Science and Medicine in Imperial China—the State of the Field', in *Journal of Asian Studies*, 47, 1, pp. 41–90.

Sorböm, Per (1984), 'Om spridning av spridningstheorier: Exemplet kompassen', in *Lychnos: Lärdomshistoriska Samfundets Arsbok*, pp. 27–65.

Spence, Jonathan D. (1992), 'The Dialogue of Chinese Science', in *Chinese Roundabout: Essays in History and Culture*, Norton, New York, pp. 141–54.

Teich, Mikulás and Robert M. Young (eds)(1973), *Changing Perspectives in the History of Science*, Heinemann, London.

Thorndike, Lynn (1923–41), *A History of Magic and Experimental Science*, 6 vols, Columbia University Press, New York.

Werskey, Gary (1988), *The Visible College: A Collective Biography of British Scientists and Socialists of the 1930s*, second edn (Foreword by R.M. Young), Free Association Press, London (originally published 1978).

Whewell, William (1837), *History of the Inductive Sciences*, 3 vols, Parker, London.

Whitehead, Alfred North (1926), *Science and the Modern World*, Lowell Lectures, 1925, Cambridge.

——— (1955), *Adventures in Ideas*, New York (originally published 1933).

Williams, William Appleman (1980), *Empire as a Way of Life*, OUP, Oxford/New York.

Winstanley, Gerrard (1941), in G. Sabine (ed.), *Works*, Cornell University Press, Ithaca, New York.

Wittfogel, Karl A. (1931), *Wirtschaft und Gessellschaft Chinas. Versuch der wissenschaftlichen Analyse einer grossen asiatischen Agrargesellschaft*, Leipzig.

——— (1957), *Oriental Despotism: A Comparative Study of Total Power*, New Haven.

Wylie, Alexander (1897), *Chinese Researches*, Kelly and Walsh, Shanghai.

4

Aant Elzinga

Revisiting the 'Needham Paradox'

The Multifaceted Nature of Needham's Question

Introduction

Why did modern science and technology as we know it take off in Europe and not in China, given the latter's high level of development before the sixteenth century? This was the question posed by Joseph Needham, and in answering it he pointed to a number of different dynamic factors, both internal and external to science.

Needham's own view was shaped by several contingencies. The first was his practical bent and early experience with toolmaking while at Oundle School in Northamptonshire under its headmaster F. W. Sanderson, and later in university laboratories.[1] Another was his training as a biologist working on embryology, with particular emphasis on morphology. This allowed him to look at other phenomena, including institutionalized practices in society from an evolutionary perspective of growth. A third aspect was his attachment to Marxist thinking, which he combined with a Christian ethic. In this respect, he subscribed to many of the assumptions elaborated by John D. Bernal, for example, the notion that science as a productive

[1] 'See Maurice Goldsmith, *Joseph Needham: 20th-Century Renaissance Man* (UNESCO Publishing, Paris, 1995), p. 25. H. Floris Cohen also brings attention to Needham's experiences with toolmaking—see his *The Scientific Revolution* (Chicago University Press, Chicago, 1994), p. 421.

force in society has a dynamic which requires particular social relations and ideological conditions for its sustenance.

Central to Bernal's monograph on *The Social Function of Science*, published in 1939, was the notion that capitalism was initially a stimulus, but later it became a hindrance to the development of a science oriented to social equity and welfare, both nationally and internationally. This thesis of science being frustrated and prostituted in Western countries in the 1930s, I call the strong thesis of Bernalism. It called for a socialist revolution so that the full potential of science could be released and applied for human benefit, instead of being chiefly in the service of exploitation and oppression of working and colonized peoples.

Later, after the war, Bernalists weakened this proposition, maintaining that an introduction of planning, more efficient use of resources and competence, as well as greater public accountability would lead to a better and more democratic science. Indeed, an introduction of reflexivity, in the sense of subjecting science itself to steering on the basis of knowledge in a 'science of science', was called for.

As it turned out, this call was as scientistic as the narrower minded non-responsibilist view it was meant to replace, something Needham was quite aware of in the late 1960s. He was led to adopt the idea strongly propagated by the radical science movement, viz. that science itself also required a fundamental change of power relations in its institutional structures. This belief was reinforced by his positive stance regarding the Chinese cultural revolution initiated by Mao Zedong.

The contingencies shaping Needham's early thought, and his formulation of the question about the historic scientific revolution in Europe are important for analysis. Both the context and the formulation of the question were permeated with elements of the same evolutionism that he otherwise opposed, especially in the traditional historiography of science, which he found to contain a hidden Eurocentric bias.

Implicitly referring to Rupert Hall's position in the 1960s and early 1970s, he points out how that scholar, like most contemporary professional historians of science in Britain or the USA, might be interested in Greek science, but rarely Arabic or other civilizations' medieval science.

The last thing they want to hear about is non-European science, and that partly because distinctively modern science originated only in Europe, the only interesting ancient and medieval sciences must have been European. This glaring non-sequitur still dominates the intellectual outlook in the West inspite of all that has been done by enlightened comparative historians of technology.[2]

On the other hand, affirming human progress in the enlightenment mode, with science, medicine, and technology as the most obvious manifestations of this progress in one direction, that of a common progressive heritage of humankind, the same Needham writes:

Here ... humanity has always been marching in one column, and though the particular systems of natural philosophy may have been ethnically bound, ... the actual understanding and control over Nature has been passed from mind to mind across all barriers to build up what the early Royal Society- called the great edifice of 'true natural knowledge'.[3]

Not only science, but the political organization of society, too, is seen as a manifestation of human progress. After biological evolution comes social and institutional evolution and within this a reflexive consciousness called science, which may also be turned upon itself for the improvement of its own growth conditions. This conviction, in line with the Bernalist notion of science being frustrated and prostituted under certain types of social relations, is largely maintained into the 1960s, whence the radical science movement criticized the scientism implicit in it, pointing to class and gender differences in the laboratory, and so the need for concomitant struggles at social and ideological levels, in such sites too.[4]

This new consciousness provided one impulse to the social studies of science. The other impulse, more internalist, came from the social turn in the historiography of science, which in turn tends in retrospect

[2] J. Needham, *Science in Transition* (1981), cited after Floris Cohen, p. 425.

[3] J. Needham, *Science in Traditional China*, cited after Floris Cohen, p. 423.

[4] See Hilary Rose and Steven Rose (eds), *The Political Economy of Science* (MacMillan, London, 1976) and its sequel, *The Radicalisation of Science* (MacMillan, London, 1976); Hilary Rose (1994), *Love, Power and Knowledge: Towards a Feminist Transformation of the Science* (Indiana University Press, Bloomington and Indianapolis). Rose recalls and associates herself with this more radical stream of science studies, that has in part been eclipsed by the new quest for professionalization, as well as the newer field that was once set up in opposition to Bernalist externalism and academic internalism of history and philosophy of science departments.

to be attributed to the introduction of Thomas Kuhn's notion of the paradigm or disciplinary matrix which encompasses not only cognitive and epistemic but also social dimensions in the growth processes of science, albeit conceived internalistically. As more recent scholarship shows, this view also resonated with a broader cultural narrative, associated with the rhetoric and reality-defacing imagery of science in society generated in the cold war era.

More recently there has been the cultural turn in the historiography of science, and a linking of the project of comparative cross-civilizational studies of science to a much more radical critique of modernity, one that focuses on the mode of representing basic categories and on a politics of culturally situated identities. In such discourses that emphasize the locally situated nature of knowledge, playing up the intersection of culturally based representations and identities, the very basis of the question at the core of the 'Needham paradox' becomes absurd. One does not ask why and how what was to become scientific knowledge in different civilizational contexts came to be more successful in certain of these contexts and less so in others. Thus, the very basis of Needham's question, and the evolutionism and residual scientism upon which it was predicated, is dissolved.

What, however, remains, more important, is the ethos and activism that fired the broader humanistic basis of Needham's project. In his estimation, science is the common heritage of all humankind. Modern science too cannot be identified as European ethnoscience, but only as the modern amalgam of the sciences of various different civilizations. Therefore Eurocentrism was to be opposed and other traditions were no longer to be treated as 'backward', awaiting only the embrace of advanced Europe, for, 'we are living in a dawn of new universalism, which ... will unite the working people of all races in a community both catholic and cooperative .... The mortar of this edifice is mutual comprehension'.[5]

Shifts in the Comparative Studies Agenda

This erosion of Needham's symbolic question 'why was Galileo Galilei born in Europe and not China' already started in the late 1960s, with the social turn in the historiography of science, which preceded the present culturally reflexive turn. Indeed, Needham's original question

[5] Preface to *SCC*, vol. 1, p. 6, cited after Goldsmith, p. 107.

has generated some heat. Several scholars, like Nathan Sivin and others,[6] have branded it a non-question.

In the civilizational study of science and technology several shifts of agenda are discernible. There was, to begin with, the positivistic enterprise of Sarton and Singer, and, to some extent, of the generation after them, who instituted science historiography as a discipline in 'Western universities'. Needham sought to provide a corrective to this tendency, and offered a critique of the latent Eurocentrism of the second generation internalism, posing against it an exaggerated celebration of Chinese 'Triumphs'. In a new social turn, Nakayama, Sivin and others, building on Needham, make a sympathetic critique of his approach, and articulate an opposition to the radical reflexive cultural turn that tends to judgmental relativism (as distinct from mere acceptance of contingency) in historiography.

In considering these three shifts, one can ask, what the key points of methodology are at each of these conjunctures when it comes to cross-cultural comparison of science in those broad contexts referred to as 'civilizations'. Indeed, the very notions of 'development', 'growth of science', 'universally true knowledge' and the like are no longer viable. Certainly they are no longer the culturally biased notions used at the outset with Sarton and Singer. Simultaneously we are also pushed to question the Marxist evolutionist optimism assumed by Needham.

In fact, now texts with such concepts are in themselves problematic in as far as they trigger rhetorical resources associated with evolutionist points of view. The claim is that such concepts privilege the view of the European colonizer while erasing the traces of this colonizing act by a self-assumed universalization of its own vantage point.

I shall come back to this problem at the end of the essay, and point to certain pitfalls regarding the cultural turn, suggesting that in certain respects Needham's work still contains important corrective elements that can be translated in a discourse of reflexive modernity. That means they may, perhaps, be turned to advantage in drawing a line vis-à-vis a paralysing regress of hyperreflexivity.

In what follows, I shall first outline Needham's position in some detail. Thereafter, I will comment more generally on the social

[6] Ho Peng Yoke, 'Chinese Science: The Traditional Chineses View', in *Bulletin of the School of Oriental and African Studies*, LIV, 3, 1991, pp. 506–19.

epistemology of frames of academic observation and analysis, and the post-Second World War sources of the Eurocentric skew that has now become the subject of historical soul-searching and self-criticism within some historiographical communities in the West.

We thus locate Needham's efforts to counter internalist Eurocentrism in the broader context from which he derived some of his inspiration in the years immediately following the Second World War and the onset of the cold war. It will be suggested, as Nakayama has done elsewhere,[7] that 'where Needham comes from'—his own socio-political context—is significant for gaining an understanding of his project and his passionate desire to put ancient and medieval science (mostly symbolized by technological achievements) on the world map.

Needham's Question in More Detail

In brief, Needham's claim is that, on the one hand, there were external factors that inhibited a takeoff of science in the modern sense in China and on the other hand, there were internal factors inherent in the style of Chinese thought which hindered the development of abstraction in a rationalistic deductive means–ends or cause–effect mode of explanation at the philosophical level.

Chief amongst the external factors was the existence of an Asian bureaucratized social formation and the consequent lack of a bourgeoisie and mercantile class in the modern European sense. The latter was motivated by the making of profit as a powerful individual and institutional economic incentive. Linked to nascent capitalism in Europe was the character of class struggle, with the opening of a secularized political discourse and the reliance on the empirical evidence of the senses, an orientation which allowed itself to be generalized. Needham sometimes refers to it as the discovery of the method of discovery.[8] Socially it was found to be borne by a stratum of craftsmen and technicians in the interstices between traditional scholars and mercantile and political patrons. Here he follows the studies of Edgar Zilsel. Similar elaborations on science in seventeenth-

[7] S. Nakayama, 'Joseph Needham, Organic Philosopher', in Nakayama and Sivin (eds), *Chinese Science: Explorations of an Ancient Tradition* (Cambridge, Mass., MIT Press, 1973), pp. 23–43.

[8] A.C. Graham, 'China, Europe, and the Origins of Modern Science: Needham's Grand Titration', in ibid., pp. 45–69.

century Europe may be found in the writings of another Marxist, the historian of mathematics, Dirk Struik. The point is that abstracted from its local situatedness this method—experientia ac ratione, as Christian Huygens called it—turned out to be a powerful algorithm for 'truth-making'.[9]

On the other hand, what we now would call the transition, from local situated knowledge in practice, to claims of generalizability in theory and application is seen to be hindered in medieval China by the absence of a culture permitting arithmomorphic thought. The application of the idea of veracity, tied to procedures of testability, subsumes a notion of an intelligibility of nature. In the Judeao-Christian tradition, with its monotheistic notion of a lawgiver god, Needham suggested, the notion of 'law' was encoded into culture, projected into natural philosophy as a concept of 'law of nature'.

Here, again, he followed Zilsel and his study of the emergence of the concept of scientific law as having roots in the realm of religious thinking. Needham devoted considerable effort to demonstrating how the notion of a law of nature was missing in Chinese culture.[10]

According to Needham,

> two things are clear, (a) that the depersonalization of God in ancient Chinese thought excluded the conception of a divine celestial lawgiver imposing ordinances on non-human Nature; and (b) that the highest spiritual being known was never a Creator in the sense of the Hebrews and Greeks. It was not that there was no order in Nature for the Chinese, but rather it was not an order ordained by a rational personal being, and hence there was no guarantee that other personal rational beings would be able to spell out in their own earthly languages the divine code of laws which he had previously formulated. There was no confidence that the code of Nature's laws could be unveiled and read, because there was no assurance that a divine being, even more rational than ourselves, had ever formulated such a code or made it readable. One feels, indeed, that the Taoists, for example, would have scorned such an idea as being too naive to be adequate to the subtlety and complexity of the universe as they intuited it.[11]

[9] Aant Elzinga, *On a Research Programme in Early Modern Physics*, (Akademiieforlaget, Goteborg, 1972).

[10] Joseph Needham, 'Human Laws and Laws of Nature in China and the West' (I), *Journal of the History of Ideas*, vol. XII, no.1, January 1951, pp. 3–30.

[11] Joseph Needham, 'Human Laws and Laws of Nature in China and the West, *Journal of the History of Ideas*, vol. XII, no. 2, April 1951, pp. 194–230; quote from p. 229.

To establish this, Needham analysed in detail the notion of *li* which he equated with 'natural law'. He saw this as a product of the Confucians, who affirmed the body of ancient custom, usage and ceremonial, which included all those practices, such as filial piety, which were observed by the Chinese people as intuitively right for a millennium and more. Through a close study of the etymology of this and other words, among them certain ones central to the legalist and other traditions, he finds that there is a no-man's land in which *li* creeps over from juridical, i.e. natural law to scientific law ('Laws of Nature'), but in the end it never really makes it. With the Neo-Confucians *li* becomes associated with pattern or organization, as a natural and inescapable law of affairs and things, and in one sense it is said to permeate non-human things, as distinct from another sense (*hsing*), which is the *li* that permeates humans themselves. From this, Needham argued that it is, therefore, possible to delineate *li* as the 'principle of organization' in the universe. 'I conclude, therefore', he writes, 'that "law" was understood by the Neo-Confucian School in a distinctly Whiteheadian organismic sense', a sense of parts falling precisely into place with other parts in the whole organism they compose, without the slightest excess or deficiency.[12] Thus, 'while one could not say that "law" in the Newtonian sense was completely absent, it played a relatively minor (perhaps even very minor) part ...'[13] Sometimes Needham suggests that if a scientific revolution had occurred in China, it would have been within the framework of an organismic and not a mechanistic world picture. On the other hand, he believed that the mechanistic world picture was a necessary precondition for such a revolution, as in the West in the sixteenth and seventeenth centuries.

Intelligibility of the Cosmos: Europe

Nathan Sivin has pointed out that the main feature of Chinese thought was the belief that the cosmos is too large, and the texture of nature too fine, too subtle, too closely intermeshed for phenomena to be fully predictable. This proposition denies that the physical world can be fully penetrated by study as fully described in words or numbers.[14]

[12] Ibid., p. 219 and the passages previous to this on p. 218.

[13] Ibid., p. 219.

[14] Nathan Sivin, 'On the Limits of Empirical Knowledge in Chinese and Western

The Chinese did not even make a distinction between epistemological and ontological indeterminacy. One Chinese writer is quoted as saying that in order to know, we break the continuity of nature into blocks of time, which we treat as though each were uniform.[15]

This was indeed the strength of the mechanistic world picture as it developed in seventeenth-century Europe—its assumption of cuttable repetitiveness. It is the idea that lies at the base of the study of harmonic motions, capable of being grasped mathematically by the calculus of Newton and Leibniz, and successively elaborated in theoretical mechanics from Laplace to Fourier's analysis, Hamiltonians and eigenvalues expressed in tensor analysis. Non-linear dynamics, central to the so-called chaos theory in our time, only succeeded much later, opening up today, new avenues for quantification of the non-repetitive or irreversible processes. Perhaps it was this early focus on repeatable processes, abstracted in terms of idealized 'systems', with a basic distinction between primary and secondary qualities, and analysed with a mathematics of harmonics, from a 'view from nowhere' which annihilated real time, that constituted part of the success story.[16] Alfred North Whitehead speaks of the abstraction involved as the fallacy of misplaced concreteness, i.e. the mistake of taking mathematical abstractions and idealizations from the real thing, as the concrete.

Here then we have another side of the paradox. By dividing subject and object and reifying the abstract one gets on the one hand the abstraction of isolated systems which was enormously successful as a research programme. But by confusing the pseudoconcrete in the imagination with the real concrete world, natural philosophy in the West was perverted too, into a mode of domination-of-nature thinking. Potentiality was rejected and only actuality—being—was left, as isolated systems were hypostasized. Some philosophers started to think of the universe as an ideal mechanical system, reifying space and time as absolute relations. Others accepted the relativity of space

Science', in Shlomo Biderman and Ben-Ami Scharfstein (eds), *Rationality in Question: On Eastern and Western Views of Rationality* (E.J. Brill, Leiden, New York, etc. 1989), pp. 165–89; quote from p. 165.

[15] Ibid., p. 174.

[16] See further, Aant Elzinga, 'Christian Huygens and the Elimination of Time', *Bull. de la classe des Sciences*, 5e serie, Tome LXXII (1987–10), Ac. Royale de Belgique, pp. 394–404.

(and maybe time), but still needed an outside power to guarantee the continuity of everything. This gave rise to a host of epistemological and metaphysical problems. Among other things it meant an opposition between the real time of living and the abstract anonymous time of the mechanistic system.

This, then, was something of the cognitive texture that allowed natural philosophers in seventeenth-century Europe to plug in algorithms with associated boundary conditions in nature and derive consequences which again could be interpreted and attributed to nature. The mechanistic world picture, the splitting of subject and object, and attribution of intelligibility and calculability to all domains of nature was both a strength and a weakness. The weakness was the legitimation this came to lend to an exploitative domination-of-nature ethic, upheld by the constant need to draw boundaries between humankind and animalkind, mind and body, cognitive and social, etc.

Being in Tune with Things: China

A more recent elaboration of the theme suggests that Needham's focus on *li* (organizing pattern) is less useful than a focus on *shi*, which translated is roughly comparable to 'propensity of things'. Needham tries too much to identify points of congruence with natural philosophy in the West, Newtonianism for example. Another approach constitutes the theme of a book by Francois Jullien.

He argues that the originality of the Chinese lay in their indifference to any notion of telos, a final end for things. They sought to interpret reality on the basis of itself, from the perspective of a single logic inherent in the actual processes in motion. Their notion of a potential born of disposition (*shi*) is clearly displayed in early military strategic texts, where it is employed to discuss the way the configuration of things (the site, characteristics of the terrain, the insight, skill and competence of the enemy, the placement of one's own troops, etc.) must be harnessed in such a way that subsequent actions flow in tune with the evolution of events which are seen to be objective. 'The perspective operative here is of a process that can evolve to our advantage if we make opportune use of its propensity'[17] and do so with efficacy (i.e. with the least

[17] Francois Jullien, *The Propensity of Things: Towards a History of Efficacy in China*, (Zone Books, New York, 1995), p. 34.

expenditure of energy).[18] Even Mao Zedong resorted to this concept to explain the tactics most appropriate to a war of resistance, where a head-on engagement with the enemy is minimized and the enemy's morale and capacity for battle is gradually weakened through elusive guerilla tactics and encircling movements.[19]

This may be contrasted with Clausewitz's doctrine which calls for decisive and direct confrontation in battle, whence the mobilization of superior force at a given moment is used to achieve a predetermined end. This doctrine involves the traditional idea central to Western philosophy, that of a relation between means and ends. Where the Chinese regarded the moral qualities (e.g. courage) essential to the successful conduct of warfare as simply being intrinsic to the situation, Clausewitz understands these as factors in themselves.

Moreover, the theory of friction, played an important role for Clausewitz, as it referred to the gap or disparity between the plan drawn up in advance, which is of an ideal nature, and its practical implementation, which is subject to chance. 'The Chinese concept of shi', Jullien maintains, inserts itself 'into the distinction between what Westerners have opposed as "practice" and "theory", ... thus collapsing that distinction; [it] shifts "execution" toward something that, given the propensity at work, operates of its own accord and excludes any uncertainty or inadequacy; neither deterioration nor friction is involved".'[20] It is a matter of the right or appropriate way of conducting oneself in the situation at hand.

In sum, the contrast between the Chinese concept *shi* and the Western categories predicated on means and ends, results in a fundamental structural difference in thought, which also reflects itself in the domain of natural philosophy, where pronouncements are made on the nature of nature. The Chinese ancient and medieval interpretation of reality in any realm, even in the most speculative

[18] Ibid., pp. 39 ff.

[19] Mao Zedong, Military Political Writings, cited from 'Problems of Strategy in China's Revolutionary War', December 1936, *Selected Works*, vol. I, (Foreign Languages Press, Peking, 1964), pp. 179–200, also 'Problems of Strategy in Guerilla War Against Japan', May 1938, *Selected Works*, vol. II, (Foreign Languages Press, Peking, 1965), pp. 81–94.

[20] Jullien, *The Propensity of Things*, p. 38.

mode, appears, then, to proceed through an understanding of the disposition of things. Instead of the explanation of causes, we have the implication of tendencies, and at an early date Chinese civilization moved 'away from religious feeling toward the sense of universal regulation'.[21] Tendency and logic are inseparable, whence 'it would be a crude mistake to believe that "when the world is well governed" it obeys only "logic" (or "pattern", *li*), whereas "when it is badly governed," it obeys only "tendency" (*shi*), for in both cases logic and tendency are indissociable'.[22] The Chinese approach can be deemed neither mechanistic nor finalistic.

Reconstructive Critique of Theoretical Genealogies

'A methodological problem consequently arises if we are to press on with the comparison of the two cultures',[23] Chinese and European. Jullien suggests retracing the process that established our own conceptual frameworks, or cultural narratives, to identify a split, in this case say between Greek physics and the Chinese concept of process, to see how a difference emerged and try to reconstruct it by relating this to possible agreement before the split. This is a work, we are told, not of realistic historical reconstruction but of a theoretical genealogy. Thus, cross-civilizational comparative methodology should involve a heavy dose of reconstructive critical philosophical analysis. The trajectory of European thought from Aristotle onward involved particular choices, which may be more easily teased out by comparison with another genealogy of concepts, say the Chinese one, which is assumed to have a similar point of departure in a point before the 'split' from pre-philosophy. In European civilization once 'physics' was founded on substance, the static order could not adequately explain the dynamic order, requiring the introduction of a move. Conversely, Chinese thought, which dispensed with the need for a subject, could with equal logic dispense with external causality.

The conclusion Jullien arrives at seems to confirm the line of analysis from Needham to Sivin, even if the point of departure and method of analysis it illustrates is different. As a research programme it however meets with the same difficulty that it appears to privilege

[21] Ibid., p. 225.

[22] Ibid., p. 233.

[23] Ibid., p. 249.

certain categories in European philosophy in setting up the reconstructive meeting with Chinese philosophy. Still, the ultimate claim of this programme is limited. Compared to the cultural turn discussed below, the author stands out as rather rationalistic and optimistic, in as far as we are led to believe that it is possible, by reconstructive critical philosophical, or theoretical genealogical analysis, to 'decipher from the outside our own intellectual history' and come closer to seeing it 'as it is'.[24]

Cultural Narratives

'Invasion' or 'Discovery'?

Systematic observation and analysis within academic frames sometimes, as already indicated, reveals characteristics that show explanations in part to have been shaped by and forming part of cultural narratives of broader speech communities.[25] As such, they have their anchor in the nation, a country, a people, or even a region. Distinctive for the discourse of modernity has been that, historically, it has its roots in what are claimed to be significant evolutionary processes that took part in Greater Europe from 1492 to the Second World War.[26] It is further claimed that these evolutionary processes have been responsible for bringing the fruits of progress to non-Europe, through colonization, which linked up local and regional events and intellectual activities into a global structure.

Now, of course, the assumption that colonialism itself was the source of all progress has been shown to be patently mistaken. There is a wealth of critical scholarship to this end, showing how there was, in fact, a whole lot in the way of appropriate technologies and situated

[24] Ibid., p. 18.

[25] For the concept of 'cultural narrative', see Donna Haraway, 'The Bio-politics of a Multicultural Field', in Sandra Harding (ed.), *The Racial Economy of Science,* (Indiana University Press, Bloomington and Indianapolis, 1993), pp. 377–97. Haraway deals with the reflection of a non-Western perspective in Japanese primatology, where 'the mask is a figure of the two-sidedness of the structure of life, person, and society', as distinct from dualities of man–animal, culture–nature, and mind–body so crucial to Western Greek and Judeo-Christian mythology and to 'derivative ideologies of scientific objectivity' (p. 379).

[26] J.M. Blaut, *The Colonizer's Model of the World: Geographical Diffusionism and Eurocentric History* (The Guilford Press, New York/London, 1993), esp. ch. 1.

specialized knowledges in other parts of the world prior to colonization. Some of these have survived in spite of the economic, socio-political and culturally dismantling effects of colonialism which tended to discard the indigenous and traditional as irrational while crediting Western-driven development as progress. Vandana Shiva sharpened this point for science studies when she stated—in her keynote address at the joint 4S/EASST conference in Goteborg,

> Between 1492 and 1992, Europe's meeting with the non-European cultures was actually no real meeting. The interaction by the colonized was always experienced as invasion, and by the colonizers as discovery. The experience of invasion as discovery has been facilitated there through which European men had constructed a world in which evolution was understood to have created two separate minds—one for themselves and one for all others.[27]

Much of mainstream economic theory and political science in the West, within which the notion of an objective and neutral science is entrenched, represents relatively autarchic local development in pre-Columbian and other traditional societies as underdevelopment, and emulation of modern European and North American trajectories as development. Even when we try to be critical and redefine terms we have a problem in that the point of reference in the literature against which we polemicize still tends to influence our own reformulations. This, I think, was also in part Needham's problem when he introduced his particular mode of countering the then prevalent Eurocentric cultural narrative in the historiography of science.

Reflection into the History of Science

The Eurocentric view has also informed the historiography of science, especially in the founding of this field as a professional academic endeavour. In its early days, the history of science was, by and large, undertaken by retired scientists, just as military history was written

[27] Cited in Aant Elzinga, 'Traces of Eurocentrism in Curent Representations of Science', *VEST* vol. 8, no. 4, 1995, p. 85 (this is a thematic issue of the journal *VEST*, around the work of Martin Bernal and his *Black Athena*, which was the subject of a symposium organized at the University of Goteborg 25–26 May 1994, with the participation of Bernal, E. Mendelsohn, S.E. Liedman, R. Ambjornsson, F. Josephson, P. Astrom, J. Barmark and others; the entire address is translated into Swedish and published in the same journal as Vandana Shiva, 'Vetenskap, teknik och 500 ar av kolonisationerz' (Science, Technology and 500 Years of Colonizations), Tidskrift for vetenskapsstudier *VEST,* vol. 6, no. 2, 1991.

by retired generals, mostly representing themselves as the 'victors' in an historical chain of events, to which is retrospectively added a telos of leading up to the present-day textbook interpretations of the state-of-the-art knowledge in given fields.[28] Just as the history of nations was supposed to be the work of kings and their leading generals and diplomats, so the history of science was regarded as a record of the brains of great men of genius. European achievements of science at the same time served as so many benchmarks of superiority, being a measure of the way other civilizations and peoples lagged behind. Endemic to this self-congratulatory historiography was the valorization of science as the epitome of knowledge. With positivism in its various forms, from Comte to Carnap, and to Sarton and Singer, this was encapsulated in the code of scientism.[29]

The opposite standpoint, tending to romanticism, blurs the distinction between scientific and other ways of knowing. This is a tendency that flourished at various times in Europe, as in the *fin de siecle* pessimism of the previous century, again in the dark period of fascism in the 1930s and early 1940s, and there are strains of it in some pretenders of post-modernism today. Romanticism denies the basic assumptions of scientism and challenges them. Indeed, in some cases the very idea of progress in and through science is denied. Being a child of a reaction to the industrial optimism of an earlier century, romanticism promotes intuition as superior while scientific reason is equated with an advance of mechanization in human life. Also, the irrational is embraced as a possible road to truth. In the extreme this criticism of science approaches a critique of rationality.

What I am concerned with here are two images of science, one linking scientific advance with social progress on the assumption of a principle of progress through science, the other tying scientific and

[28] For a concentrated criticism of this, see J. Agassi's 'Towards an Historiography of Science', *History and Theory*, Beiheft 2, 1963 (a now forgotten text written parallel to Kuhn's *The Structure of Scientific Revolutions* of the previous year).

[29] By scientism I understand (a) a view of science as a historic socially neutral enterprise; and (b) the general philosophy which claims that all aspects of the universe are knowable through science and that science as such is the only reasonable, adequate and useful mode of cognition, superior to all others. Other ways are degraded as second rate, or useless—e.g. metaphysical nonsense, as in the language of logical empiricists, who devoted so much of their energies to finding some criteria for demarcation between science and non-science.

technological progress to social regress. In both cases science is mystified as a kind of independent force with its own irrepressible imperative—the difference is that in the technocratic assessment this force is uniquely positive, while the romantic evaluation casts it in the negative. Now, of course, one cannot equate romanticism with anti-science. It is only in the extreme that one reaches this extrapolation. Exploring the contextual embodied and socially contingent nature of the shaping of scientific knowledge claims should not be confused with anti-science, as has been done by some defenders of science in their misconstrued critique of the sociology of scientific knowledge and the work of feminist and other science studies scholars, who seek to demonstrate the local situated character of scientific knowledge in its origins, be it in the laboratory, field observation or measurements of traces of climatic change carried out within the institutional framework of some national or multinational laboratory.

An important step then was the professionalization of the subject, which involved a narrowing of its focus, and attention to internalist detail of intellectual contents using text-critical methods. Meanwhile, the issues of social, economic, cultural and other interests tended to fade out of sight. However, by the mid-1970s, after the introduction of the social turn in the history of science, which brought into focus the social history of science, first internalistically and then more broadly, Nathan Sivin was able to write:

> The generation that narrowed its conception of the history of science in order to build a profession has attained that end. The next generation, which can take professional status for granted—and choose, if need be, the freedom of the amateur—appears ready to act upon a much more general curiosity and concern for issues ...[30]

Sivin's point is that internalism and the relative absence of a comparative study of histories of science across civilizational boundaries over a considerable period of time was due to the historian's quest for professionality. In this context Needham remained the exception—'happy-amateur'.

[30] Nathan Sivin, 'Introduction', in Nathan Sivin (ed.), *Science and Technology in East Asia* (Science History Publications, New York, 1977), p. xxii. This volume celebrates the fiftieth anniversary of the founding of the History of Science Society in the US, and assembles a large number of articles that were published in *Isis* over a period of 60 years, dealing with the study of past Chinese, Japanese and Korean science and technology.

Professionalism tacitly reinforced Western or Eurocentric values. We see this codified in Parson's sociology and Merton's proposal of four norms pertinent to the ethos of modern science, norms which, according to his younger colleague Bernard Barber, happened to coincide with the democratic norms central to the ideal of the liberal or open society as promulgated in the West (against the Soviet 'East') after the Second World War.

Rather than drawing up a comparison with a view that identifies the factors facilitating and inhibiting the emergence of an idealized science when viewed through Western lenses of progress, Sivin in the mid-1970s suggested that the task for comparative studies now should be much more complex and challenging.

Sivin's own work at the time represented the introduction of the social turn in the study of science and civilization, informed both by the work of Needham and that of the professional internalists he criticized. In this turn, social and, especially, anthropological studies of science are advocated. What becomes important is:

(1) Understanding the mode of differentiation (modernization) whereby different domains of practice and concomitant cogitation emerge, each with its own criteria of evaluation of, respectively, the good, the beautiful and the true. This comes close to Needham's project, but gives it, what we might call, a Luhmanian and Habermasian twist.

(2) Considering other ways of how science, technology and society are intertwined (as a seamless web) at the cognitive as well as social and organizational/institutional levels.

(3) Considering the styles and values in science as important as their conceptual content and their veracity.

(4) Looking at'other civilizations' encounter with the West and its science and technology. This is something for which Lewis Pyenson has developed a perspective, methodology and several volumes on science and imperialism. I shall come back to this to note the realist flavour with which this consideration was introduced here.

(5) Comparing the way the kind of knowledge we call science, quantitative and qualitative, is classified within and by other cultures, and on what principles.

(6) Comparative study is perhaps the most powerful tool for examining our assumptions about those aspects of today's science

> that are universal—constrained only by the objective reality of nature—and those that reflect the values and perceptual habits of the culture in which they originated.[31]

Here we see the assertion, finally, of a realist epistemology, which is no longer present in the later 'cultural turn'.

By comparison, contemporary scholars within Science, Technology and Society (STS) or Sociology of Scientific Knowledge (SSK) would claim that the foregoing does not go far enough. Local cultures of scientific practices are taken to shape the cognitive content in a more radical way, whence universalization, the process whereby one proceeds from the local to the general or globally accepted (universal) knowledge claims becomes problematic. Thus we have a relativization of the universal which is not there in Sivin's social turn. He admits the entry of sociology and anthropology of knowledge and how this led to scholars seeing the worldview or cosmology, or science of 'people in the past', or elsewhere, but also in the West, as simply a portion of the single reality that people constructed as they grew up.[32] With SSK, this postulate was formulated in Durkheimian terms, according to which natural order replicates social order. More recently we talk about the co-evolution or co-production of social and natural order, or of knowledge and social order, or better, of epistemic claims and socio-political regimes.

Another dimension, even more important, is the normativity inherent in Needham's analysis. His work is at the same time permeated with an ethos of activism which calls attention to its own situatedness as part of a collective knowledge-producing process, linked to a sense of moral order, that of claiming responsibility for promoting science so as to enhance the welfare of humankind.

Already, with Sivin the distinction between the social, spiritual and physical worlds, taken for granted by scholars of the two earlier phases of historiography, then becomes subject to deconstruction and relativization. The distinctions we, or historians before us, make are recognized as not necessarily subsisting in the minds of those we/they study. In as far as we attribute similar categories and distinctions to the 'other' it is because we project our own categories into historical material, using these categories to structure and make sense of this material.

[31] Sivin, *Science and Technology* 1977, p. xv.

[32] Nathan Sivin, 'Over the Borders: Technical History, Philosophy, and the Social Sciences', *Chinese Science*, vol. 10, 1991, pp. 69–80.

To be sure, the need to impress people with the fact that the Chinese had a lot of science is long gone. Now it is common knowledge in most parts of the world, says Sivin, even amongst school children and on TV. 'There is no point in striving to prove it now. The game has long since been won.'[33]

As for the question of achievements and ranking, or the problem of the 'scientific revolution', these have at this stage become stale issues. One is urged to move on to more productive but neglected research issues, ones prompted by the social and cultural studies of science. What matters now is that we are able to begin comparing several strong traditions of science and technology based on the ideas and social arrangements of different civilizations. All of them must be closely studied if we want to understand the general relations, through history and across the globe between science and culture, science and society, science and individual consciousness. Without that understanding we will remain trapped in our own parochial view. Historians must have more urgent work to do than trying to prove the inferiority of every other culture to the one in which they specialize.[34]

For Sivin, as distinct from Needham, the privileged position of the West comes not from some mystical dynamic like technological determinism, but from a head start in the technological exploitation of nature and the political exploitation of societies not technologically equipped to defend themselves.[35]

Coming back to a couple of the points above, a few more things may be said, especially apropos of the fourth point, which has been explicated by Lewis Pyenson in a whole series of works. Pyenson has noted how different large European powers had different styles and strategies in their colonization, and how this has had a bearing on the shaping of scientific discourses in colonial countries, discourses which, at the same time, amalgamated impressions from the local context. The fact that science or scientific research practices in a colonial country, qua cultural imperialism, is a vehicle or beach-head in the overseas entrenchment of metropolitan domination, does not make

[33] Sivin, 'Over the Borders', 1991, p. 79.

[34] Nathan Sivin, 'Why the Scientific Revolution Did Not Take Place in China—Or Didn't It?', *Chinese Science*, vol. 5, 1982, pp. 45–66, esp. p. 48.

[35] Ibid., p. 52.

it less viable as a mode of scientific knowledge, albeit one rated less advanced by scientists in the metropolitan centres. Indeed, it is thanks to science's feature of doing the tasks assigned to it, be it inventory of natural resources, cartography, or experiment and application of specialized knowledge for the institutionalized motive of commerce, military conquest or something else, that scientific specialists of various categories were able to function as agents for the propagation of the metropolitan culture of their home country, just as colonists and high commissioners, governors and judges as functionaries were agents of social and economic expansion.[36] In the same way, one finds that science has been used as proxy for political presence and large power prowess in real 'terria nullius' terrain like Antarctica. Here the exercise of scientific research is at one and the same time an act of science and politics. In other words, science functions as a continuation of politics by other means, and is able to do so just because of its ability to contribute to a nation's accumulation of exact knowledge.

With respect to point five above, we note how in the West an influential scheme is the positivist one, coming out of the emancipation from feudal hegemony and institutionalized religious thought. Thence, as in Comte's scheme, the positive territory to which a science refers is taken as definitive for its boundaries. Marxism too, in Engels' (realist) concept of irreductive materialism, takes a similar approach. Later, method became an important boundary-managing device, as with Dilthey, who set apart the Geisteswissenschaften from the increasingly successful and powerful Naturwissenschaften in the state-steered realm of German scholarship in the nineteenth century. This prompted the formation of a whole philosophy of science (hermeneutics of the humanities) as distinctively different from and counterposed to that of the natural sciences (allegedly, positivism). At a later stage, after the Second World War, we have seen tendencies of defining new disciplines on the basis of their social function, professional occupational links, or the anti-authoritarian and anti-establishment ethos of critical social movements. The notion of 'area studies' is an example. 'Science studies', or STS, itself may be interpreted as a

[36] Lewis Pyenson, 'Pure Learning and Political Economy: Science and European Expansion in the Age of Imperialism', in R.P.W. Visser *et al.* (eds), *New Trends in the History of Science* (Rodopi, Amsterdam/Atlanta, 1989), pp. 209–78, esp. the concluding section pp. 274 ff.

manifestation of a field thus shaped at a particular historical conjuncture. Lyotard thus refers to this field as an expression of a new reflexivity emerging in a post-modern age.[37]

Taking Sivin's premise that cultural factors contribute to the kind of demarcations that are made in science, or, in other words, that the general concepts of different branches of science also are culturally determined, as our point of departure we come back to the question of what role a scientistic self-understanding plays in this respect. Now, we can ask, with the onset of imperialism and neocolonialism, to what extent can one find similar or comparable trends in other civilizations, influenced by events in the (thereupon defined) metropolitan countries. This question brings in dimensions of world picture and philosophy.

This question is still posed from a moderate programme of contextualization, which does consider the more radical activist normative quality and question of situatedness as posed by, for example, Donna Haraway.

When it comes to defining a speciality by its 'intent' or use context, the example of geomancy in traditional China is a good one. Geomancy is the specialist study of wind and water, which decides the auspicious placement of houses and tombs with respect to features of the landscape, like aesthetics of land use, attention to filial piety and tradition, avoiding violation of the 'spheres' of deceased elders. Geomancy contains the typical expression of what has been called the concern with 'the propensity of things', an idea quite different, as a constitutive element of knowledge, from the principles of cause and effect or intelligibility at the core of modern science.[38] 'Mathematics, geomancy, and so on, each comes equipped with its own social meaning.'[39] There was no common meaning in line with our meaning of science, or even that of natural philosophy in previous centuries in

[37] J.F. Lyotard, *The Post-modern Explained* (University of Minnesota Press, Minneapolis, 1993), cf. p. 20.

[38] Jullien, *The Propensity of Things*, p. 92.

[39] Sivin, 'Why the Scientific Revolution Did Not Take Place in China', p. 49. On the geomancy tradition, still very much alive in China today, see also Stephan D.R. Feuchtwang, *An Anthropological Analysis of Chinese Geomancy* (Ventiane, Laos: Vithagna, 1974). Jullien, 1996, also refers to Ernest J. Eitel, 'Fengshoui ou Principes de science naturelle en Chine', *Annales du musee Guimet* 1(188) and J.J.M. de Groot, *The Religious System of China*, (E.J. Brill, Leiden, 1967), vol. 3, ch.12.

Europe. Nor were there individual sciences influenced by philosophy in China. Philosophy just did not provide a unitary amalgam. Thus, to try and assess to what extent such thought measured up to 'science' in the modern European meaning just does not make sense, unless of course one falsely wants to claim the superiority of one over the other. The 'scientific revolution problem', Sivin concludes, is, nevertheless, heuristically fruitful, even if it does not lead to any answers with explanatory prowess. It certainly helped open up and sensitize us to new dimensions in our own history.

Where does Needham Come From?

All too frequently, the appreciation and critique of Needham's work focuses narrowly on the quality of his scholarship and fails to bring out sufficiently the circumstances that contributed to driving his normative and moral activist (ecumenical) passion to set the record straight. Rather than rehearsing his major claims regarding internal and external factors in the development of science and technology in China prior to the sixteenth century, my focus here will be on the wider context that gave meaning to his life and lifelong project.

The ultimate aim of Needham's monumental project *Science and Civilization in China* was to promote mutual understanding among different cultures.[40] The turn to things Chinese came in 1936, when he was an active member of the movement of natural scientists against fascism and for a socially responsibilist stance in research. As such, he is usually counted as one of the inner group of Bernalists. Personal friendship with three young Chinese scientists, Lu Gwei-Djen, Shen Shi-Chang and Wang Yi-Lai, who came to work in the Cambridge laboratories at that time, was an important catalytic factor. As he used to say, 'the fact that as scientific minds they were so much like my own raised vividly in my consciousness the historical problem of why modern science had originated in Europe alone, and not in China or India'.

In 1942, Needham went to China to head the British mission to Chaing Kai-Shek's Koumintang government, and went on to become a science adviser. It was this experience in organizing the British Scientific Office in Chunking, one of the most successful wartime scientific cooperation offices, that became the model for a proposal

[40] Goldsmith, *Joseph Needham*. p. 146.

to establish an international science cooperation service, meant to function in parallel with other international organizations, such as the International Labour Office, the Permanent Commission for Food and Agriculture, the International Health Organization and the like, to promote peace and reconstruction in the post-war era. Even if such an organization never did emerge, Needham, through the circulation of a number of important memoranda and through his participation, as an expert, in the talks amongst Allied countries on creating a supranational intergovernmental organization for culture and higher education was instrumental in seeing to it that the 'S' for science got into UNESCO when this body was created in 1945.

In one of his memoranda, Needham argued that international collaboration in science, if it is to benefit all nations, cannot be left to the spontaneity of market forces. The fundamental error of believers in *laissez-faire,* he said, is that they look at the scene too exclusively from the EuroAmerican point of view, that is to say, they think of oscillating between Paris, Brussels, London, New York, Washington, Montreal and the like. They do not realize that the picture of the world looks very different when seen from Romania, Peru, Java, Siam or China. For historical reasons, since modern science grew up in Western Europe, there is a 'bright zone' covering Western Europe and North America, where all sciences are very advanced and industrialization is highly developed. It is particularly the scientists and technologists in the far larger regions of the world outside the 'bright zone' who need the helping hand of international science.[41] Against the '*laissez-faire* principle', Needham thus mounted what he called the 'periphery principle', to concentrate on scientific and technological aid to what we today would call the Third World.

From the Creation of UNESCO to the Cold War

Then UNESCO was created in 1945. Its first Director General was Julian Huxley the biologist, and Joseph Needham was made the first Head of the Natural Sciences Section. In this capacity he was instrumental in supporting history of science as an academic discipline, doing so with clear externalist ambitions which coincided with his

[41] Joseph Needham, *The Place of Science and International Scientific Cooperation in Post-World War Organization*, Memorandum III, 15 March 1945 (UNESCO Archives).

concern to understand both external and internal 'determinants' or regulatives in the growth of science. From the archives of the International Council of Scientific Unions (ICSU) and UNESCO it is evident that this coincided with insights that had their origins in the debate on the social relations of science in the 1930s. The threat of fascism, events in Italy, Germany, China and elsewhere had a consciousness-raising effect on many scientists, especially natural scientists, during the 1930s and early 1940s, in the latter instance also as advisers and experts in the Allied war effort. Bernalists in Britain are well known, but in other countries too there were similar anti-fascist groups of scientists and intellectuals.[42]

It is interesting to note how the emphasis was on the impact of science on society, and not the other way around, what we today would call the social shaping of science. The Dutch Royal Academy of Sciences was one of the main advocates of this effort, but their more radical formulation was toned down in ICSU.

Huxley and Needham later left UNESCO to go on to do other things, and the social and external history of science approach lost important supporting forces. The science and society effort in the historiography of science turned out to have been promoted from outside the community of scholars. Amongst historians of science organized by the International Union of History of Science (IUHS), professionalization of the discipline and, with it, a turn to internalist intellectual history became more and more prominent, reinforced, during the cold war, by a concern for disengagement from politics and ideology. In reality this turned out to be a harmonization with the politics and ideology of the 'Free World', the West, under the cloak of a professional ethos of academic purity. Other committees within the IUHS, dealing with bibliographing internal historiography, or recording the development of ancient science in West Asia and Greece overtook the Commission on the Social Relations of Science (CSRS) in terms of funding and intensity of activities. The documentation of ancient science in West Asia, of course, was significant for future comparative studies, but otherwise, strictly speaking 'externalists' as they were later called, after the critical reception of Lilley's essay, were externalized (marginalized).

UNESCO, on its part, continued to support a more scientistic variant of social studies of science through its special journal, *Impact of*

[42] For a detailed discussion, see Petitjean's essay in this volume.

Science on Society, published and distributed from Paris, first appearing in 1950. The first issue still bears the imprint of the socially responsibilist concerns, assembling, as it did, a vast bibliography on science, its social roots and its social impact, a project started by the CSRS of the IUHS. Well into the cold war, and with the entry of the USSR as a member of UNESCO after Stalin's death, there was a remarkable depoliticization of *Impact of Science on Society* too, followed by a more technocratic orientation around the systems-theoretical 'science of science' approach that later emerged in the Soviet Union and Eastern Europe, coinciding with a reformist (the weaker) version of Bernalism.

What was the Externalism–Internalism Debate About?

Steven Shapin some years ago pointed out how the old internalism–externalism controversy regarding historiography has been left behind without really ever having been resolved or played out properly.[43] The relationship between the two sets of factors, internal and external, he notes, is asymmetrical, in the sense that internalist historiography can flourish on its own, while externalism cannot. It is parasitic on the intellectual content of whatever is under consideration from a contextualizing point of view. A total externalist historiography would make no reference to scientific ideas in their own right, but reduce everything to the cultural or societal dimension, which becomes an empty absurdity. Thus, internalism is always posed as primary, while externalism is supplementary.

A second point is that as theories, internalism and externalism were never very well characterized in their own terms. It was more a matter of a figure of speech, a thought figure, or two rather broad approaches. The debate never focused sufficiently on the conditions under which an external factor could be taken to be sufficient cause for a specific kind of cognitive development in history. Also, it remained largely unclear what internalism–externalism really represented. Were there rival theories of scientific change, research programmes, heuristics, or what?

Once Samuel Lilley's now classical essay and similar work had been classed as extreme (or Marxist—in turn equated with mechanical)

[43] S. Shapin, 'Discipline and Bounding: The History and Sociology of Science as Seen Through the Externalism–Internalism Debate', *History of Science*, vol. 30, 1992, pp. 334–69.

externalism and rejected by most 'serious' (i.e. 'professional') historians of science by the late 1950s and early 1960s, the internalism–externalism opposition was accepted as something to which lip-service was paid in the instance of either a purely internalist pursuit or a balancing of this with eclectic references to externalist dimensions. In other words, externalism was replaced by eclecticism, an eclecticism that made use of the best of the real externalism that had emerged in the 1930s and immediately after the end of the Second World War, but without admitting this and therefore recognizing externalism as a viable historiographical programme.

So-called extreme externalism was a strawman, a bogus construction erected by professional internalists to legitimate their own turn away from the social responsibility of science perspective that had once fed the most avid externalists. This rejection was, in essence, a boundary management exercise, at the two levels of marking out ideological identity during the cold war and of staking out a distinct discipline. Professional ethos, liberal ideology and the quest for a disciplinary identity coincided and tended to reinforce one another in academe.

During the cold war, in the West many a researcher in the scientific community, historians of science included, increasingly sought to portray an ideal picture of science as a disembodied, factually oriented, impartial, neutral/objective pursuit. Indeed 'objectivity' came mistakenly to be conflated with the absence of social partisanship ('neutrality'), which is actually not the same thing. In Needham's ideal of science, for example, the striving for objectivity is closely linked to his sense of working in an ecumenical moral order, which feeds his normativity and humanistic activism. If we look more closely, the two can and do interact, and need to do so, dialectically—objectivity being a relationship in the epistemic domain, where objects of research and procedures for validation are constituted as elements of a scientific discourse, while partisanship is the conscious expression of the broader institutional motive of research work in a societal utilization, or a moral context, actual or intended.[44] A totalization on the one side leads to objectivism and pins the problematique to epistemology, whence realism–relativism dichotomy becomes a cul de sac, and ideology is at par with scientism. A totalization on the other side

[44] Aant Elzinga, 'Objectivity and Partisanship in Science', *Ethnos*, nos 1–4, 1975, pp. 406–27.

introduces subjectivism, feeding into romanticist ideals of science.

Some of the confusion in the internalism–externalism discussion also comes from the equation of the external with the social, as opposed to the ideational as cognitive and internal. In fact, as Thomas Kuhn, as much an internalist as anyone else, demonstrated in 1962, social factors are very much part of the story of disciplinary development.

Recent scholarship, after the end of the cold war when archives from 1945 to 1949 are being opened up, reveals that scientific communities in those days became very concerned with their professional self-image. Moreover, it appears, this image moved from externalism to internalism, thereby tending to reinforce the ideal of scientific communities as factual, neutral, objective and impartial, all the more as the reality of science *groso modo* tended to be just the opposite. The major dynamic force for physics at the individual level might have been intellectual curiosity, but at a collective level of institutional motives it was something else, viz. the growing military–industrial expansion of Big Science. This intermeshing of post-war physics with military and military–industrial goals has been carefully researched by Paul Forman, among others. He has shown how the defence effort in the United States, for example, set the agendas of basic research in certain branches of physics.[45]

Steven Shapin suggests that the internal–external boundary-drawing process, of which UNESCO and Needham were an institutional part on the losing side, did not have much to do with historiography. It had more to do with the cultural narrative of the cold war politics of representing science as an impartial neutral affair. To this may be added Nathan Sivin's point about historians' concern with professionalism. The two agendas appear to have been mutually reinforcing.

Steve Fuller, in connection with work on the history of Thomas Kuhn and his paradigm theory, has looked into the archives at Harvard, where he finds how early history of science course proposals just after the war made ample reference to Bernalism, social class and the social function of science. These were, however, screened out to promote a more palatable image of the detached expert, depoliticized and neutral—the puzzle-solving professional who will fill an important

[45] Paul Forman, 'Behind Quantum Electronics: National Security as Basis for Physical Research in the United States 1940–1969', *Historical Studies in the Physical and Biological Sciences*, vol. 18, pt 1, 1987, pp. 149–229.

slot in industry. Neither the social nor the technological base of science, nor the fact that science costs money was discussed. In other words, an image of science disembodied, floating in a world of ideas in the air, above society was projected. Fuller thus argues that Kuhn's work, started at Harvard under J.B. Conant, as part of a programme for liberal arts and scientific literacy for veterans returning after the war, was also part of a tacit design—the fortification of an internalist self-image of science, a kind of public relations strategy to promote science in a business-as-usual mode.[46]

After the war, this image of purity was further perpetuated in the image of the strict separation of basic from applied science, and of science from society in Vannevar Bush's influential science policy treatise, *Science—The Endless Frontier* (1945). Here we find the disembodiment of science abstracted into a linear model of the innovation process. In this there is a successive chain of events that leads from discovery or some basic scientific idea, hatched in the mind of a researcher, to the maturing of concepts, to a possible application and finally the development of a product or process for the market, i.e. an 'an innovation'. Note how the very definition of 'innovation' was given a tacit social (commercial) connotation. This linear innovation chain model came to be a powerful metaphor which, because of its simplicity, still guides the thinking of many a science administrator or policy-maker; this metaphor conveys the image of 'purity' speaking 'truth' to 'power'.

It is only today, with the advent of high technology, strategic research, the end of the cold war and current globalization processes that it is being challenged. Don Price early on pointed out how this model with its dichotomization of science from society was congenial for both scientists and politicians. It suggested that the greater the purity of universality in scientific knowledge claims, and hence of the disembodiment from the socio-historical contexts which gave rise to them, the greater also the clout for political action or technological trajectories that were legitimated by such claims. Something similar is still at work today, especially with the increasing frequency of an

[46] Seminar discussion at the Department of Theory of Science, University of Goteborg, Spring 1995; cf. Steve Fuller, 'Being there with Thomas Kuhn: A Parable for Postmodern Times', *History and Theory*, October 1992, pp. 241–75; also Steve Fuller, 'Teaching Thomas Kuhn to Teach the Cold War Vision of Science', *Contention* (UCLA, 1994).

element of 'risk' and uncertainty surrounding, what Funtowicz and Ravetz have called, post-normal science.[47] This latter concept counterposes scientific knowledge and technological projects with a high degree of uncertainty and risk respectively, under duress of political pressure, with 'normal science', in which knowledge claims are ideally very certain and political pressures or partisan ideological influences are absent. Now, as then, it is in the interest of both scientists and politicians to keep the boundaries between science and society clean and clear, at least in popular and textbook perceptions. Thus, the imagery at the core of the 'professional' side of the 'internalism'–'externalism' discussion seems to be constantly reproduced socially and culturally in present-day society, despite all the talk to the contrary, for example in the guise of the post-modern.

The Cultural Turn and the Notion of Modernity

From the foregoing we can see that Needham was a modernist, who, like Montesquieu, believed that ethnocentric prejudices are undermined by the experience of otherness. Further, it is only possible fully to understand the self after encountering and experiencing the other, and only in this way, by beginning with the attempt to understand the other, will knowledge approach universality and escape ethnocentrism.[48]

Richard Devetak, in a Habermasian critical reconstruction of a 'renewed project of modernity', has pointed out how the spirit of cosmopolitanism, and the question of 'community', with universal emancipation as a regulative ideal is a central tenet in such an interpretation of modernity, which rejects the lack of reflexivity and naivete present in social evolutionism. Another tenet is the ethos of critique. 'Together, the ethos of critique and the spirit of cosmopolitanism constitute the central features of modernity. They are thought to make emancipation possible.'[49] In a way, we can say, Needham in the course of his lifetime moved in this direction, of a

[47] Silvio O. Funtowicz and S. Ravetz, 'Science for the Post-normal Age' *Futures*, vol. 25, no. 7, 1993, pp. 739–55.

[48] Richard Devetak, 'The Project of Modernity and International Relations Theory', *Journal of International Studies*, vol. 24, no. 1, 1995, pp. 27–51.

[49] Ibid., p. 35.

more hermeneutically reflexive modernity, even if he probably never reached it in the way Anthony Giddens and Ulrich Beck have elaborated such views.[50]

It is interesting to note also how Needham's defence of modernity as emancipatory and democratic was adamant in the need to criticize fundamentalist views that call for, say, the Islamization of science, as well as the other side of the coin, the pitfalls of cultural relativism symbolized by the work of Spengler. In the fifth volume of *SCC*, in the 'Author's Note' there is an interesting polemic. He warns against an absolutization of the notion of ethnoscience, used to underline the achievements of eastern cultures or those of other civilizations, like ancient Egypt, India and China. That would be to deny the universal validity of certain scientific findings independent of geographical and cultural meridians. Relativism on the one hand, and the claim for a need for a fundamental resacralization of science, in tune with the values of some other civilization made absolute, Needham, it appears, saw as twins on the tree of ethnocentrism, where Eurocentric scientism is a further branch, also to be rejected. A distinction is made between values and science, and between scientism and science. Thus he writes:

> There is a danger to be guarded against, the danger of falling into the other extreme, and of denying the fundamental continuity and universality of all science. This could be to resurrect the Spenglerian conception of the natural sciences of the various dead (or even worse, the living) non-European civilizations as totally separate, immiscible thought patterns, more like distinct works of art than anything else, a series of different views of the natural world irreconcilable and unconnected. Such a view might be used as the cloak of some historical racist doctrine, the sciences of pre-modern times and the non-European cultures being thought of as wholly conditioned ethnically, and rigidly confined to their own sphere, not part of humanity's broad onward march.[51]

For Needham it is clear that extreme cultural relativism makes nonsense of any idea or attempt at cross-cultural comparison, or dialogue. The comparative study of technology and culture as a project falls apart, or it becomes an idiosyncratic absurdity. Needham argues that the

[50] Ulrich Beck and Anthony Giddens, *The Consequences of Modernity* (Polity Press, Cambridge 1990).

[51] Needham, *SCC*, vol. 5, p. x.

US-based Persian historian of science, Seyyed Hossein Nasr's gnosis-oriented call for a resacralization of science is guilty of a falsified reconstruction of the history of science. Moreover, Hossein Nasr's own view is seen to be anti-democratic, since it involves a denial of the equality of forms of human experiences, while privileging the particular Islamic one he supports. This is in conflict with the basic democratic ethos of modern science, interpreted not as a European virtue but as ecumenical good. Gnosis as distinct from episteme leads us onto a road of obscurantism, not enlightenment and emancipation.

Hossein Nasr continues to this day to maintain his critique of modern science, founded on what he calls its immanent scientistic value base, which he equates with its European heritage, and asks for a retranslation of science into the value base of Islam. He is one of various voices in a discussion concerning the Islamization of science, an issue on which there is considerable diversity of opinion as to basic view and strategy. In a recent paper on Islam and Science, delivered in Islamabad (1995), Hossein Nasr points out the political importance of this question in a world where globalization has become a catchword and the Gulf War symbolizes high tech.

> Of course, one can understand why the blind praise and almost 'worship' of modern science and technology are manifested everywhere among sectors of Islamic society with profound political, theological, and juridical differences when one sees a war in which some sixty people are killed on the other side, and nearly three hundred thousand Muslims with others from Bosnia to Chechnya to Kashmir to the Philippines, the Muslims are always on the defensive and the ones to be massacred and 'cleansed'. Yet this desire to adopt western science and technology blindly and without critical appraisal, no matter how understandable on the emotional level or that of political expediency or even necessity, cannot remain heedless of truths that underlay the profound issues of the relationship between the Islamic world view and modern science.[52]

In Hossein Nasr's view, modernization in the Western sense, and with it self-congratulatory Eurocentric historiography of science and technology, have helped create a mirage which must be revealed for what it is.

[52] Seyyed Hossein Nasr, 'The Islamic World-view and Modern Science', *The Islamic Quarterly*, vol. 39, no. 2, 1995, pp. 73–4, quote from p. 74.

From the foregoing it is evident that the issues of the debate in which Needham was involved through the years have not abated. If anything they have intensified, particularly when we consider the post-cold war world order and the articulation today of so-called post-colonial voices, or the concern with how statements, such as Needham's and Hossein Nasr's, are loaded with a particular rhetorical flavour meant to enhance 'other' cultural identities.

Multiculturalism

With the reflexive cultural turn, we find ourselves in a new discourse in which the project of comparative studies obtains a new meaning. Now the rhetorical dimension of accounts about science and their role in the promotion of particular identities, be it one of expertise, or specialist disciplines is subjected to cultural analysis. In the context we are discussing, the main question is: what is the process parallel to the one Edward Said and Martin Bernal have described as the socio-historical shaping of 'orientalism' as a discipline pursued as unbiased professional academic study, which discounts and hides its own filiation with an overriding Eurocentristic social evolutionist cultural narrative? In our review of the closure of historiography around 'internalism', in tandem with the strictures of the cold war, we have already indicated how the politics of identities, even in historiography, can change with time and societal context. At the same time, Needham's work remains a monumental benchmark that has survived through this era. Instead of its contradiction with traditional Eurocentrism today, the significance of Needham lies on the other side, as a demarcation vis-à-vis the incommensurability present in the kinds of fundamentalism and hyperrelativism that he argued against in the cases of Hossein Nasr and Spengler, who were designated as two of a kind.

In order to try and grasp what this new discourse might imply, let me make an entry into it with Edward Said's *Culture and Imperialism*.[53] Said describes his method as focusing, as much as possible, on individual works of literature, 'to read them first as products of the creative or interpretative imagination, and then to show them as part of the relationship between culture and empire'.[54] This is a kind of

[53] Edward Said, *Culture and Imperialism* (Alfred A. Knopf, New York, 1993).

[54] Ibid., p. xxii.

bifocal view, shifting perspective from first content then context. Its aim is to demonstrate diversity and complexity of human experience. Binary oppositions should be avoided, and alignments, intersections across cultural borders and how these are constituted should be brought to the fore instead. Partly because of empire all cultures are involved in one another, none is single and pure, all are hybrid, heterogeneous, extraordinarily differentiated, and unmonolithic. Said pleads for the bringing into evidence of rupture, difference, as reflected in counter-narratives, those of the nomad intellectual traversing boundaries in many dimensions, to display multiculturalism, fed by a non-adversarial community of awareness and anti-imperialism.[55]

This is in an effort to prevent closure around totalizing features that are mistakenly taken to be the essence of identities. In other words, it becomes a matter now of rhetorically binding the European and non-European together in a kind of new universalism. In this project, when compared with Needham's, the emphasis is more on the cultural dimension, on questions concerning how people perceive and understand themselves, more than the social and material substrate pointed to in traditional Marxist analysis, where technology is merely a force of production, an instrumental tool. Now technology becomes culturally loaded, recognized as the repository of a multitude of possible self-interpretations, which one being assumed will depend upon context, tradition, and other contingencies. '... We must all write our histories and cultures rescriptively in a new way; we share the same history, even though for some of us that history has enslaved. This, in short, is writing from the colonies coterminous with the real potential of post colonial liberation ...'[56] 'It is a question of mocking the linear (social evolutionist) cultural narrative by embracing the hybrid, the mixture of real and surreal in all experience.'[57] Failing this we end up in binary logics that affirm the hegemonic symbolism of say the Gulf War and the imagery of 'surgical' military–cybernetic technologies.

The programme that follows has several points: First, by a new integrative contrapuntal orientation in history that sees Western and non-Western experience as belonging together because they are connected by imperialism. Second, by an imaginative, even utopian

[55] Ibid., p. 274.

[56] Ibid.

[57] Ibid., p. 276.

vision which reconceives emancipatory (as opposed to confining) theory and performance. Third, by investment neither in new authorities, doctrines, and encoded orthodoxies, nor in established institutions and causes, but in a particular sort of nomadic migratory, and anti-narrative energy.[58]

What this seems to mean is a methodology that would juxtapose things from different cultures instead of striving for purity, to counter ethnic purification at the cultural level. It means playing up the multivalence of identities rather than putting single labels onto people and things. As a movement this is held to be one of audacity and generosity, that looks to humankind, and not to digging oneself into (ethnic) trenches of spite and hate, fired in a binary way.

In this respect Said would probably agree with Needham against Hossein Nasr and other totalizing narratives, whether they come from East or West. In the politics of civilizational identities today, Islamization is a strategy that plays into the hands of its binary counterpoint, the new hegemonic agenda of imperialism under the mantle of globalization. Instead of opting out of globalization Said seems to opt for a strategy of supporting local counterpoints of resistance in a new mode. I guess he would agree with Everett Mendelsohn also, who observes in an article on religious fundamentalisms, how the 'ultimate subordination to sacred authority ... marks fundamentalists' approaches to modern science and technology'.[59] All of these writers who are concerned with a common aim of promoting mutual understanding prefer the secular mode, well aware that this may be construed to be tainted with Western repression, but believing that such a simplistic view must also be rejected since it fails to see what Needham already tried to make clear, viz. that 'the modern' is no longer 'the European', nor does Europe have a monopoly on modernity.

Critics of Said's previous book, *Orientalism*, have held that he overemphasizes the cultural dimension, to a point where 'racism' is identified as a trope in modern times only to be projected backwards into history to the times of the Greek *Iliad* and Homer. Therewith

[58] Ibid., p. 279.

[59] Everett Mendelsohn, 'Religious Fundamentalism and the Sciences', in Martin E. Marty and R. Scott Appelby (eds), *Fundamentalisms and Society* (University of Chicago Press, Chicago and London, 1993), pp. 23–41, p. 25.

the concept of racism loses its connotation as a category associated with a particular mode of economic householding that has been exploitatively globalized via imperialism. Exaggeration of the cultural dimension and concern with the politics of identities also deflects attention from the kinds of external factors Needham points to as important in the development of modern science, for example the space created for investments in science by virtue of accumulation of surplus capital in the process of exploitation of both labour and the colonies. This is still a distinctive feature of the history of Western science, especially its seventeenth-century revolution.

In other words, one has to be careful not to put the burden on representations and literary and other imagery per se, reconstructing these and then pushing them endlessly back into history as specific tropes. Otherwise we open up a kind of orientalism-in-reverse, says Sadiq Jalal al-'Azm in an article with this title.[60] According to him, playing up the cultural dimension, without cognizance of the material economic one, suggests a reversal of the real relationship, to one where academic orientalism is supposed to have created institutional orientalism, and not just part of its legitimatory cultural goods or deeper narrative. It seems to assume the textual and the contemplative qua orientalism (the discipline) preceded and shaped the administrative, economic and even military orientalism, which definitely came much later than the minds, texts, aesthetics, representations, lore and imagery of Westerners that go as far back as Homer, Aeschylus and Euripedes. This is a pitfall inherent in a textual attitude to reality. Orient and occident are fundamental ontological categories and classificatory schemes, coproduced in an historical process. Pushing the discovery of its character as a social construction in space and time to an ideological ontologization back into history before the time of its actual emergence, al 'Azm holds, is tantamount to a new kind of essentialism.

[60] Sadiq Jalal al-'Azm, 'Orientalism and Orientalism in Reverse', *Khamsin,* vol. 8, 1981, pp. 5–26. This is a shortened version of an earlier paper in Arabic. The article has since then been reprinted in an anthology, but I have no reference available for this. See, however, Emmanuel Sivan, *Interpretations of Islam: Past and Present* (Princeton, 1985). Further discussion may be found also in A. Rogers, 'The Boundaries of Reason: The World, the Homeland and Edward Said', *Environment and Planning: Society and Space,* vol. 19, 1992, pp. 511–26. Also see Sadiq Jalal al-'Azm, 'Islamic Fundamentalism Reconsidered: A Critical Outline of Problems, Ideas and Approaches, part II', *South Asia Bulletin*, vol. 14, no.1, 1994, pp. 1–27.

There is a problem here, certainly. On the view that the orient is only and can only be a 'representation' in the mind of the culture of the occident, orientalism becomes the specialized repository for this genealogy of representation. However, orientalism and the notion of the 'orient' as a representation have to be distinguished. The former is an academicized institution whose fortunes were linked to the process of European colonization and empire, which provided its institutional motive and narrative underpinnings in European culture. The latter, in Said's analysis, is a representation that involves several things: (a) the domestication of the alien, (b) a complete transformation, an interpretative package of concepts linked to the interests of the receiver, (c) reinterpretation of the 'raw reality' as the orient so as to make it understandable in the matrix of the occident, to change its elements from free-floating objects into units of knowledge, and (d) it follows the natural bent of the human mind in resisting the assault on it of untreated strangeness. The 'other' is retranslated into an image closer to the interpreting self. This is a grossly hermeneutic problem, and I shall not endeavour to go further into it here. It may well be that Said gets out of the pitfall by making clear the bifocalism in his point of departure, first content then context, especially if context is taken to include socio-economic materialities.

We find a similar tension between emphasis on the cultural contra the spatio-temporal material dimension in the dialogue between Donna Haraway and David Harvey, where the latter is more concerned with the circulation of money and its relation to the circulation of capital when it comes to 'situatedness'.[61] In this case, the circulation of imagery, texts, aesthetics, representations, lore, etc. as well as the circulation of money and (institutional) capital becomes interesting. As already indicated, it was the latter and its expansion in terms of surplus, together with a 'circulation' of instruments and techniques in the mercantile cities of Italy, and later on the Netherlands, that contributed substantially to the opening of a 'space for science' in the modern sense.

At the same time, Haraway reminds us that contingency and relationality are not equal to relativism.

[61] 'Nature, Politics, and Possibilities: A Debate and Discussion with David Harvey and Donna Haraway', *Environment and Planning: Society and Space*, vol. 13, 1995, pp. 507–27.

Radical historical contingency is about understanding the modes of materiality in the world; that understanding issues of possibility or understanding issues of apparatuses of production is materiality; and that it's semiotic all the way down. There is no gap between materiality and semiosis; the meaning-making processes and materiality of the world are dynamic, historical, contingent, specific; that bodies and institutions and machines are made, not made up.[62]

We may ask 'What kinds of connectivity globalize' and to what end?[63] At the same time, it is important to bear in mind that the concepts we use to point to transboundary entities, like the cyborg, are after all at one level a 'configurational move' in a politicized narrative, and not a literal statement about the world. The trouble is that 'some folks read it that way too, using it in a celebrational mode, thus missing the argument that the cyborg issues specifically from the militarized, indeed a permanently warbased state based, industrial capitalism of World War 2 and the post World War 2 Cold War'.[64]

In much the same way as Aijaz Ahmad argues, it seems that the terms 'colonial', 'post-colonial', and 'post-coloniality' too have been stripped away from their original connotation, as terms referring to a periodization of history as regards the rule over peoples of particular countries by ruling classes of other countries, to push them back to such configurations as the Incas, the Ottomans and the Chinese, well before the European colonial empires began.[65] This construction of a global transhistoricity of colonialism, it is argued, evacuates the very meaning of the word, dispersing it so wide that we can no longer speak of determinate histories of determinate structures, such as that of the post-colonial state, or of such states as vehicles of a compact between national and imperialist capital. And, we might add, it delinks from the activist ethos that nevertheless may be found amongst those who, like Needham, seem to represent a possible extension of the notion of modernity.

Hybridity, ambivalence and contingency identified by cultural analysts are important, but they should not be essentialized. Such

[62] Ibid., p. 509.

[63] Ibid., p. 511.

[64] Ibid., p. 514.

[65] Ibid., Aijaz Ahmad, 'The Politics of Literary Post-coloniality', *Race and Class,* vol. 36, no. 3, 1995, pp. 1–20, especially p. 6.

concepts become powerful tools of analysis only when they are incorporated into the concrete analysis of specific materialities and conditions within a frame that also can say something about the structural dialectic of imperialism. Aijaz Ahmad concludes: '... The celebratory invocations of a uniquely postmodern hybridity, uniquely available to migrants residing in metropolitan cities and emanating from there to the rest of the globe, is, in a profound sense, simply the opposite of cultural differentialism.'[66]

The latter trades on a binary finalist notion of going-back-to-the-roots in a mythic past which is posited as the only true moment of cultural authenticity, as in religious revivalism and protofascist nationalisms. It abstracts from this to hypostasize an ever-present unique ethnic authenticity which is anti-historical and ever-divisive. Post-modern hybridity on the other hand, countering myths of origin and myths of 'deep nation' and the 'long past', etc. tends equally into anti-historical abstraction, giving semblance to the myth that one is free to invent oneself and one's community in a constantly changing mirror of a multiplicity of cultural others. The notable thing is that post-structuralist thought lends itself equally to both stances, 'the identitarian-particularist anti-history as well as the globalising-hybridist anti-history'. For its part, cultural analysis which exclusively focuses on hybridity, in playing up contingency ends up denying structuration and institutional regulatives in the distribution of powers and privileges. This is always convenient for the powers that be, since it does not link the more substantial underpinnings to their tacit legitimations of this economic, political or professional privilege that meets our eyes and ears more superficially at the textual level.

On the one hand, as Needham pointed out vis-à-vis Spenglerian relativism, differences between cultures may be overplayed to the point that expression of the position of one in terms understandable to the other may seem to be in danger of being ruled out in principle. The assertions of alleged differences become self-defeating since if the claims are true, they cannot be known as such. On the other hand, sensitivity to hybrid alliances presupposes a recognition of difference too. In both cases we end up with the problem already discussed in a more concrete fashion above, in the comparison of putatively Western

[66] Ibid., p. 18.

mechanistic and Eastern organismic world pictures. At bottom we are dealing with concepts that apply to a whole set of phenomena which separate science from other cultural manifestations. Just as in the case of the concept of 'religion', which comes out of our culture, other civilizations may not have these separating concepts, accepting them only later in the meeting with Western civilization, whence they are grafted on.[67]

How then do we get out of this blind alley in which cultural analysis of a hyperreflexive kind leads to an infinite regress, disabling us from doing anything but comparison as a work of bric-à-brac juxtapositions? One way, it seems, is to go back to Needham's point about European science also being an ethnoscience, which in confluence with other ethnosciences brings us to a 'modern science', ultimately ecumenical, when seen as a global system with a more equitable potential than the present one. The ethos attributed to this science viewed disembodiedly may then, together with its truth claims be contextualized, shown to be a product of a co-production of ideas and social order in a particular civilization that exports this as universally true and good. In other words, seemingly absolute universality is shown to be located or situated universality.

As Sandra Harding has argued recently, mainstream science studies have tended to leave intact 'the science vs. ethnoscience construct, and the Eurocentric, unbalanced valuing of modern vs. other traditions

[67] For the discussion of the concept of 'religion' and the presupposition of its differentiation as a specific realm, see Ursula King, 'Historical and Phenomenological Approaches', in Frank Whaling (ed.), *Contemporary Approaches to the Study of Religion, vol. 1: The Humanities* (Mouton Publ., Berlin, 1983), pp. 29–164, cited in David A. Paulines, 'Response to S.N. Balagangadhara's Study of the Understanding of 'Religion', *Cultural Dynamics,* vol. 8, no. 2, July 1996, pp. 171–87. This is one of several articles dealing with S.N. Balagangadhara, 'The Heathen in His Blindness ...', *Asia, the West and the Dynamic of Religion* (E.J. Brill, Leiden, 1994). Several of these articles touch on questions pertinent to the present discussion, including the one of how the scientific enterprise may originally have been grounded upon a particular religiously grounded understanding of the world as intelligible, whose emphasis on truth ultimately through scientism led to a cannibalizing of its own religious or metaphysical foundation. In the discussion in the journal, much is made of Balagangadhara's distinction between theoretical and performative cultures, which one author finds to be reified in the hands of that scholar, see David Loy, '... While the Scholar in His Wisdom Bows Down to Truth: Asia, the West and the Dynamic of Truth', *Cultural Dynamics,* vol. 8, no. 2, pp. 147–60.

that it legitimates'.[68] This is because, with their focus often on the laboratory or micro-level construction of knowledge, they privilege this activity, not explicitly challenging its implicit claim to universality in principle, and superiority vis-à-vis competing theories that might have emerged in other cultural traditions. They do show 'the cultural fingerprints on modern sciences, but do not confront the conflicting assumption that modern sciences are nevertheless special because of their internal features'.[69] In order to confront this question, which involves analysis of the historical shaping of the very claim to universality within a Western context, Harding points to the work of post-colonial 'development' theorists which might be fruitfully integrated with science studies so as to deal with questions regarding the role of European expansion in the advance of European ethno-science, one of the tributaries in Needham's 'modern' science, or the role of the self-assumed notion of 'intelligibility' of the cosmos or 'neutrality' qua 'objectivity'.

Donna Haraway also argues for greater sensitivity to our own situatedness, but appears to do so in a somewhat different way. Harding and Haraway draw, it seems, quite different conclusions from this awareness of the situatedness of knowledge. The latter assumes a more activist stance that appears to dissolve the boundaries between science and politics to a much greater degree. The differences and their implications are worth further discussion.

There is an important political dimension here too. As Susantha Goonatilake has argued, a multicultural polyvalent approach allows us to recognize what modernizing Western science has borrowed from its surroundings, including other cultures' knowledge traditions. Needham's corrective against the pitfalls of hyperreflexivity also lies here, obliging us at the moral and political levels to recognize the contributions of other peoples and cultures, and thence current knowledge as the common heritage of humankind. This principle is in jeopardy when multinational companies send out their experts to bioprospect ethnobotanical knowledge that is part of other traditions

[68] Sandra Harding, 'Is Modern Science an Ethnoscience? Towards a Postcolonial Epistemology for Science Studies', contribution to the ORSTOM–UNESCO Conference on Twentieth Century Science: Beyond the Metropolis (Paris, September 1994).

[69] Cited after ORSTOM-paper, p. 6.

of materia medica in order to incorporate this into the cycles of money-making and capital in the North, thereby expropriating the original innovators who then find themselves in the situation of having to buy their own therapeutic products repackaged on a Western-dominated global pharmaceutical market.

Unless we move from textual recognition and political reading of signs and symbols, to taking a stand on property rights and multinational corporations' control over resources emanating from ethno-epistemic practices in the Third (and Fourth) World(s), we are letting global robbery pass by in silence. A recognition of the situatedness of knowledge claims must therefore be relinked to a sense of equity, justice and solidarity in the global arena. In this respect Needham's own life history and his mode of boundary-crossing will always be worthy of emulation.

5

Steve Fuller

Prolegomena to a World History of Science*

Recovering a Sense of Historicity in the History of Science

A sense of *historicity* is essential for understanding the grip that science has had on the global imagination. I mean here something more than just an awareness of the temporal ordering of achievements that have been made in the name of Western science. Rather, by 'historicity' I mean the way in which one's perception of what has transpired in the history of science shapes one's sense of the prospects for future developments. The tricky thing about coming to grips with the historicity of science is that it typically involves examining the long-term consequences of people acting on the basis of profoundly false beliefs about the history of science. While it is now widely conceded that the most pervasive accounts of the history of science—those found in science textbooks and science popularizations—are misleading to the point of being self-serving, few have stepped back to wonder what this might mean for an activity that advertises itself as unique in its quest for some ultimate, singular and unified Truth.

*Many thanks to Dhruv Raina for his critical comments, to which I hope I have done justice. This essay has since been expanded into chapters 5–7 of *Science* (Milton Keynes, UK: Open University Press, 1997).

For their part, scientists, often aided by obliging philosophers, have gone into a state of denial over this matter, arguing that if our historical understanding of science were indeed so flawed, then 'we' (mostly Westerners) would not be enjoying our current high state of civilization. However, these scientists are often the first to admit that their world-historic predecessors—Newton, Maxwell, Einstein, etc.—did not fully fathom the consequences of their world-shattering ideas and, given the chance, would perhaps disown the consequences that we now consider to have earned them their place in the scientific pantheon. From the standpoint of scientists who make such arguments, it seems epistemologically safer to say that their precursors failed to see into the future than that they themselves have failed to see into the past. Thus, time and truth are made to move in the same direction—forward. No definition of *progress* could be more concise. In a sense, the reader would be right in suspecting that little more than an exercise in semantics is at stake in trying to distinguish between, say, Newton's misunderstanding of the long-term significance of a truth he uttered 300 years ago and the latest Nobel prize-winner's attribution of contemporary significance to a statement that he falsely believes Newton to have made 300 years ago. Nevertheless, as we shall see throughout this essay, the reinforcement of this slim difference between regarding the history of science as, so to speak, a series of *unrealized truths* and a series of *realized falsehoods* is crucial for representing the sense of superiority that science is thought to confer on the West.

For example, a broad spectrum of philosophers of science have believed that the 'best explanation' of science's success lies in the fact that its most recent theories account for things better than did earlier ones. These philosophers invite us to imagine scientists' attempts to describe and explain bits of reality as a kind of linguistic archery, in which science's aim at the essence of things grows better with practice over time. The 'theory of reference' is largely given over to such considerations. However, the superficial plausibility of the archery analogy starts to fade, once we consider that all the 'contestants' are not present at the same match. Indeed, the closer we look at the history, the more we see that our scientific contestants were probably not even playing the same game. Step back for a moment and pose the following challenges to our everyday sense of the history of science: How many of the people responsible for the exemplary achievements of Western science would have approved of the directions in which their work have been subsequently taken? Would they even have been comfortable with the specific role they have been assigned in popular renditions of the progressive march towards The Truth? More specifically, would Aristotle—or Darwin, for that matter—appreciate being seen as landmark figures in a science, biology, that has increasingly reformulated its problems and insights in terms that shift the locus of investigation from the observation of nature to the manufacture of experimental effects?

Nevertheless, while the rhetorical force of these questions may be lethal for 'best explanation' accounts of the success of science, the avenues of enquiry they open up should not really surprise us. After all, the faultiness of human memory, biases of the written record, and the tendency for words to change meanings as they move across contexts all conspire to render any straightforwardly progressive account of the history of Western science patently false, yet also difficult to falsify. This paradoxical state of affairs is enhanced as our liabilities gravitate towards our interests: in a phrase, the fine art of *convenient forgetfulness*. As we shall see, in what follows, much of the historiography of science can be discussed in terms of who is and is not convenienced by a particular strategy of forgetting. An important part of this story is inscribed in the disciplinary difference between those who do science and those who engage in 'second-order' studies of how science is done. The latter group includes students of the history, philosophy, sociology, psychology, political economy, and so forth, of science. Taken as a group, with the exception of most of the philosophers, they draw attention to the fallible features of the human condition that, by implication, render the narrative of progress recounted in science textbooks little more than a myth. Practitioners of these second-order fields typically welcome their professional separation from the scientific fold because of its liberation from the scrutiny of scientists worried about how an altered view of the past might affect their legacy to future scientists. However, there is a downside to this hard-won autonomy, especially in the case of historians of science. Not only are they no longer bothered by scientists, they are ignored altogether, as scientists have virtually stopped appealing to the past to justify contemporary concerns, except in *purely* mythical terms, as in 'the leap of Thales', to quote one recent popularization (Wolpert 1992).

It is important to point out that this was not always the case.

Many of the people responsible for drawing attention to original archival materials in the history of science at the end of the last century were such distinguished scientists as Ernst Mach, Pierre Duhem and Wilhelm Ostwald. They used these materials against colleagues who championed the dominant research programmes of their day. The common strategy of these historically minded scientists was to show that whatever empirical successes could be credited to the dominant programmes were offset by conceptual problems that were never adequately addressed when first posed, often several centuries earlier but since then conveniently forgotten. It may now seem quixotic for the late philosopher Paul Feyerabend (1975) to have cited Aristotle's metaphysical conception of motion—which included both locomotion and organic change under the same rubric—as a strike against the disciplinary distinction between physics and biology, yet such a move would have been familiar to Duhem or Ostwald, who favoured the concept of 'energy' as a move back to Aristotle's unified vision. Indeed, these scientists spoke for the past much as environmentalists today speak for 'future generations' in their attempt to divert their contemporaries from potentially destructive courses of action. To the argument that the past is no longer relevant to the present, they observed that problems do not necessarily disappear simply because they are forgotten. The collective memory trace is still there to be activated. This is the spirit in which the half-century study of Chinese science conducted by the late embryologist, Joseph Needham (1901–95), should be seen. When compared with the critiques of Mach, Duhem and Ostwald, his had a depth that (one might say) corresponded to his removal from the day-to-day conduct of science. Although Needham's message is not as clear as one might like it to have been, the general tenor of his lifework was to highlight the contingent character of the West's scientific ascendancy, such that science as the West has come to know and value it may be only a transitory feature of the human condition.

Interpreting Needham Metahistorically

Needham posed his 'Grand Question': *Why did the Scientific Revolution not take place in China?* There are two subtly different ways of interpreting it that have profound implications for both how one goes about answering the question and what one takes to be the larger significance of the question. One reading presumes that the revolution would happen unless it were explicitly blocked, whereas the other presumes that it would not happen unless it were explicitly promoted. Since Needham himself seemed to play off both interpretations, it becomes especially important that we highlight the distinction between what are, respectively, *overdeterminationist* and *underdeterminationist* interpretations of the Grand Question (see Fuller 1993: 210–13).

On the one hand, one may suppose that a 'Scientific Revolution' inevitably happens to human beings everywhere, once a certain critical mass of (for want of a better word) 'social intelligence' has been reached—that is unless a locally occurrent factor prevents it. According to this overdeterminationist view of history, one would not necessarily expect China to have produced exactly the same conditions that led to the Scientific Revolution in the West—only functionally equivalent conditions, as constructed from China's own cultural and material resources. For example, instead of the Greek natural philosophy tradition, Needham pointed to Daoism as having provided comparable cultural precedent. Here the comparison of European and Chinese science can be seen as a kind of 'controlled experiment' to isolate the factors that enabled the mathematical–experimental sciences to take off in seventeenth-century Europe but not in China. History simulates laboratory conditions, in that Europe and China were sufficiently isolated from one another such that where comparable precedents failed to produce identical results, the inhibiting factor would be intrinsic to the culture, not a product of any cross-cultural interaction. The overdeterminationist position largely captures the historiographical sensibility of practising scientists. It envisages that the universality of science is represented historically in terms of several culturally distinct lines of inquiry gradually converging on a common epistemic trajectory. Thus, the overdeterminationist is prone to look for 'multiple' or 'simultaneous' discoveries of the same scientific phenomenon. Because all cultures have the potential to reach the same conclusions, it follows that a value can be placed on being the first to arrive at an intellectual destination that many others seek. But, for the very same reason, the overdeterminationist does not regard cultures as the unique possessors of scientific ideas in any literal sense that would entail the assignment of intellectual property rights. In all these respects, the overdeterminationist is one with eighteenth-century Enlightenment thought about science as a universal form of knowledge.

On the other hand, the underdeterminationist, more a creature of the early nineteenth-century Romantic mentality, comes closer to the professional historian's sensibility. Here Needham's question is interpreted to mean that the Scientific Revolution was a unique event in the history of the West that occurred at all only because certain culturally specific conditions obtained. This revolutionary moment subsequently acquired its global significance, not through independent rediscovery but rather by the West's active intervention in the affairs of non-Western cultures. In other words, China was never in a 'race' to reach the revolutionary moment, notwithstanding the superficial similarities in its cultural trajectory with that of the West. This is more in the general spirit of Needham's project, which is organized more around a holistic understanding of Chinese civilization than an equally in-depth comparative history of Chinese and European science. It is also the interpretation presumed by the sinologist Nathan Sivin (1982), who then criticizes Needham's lapses into overdeterminationism, as in his inferring a failure on China's part from a simple difference in outcome between it and the West. According to Sivin's diagnosis, whenever Needham marvelled at why the revolution did not take place in China, he seemed to be focusing on intellectual factors to the exclusion of social ones. At the same time, it might be claimed that Needham's 'lapses' into overdeterminationism were simply an expression of his approval of the West's Scientific Revolution as a desirable intellectual destination—at least when it occurred, if not in all of its subsequent developments (Cohen 1994: 479).

Considered from an epistemological perspective, the choice between underdeterminationist and overdeterminationist historiographies may depend on what needs to be explained. The former is better suited to explain the singular emergence of science, the latter to its spread once it has been introduced. In his underdeterminationist moments, Needham (1969) cast his thesis in terms of historically contingent social factors which ensured that the rise of science appeared neither miraculous nor racially grounded, two interpretations he strenuously disavowed. But in that case, history must be only moderately underdetermined, for, the more necessary a particular historical trajectory appears to have been for the rise of science, the less probable the revolution's actual occurrence then becomes. In other words, for Needham's claims about his own achievement to make sense, it must have been historically possible for the Scientific Revolution to have happened in Europe in a way other than it actually did. Needham himself suggests as much when observing that the social preconditions for the general testing of speculative hypotheses by technological means (the essence of the experimental method) had been in place as far back as the thirteenth century, when the first European universities were chartered as corporate entities on the model of guilds. This formal equality in the legal status of academics and craftsmen made commerce between them much easier than in the more class-segregated societies of the East, which presumably helped seed the academic imagination with thoughts of how their speculations could be materially realized. In that case, the question is opened: Could not the Scientific Revolution have occurred somewhat earlier than it did, given that the crucial precondition had already been in place for nearly five centuries?

But if history could have gone otherwise at too many junctures, then none of these 'turning points' can serve as genuine crossroads, in the sense of a set of outcomes that are more or less determined by the road one chooses to follow at one point. The reason is that each fork on the road would simply be an invitation to encounter other forks later on. For example, the rhetorical power of Shapin and Schaffer's (1985) *Leviathan and the Air-Pump* rests on the assumption that the debate between Robert Boyle and Thomas Hobbes over the validity of experimental knowledge was the *only* major turning point in the history of Western science, with the subsequent 350 years being little more than a ratification of the experimentalist perspective championed by Boyle. Much of the force of Shapin and Schaffer's argument would have been subverted, had they tried to explain why it took another two centuries (or more) after Boyle's victory for the experimental sciences to find a regular place in the Western university curriculum, since they would then have to reintroduce many 'Hobbesian' considerations that Boyle had supposedly suppressed for good. In particular, experimentation had to be reinvented as empirical checks on what was still essentially a theory-driven (or 'paradigm-driven') enterprise, one compatible with the traditional goals of liberal education.

It is here that overdeterminationism comes into its own, namely as an explanation of the apparent *irreversibility* (often read as the 'universality') of the spread of the Western conception of science. The standard overdeterminationist appeal to science's patent 'rationality' as a motive force is meant to mediate between the extremes of *innatism* and *imperialism:* that is between explaining the spread of science in terms of humanity's natural susceptibility to the scientific worldview and the sheer political power of the imperial carriers of that worldview. In this way, non-Westerners are rehabilitated as more or less voluntary collaborators in the Western scientific project. But the plausibility of this story starts to be stretched once it seems that any culture, regardless of its epistemic history, can converge on *the* scientific trajectory associated with the West: How can something so powerful *not* exact substantial costs from its disparate host cultures? Francis Fukuyama (1992) stands virtually alone in taking this question as being something more than rhetorical. He argues that what enables science to translate so well (i.e. 'peacefully') across cultures is that it underwrites the technological know-how needed for a society to be an efficient engine of wealth creation. On this view, science's vaunted 'universality' comes from its being a necessary means to universally desired ends, the price for which all societies will gladly pay in terms of making the requisite cultural adaptations. Of course, the alternative story is that science transforms everything in the course of subserving them to its own imperatives. A version of Langdon Winner's (1977) 'autonomous technology' thesis applied to science would thus be the negative version of Fukuyama's sanguine gloss on overdeterminationism.

Metahistoriography	Overdetermined	Underdetermined
Whose perspective	Scientist's	Historian's
History's natural tendency	Convergence upon one end	Divergence to multiple ends
Key to history	'Cunning of reason'	'Turning points'
Extreme version	Fatalism ('All cultures eventually become scientific')	Indeterminism ('It's a miracle that any culture became scientific')
Likelihood that a Scientific Revolution would occur	High	Low
What needs to be explained about the Scientific Revolution	How it was disabled (in China)	How it was enabled (in Europe)

Temporal Colonialism: Retrospected Past and an Irreversible Future

Even if we grant this division of explanatory labour between the two metahistoriographies, another question remains: How does one account for 'pre-scientific' forms of knowledge? Here we can find a backward-looking form of overdeterminationism: a *retrospective colonization of the past.* It implies that one has not gained legitimate access to, say, a form of technological knowledge unless the scientific principles on which the technology is 'based' have been mastered. If this 'basis' historically appeared after the technology was already developed and used, the proposal amounts to systematically disempowering the users and then reacquainting them with their artefacts by means of theory-driven instruction. This is not an idle possibility, but in fact a pattern has been noted historically in the 'scientization' of the professions, as craft-based training in medicine and engineering yielded to academic certification. It can also be seen in efforts to Westernize Third World curricula for purposes of rendering the natives 'governable' by making them epistemically accountable to standards that Western authorities can understand and evaluate. One of the most insidious contexts of native disempowerment occurs when Europeans rewrite the histories of non-European peoples who are no longer around to criticize the histories reconstructed on their behalf. Take the Incan settlement at Machu Picchu, Peru, which archaeologists now reckon to have been an agricultural station devoted to the effects of different soils on crop yield (Weatherford 1988). The level of soil discrimination evidenced at Machu Picchu remained unmatched by the combined efforts of botanists and chemists in Europe until the mid-nineteenth century. Indeed, European scientists had not been driven to acquire agricultural knowledge with the Incas' eye for detail until potatoes imported from the Americas helped eliminate many nutritional disorders in Europe, thereby lengthening lifespan and, coupled with a constant fertility rate, leading to an increase in the rate of population growth. Under the circumstances, it was important not simply to grow more potatoes and kindred crops, but to determine how crop yields could be increased per unit of arable land. Experimental investigations along these lines had been clearly undertaken at Machu Picchu five centuries earlier. However, because the Incas demonstrated their agricultural knowledge 'merely' in experimental practice, and not as deductions from any recognized principles of natural science, Europeans have tended to downgrade the Incan contributions as 'pre-theoretical', 'trial-and-error' ('inductive', if one is being especially polite), or 'technology' (as opposed to 'science'). Each of the highlighted expressions is meant to suggest that had the Incas 'really' understood what they were up to, they would have eventually arrived at the theories of agronomy that we now regard as providing the best explanation for the level of agricultural success they achieved.

Once again, we see the calling card of overdeterminationism, namely the assumption that all genuine knowledge eventually converges on the same trajectory—'ours'. It used to be claimed that a symptom of the Incas' lack of scientific understanding was that they did not exploit opportunities for extending their knowledge that 'we' would have under comparable circumstances. However, as military history has come to be fitted into the history of science, Europeans have reluctantly admitted that their own conquistadorial urges may have contributed to the Incas' failure to extend their agricultural knowledge beyond a certain point. A still bolder, more underdeterminationist hypothesis is that the Incas had a radically different sense of the social role of knowledge in their world, one that the visiting Europeans could not readily grasp, given their lack of schooling even in their own scientific traditions. For example, the fact that the Incas lacked a sense of the cosmos unified under a creative deity could count as prima facie evidence for their worldview being radically different from the Europeans'. But such speculation aside, historians today have no trouble asserting that when it came to acquiring the Incas' degree of agricultural sophistication, European scientists were at a distinct disadvantage, given the Western tradition of treating plants as degenerate animals and biochemical changes as 'emergent' on underlying physical processes. This is simply to say that agriculture significantly deviated from the paradigmatic cases of Western scientific inquiry, mathematical physics.

Notwithstanding its political correctness, if we dogmatically stress the radical differences between Incan and European modes of thought, it then becomes just as difficult to *credit* the Incas for having 'anticipated' European agricultural science as to discredit them for not actually having invented modern agronomy. This problem becomes especially acute once we consider forms of identity politics that turn on a culture's possession of a privileged, or at least distinct, form of knowledge. If the culture is not careful, its efforts at establishing epistemic autonomy can easily turn into epistemic isolation. Good contemporary examples of contrasting strategies for dealing with the problem can be found among *feminists* and *Afrocentrists*. Rose (1994) and Asante (1988) are exemplary works in these two movements. They represent, respectively, an underdeterminationist and overdeterminationist approach to the identity politics of science.

Feminists often stress alternative ways of doing science, ones based on intuition, empathy, caring and other 'subjective' modes of epistemic access that have been routinely devalued by the male white European scientific tradition. At their most utopian, they envision a day in which scientific practice will be transformed by the leadership of enlightened women and feminist males, perhaps even at the very earliest stages of socialization, as boys are reared to relate to the world with the same degree of receptiveness to the standpoint of 'the other' as girls typically are (Fox Keller 1985). However, there is a crucial ambiguity as to the exact identity of this utopia. Does it simply involve according subjectivity the dominant role traditionally reserved for objectivity, or is some sort of dialectical synthesis in the offing? A complicating factor is that the subjective mode associated with women's experience is neither exclusive to women (consider such 'pre-feminist' male defenders of intuition as St Augustine of Hippo, St Bonaventure, Friedrich Schelling, and Michael Polanyi) nor perhaps even a distinct form of experience in its own right. In the latter case I mean to suggest the possibility that the historical emergence of a gendered division of labour—the 'head' versus the 'heart', so to speak—has produced complementary forms of alienation for men and women. From that standpoint, to call for the replacement of objectivizing modes of experience with subjectivizing ones is a bit like saying that the cure for sadism is masochism. In other words, a close examination of all the claims that have been advanced on behalf of intuition as a source of knowledge in the history of the West would reveal that they do not share much more than an interest in compensating for the inadequacy of discursive forms of knowledge. Moreover, given the many centuries in which a gendered division of labour has shaped a bifurcated consciousness in human beings, it should not be assumed that the reintegration of the masculine and feminine poles would constitute a return to an 'Edenic' unified human psyche. Given these uncertainties about the exact character of a feminist scientific utopia, it is understandable that the practical politics of feminist science have gravitated towards more procedurally democratic claims for proportional gender representation in the scientific workplace, which renders any change in consciousness primarily an emergent feature of this institutional reform.

In contrast, Afrocentrists envision a day in which Africans will finally be given credit for discoveries that Europeans have so far claimed for themselves. Here we find the rhetoric of the 'stolen legacy' (Diop 1991). Yet, for all its surface harshness, Afrocentrist rhetoric presupposes that the Africans and Europeans have basically the *same* scientific legacy, one whose African precedents the Europeans refuse to acknowledge. Sometimes this refusal is traced back to the Greeks themselves, whose mercantile practices abroad are interpreted as cultural theft by Afrocentrist scholars. A more historically nuanced charge is associated with the 'Black Athena' thesis of Martin Bernal (1987), who argues that Prussian philologists manufactured a Greek origin for European culture in order to complement Napoleon's appropriation of ancient Rome as France's heroic mythical ancestor. As the concept of 'race' scientifically hardened over the course of the nineteenth century, the Greeks were made the source of the 'Aryan' race, whose inherent cultural superiority is still signalled whenever someone today remarks on 'how much further' a Greek was able to take a concept or technique that originated in the Near East, India, or Egypt.

Nevertheless, it would seem that compared with the Afrocentrists, the feminists are charting the more ambitious agenda, one which goes beyond redressing past injustices and moves towards a genuine alternative science. The few cases of Afrocentric inquiry in which Africans supposedly enjoyed epistemic access to something that has completely eluded European science probably constitute the most contested area of Afrocentric scholarship. A case in point is the so-called 'Melanin Hypothesis', which ascribes special psychic powers to the skin colour of pre-colonial Africans (Van Sertima 1986). Even committed Africanists often find it a little too convenient to hold European invaders responsible for preventing Africans from transmitting their pigment-based knowledge to future generations. Rather, the difference in the visionary scope of feminists and Afrocentrists is best explained as a reflection of a difference in their perception of the prospects for a political change that would benefit their respective constituencies in the present day. The feminists are more optimistic than the Afrocentrists that we can collectively move to a radically reformed world. Afrocentrists would seem to be willing to settle for a share of the credit for what has already been accomplished in the name of the West.

Whatever is ultimately made of the political and intellectual merits of Afrocentrism, it deserves credit for drawing critical attention to the forward-looking perspective that is taken for granted by the overdeterminationists' philosophical cousins, the *scientific realists* (Leplin 1984). For example, why do we automatically say that Africans, Incans, or some other non-Western peoples 'unconsciously anticipated' Western scientific discoveries? Why not say, instead, that Westerners have merely recovered or explicated the insights originally gleaned by the Africans or Incans, which because they were too organically integrated in their lifestyles—'second nature', so to speak—did not require any explicit articulation in their culture? As we have seen in the case of the Melanin scholars, the second nature can be literally inscribed on the epidermis itself. Such an historical orientation is no less 'realist' than that of the philosophies of science that usually travel under that rubric—except for the fact that it locates the sought-after reality in the *past* rather than in the present (and future) state of science.

If we have become too accustomed to envisaging disparate inquiries gradually converging on an ultimate theory of reality, we may need to recall an alternative vision, one often adumbrated by Afrocentrist and some feminist rhetoric: namely that there was a golden age in the distant past, followed by something like the Biblical Fall from Grace, from which we are now only slowly recovering. Indeed, according to this view, a premier sign of our 'fallen' state is that we require specialized discourses that are alienated from the normal run of life in order to capture what the Africans, Incans, etc. were able to embody in practices that were integrated into their everyday lives. And lest one dismiss this vision as tainted by a naive romanticism, it is worth recalling that this was largely the West's own sense of history until the middle of the eighteenth century. The key notion here is what the Renaissance Humanists called *prisca sapientia*, or 'pristine wisdom', which justified the reading of ancient Greek and Hebrew texts in their original—and presumably uncorrupted—languages. In the hands of the Protestant reformers, this became a potent weapon for challenging the religious authority of the Roman Catholic Church and the intellectual authority of its Scholastic teachers, who preferred a homogenized Latin tongue (much as English would be preferred today by professional academics as the melting pot for cultural differences). But perhaps the most striking case of the persistence of the belief in *prisca sapientia* is Isaac Newton's lifelong attempts to locate precedents for his physical discoveries in the Bible, the pre-Socratic philosopher Pythagoras, and the Hermetic Corpus of Egypt.

But given the endurance of the idea of a *prisca sapientia*, when did the West acquire its characteristically forward-looking view of where the secrets of reality lay? The answer may be found in the complicated history surrounding the 'Battle of the Books'. This battle, so-named by the Irish satirist Jonathan Swift, transpired mostly in Britain and France over the entire stretch of the so-called Age of Reason (roughly 1650 to 1800). Sometimes it is said that the two sides of the dispute—the *ancients* and the *moderns*—were contesting the relative merits of ancient (i.e. Greek and Latin) and modern (i.e. English and French) poetry. However, neither side denied the goodness of the ancient classics. A better clue to the nature of the dispute is the origin of the distinction between 'ancient' and 'modern' starting in the late thirteenth century, when the Scholastics wrangled over the adequacy of Aristotle's syllogistic approach to logic, which required that one suppose the existence of whatever one reasoned about. Here the 'via antiqua' stood for a defence of Aristotle's logic, whereas the 'via moderna', championed by the ever-sceptical William of Ockham, stood for those who wanted to adopt a proposition-based logic, one whose major premises would be formulated as hypotheses rather than straightforward existence claims. In other words, the moderns did not want to commit to the truth of the premises from which they argued before deriving their logical consequences. This detachment of substantive truth from formal validity became characteristic of the ironic stance of the Scholastics at the end of the Middle Ages. On the one hand, it enabled them to entertain the possibility of a heliocentric universe and inertial motion, both of which contradicted the received word of Aristotle but was eventually vindicated by Galileo; on the other hand, it also prevented them from committing to the truth of these hypotheses.

Similarly, in the Battle of the Books, the ancients held that the literatures of the West's founding cultures contained original truths that cannot be reduced to a formula or method whose structure later texts can elucidate. Rather, the contemporary reader simply needs to become sensitive to the subtlety of ancient expression in order to appreciate its natural superiority. In contrast, the moderns argued that the idiosyncracies of the ancient texts are little more than that, and are therefore best taken as imperfect expressions of ideas that can be brought out more clearly today and even more so tomorrow, as we hypothesize what the ancient authors meant to say and then, using evidence unavailable to them, try to say it better ourselves. At the limit, the modern would replace all the ancient texts with contemporary ones that transcended the limits, while preserving the best, of their precursors. As Gellner (1988) has observed, this 'later is better' mentality marks the difference between scientific attitudes towards writing and broadly 'hermeneutical' attitudes that encompass not only the secular classics but more importantly, from a cross-cultural perspective, the religious ones as well. If the scientist sees past texts as rough drafts of future ones, the hermeneutician sees future texts as pale copies of the originals. It is worth noting that by the mid-nineteenth century, the Battle of the Books had reached the point of literary critics wondering whether modernism taken to its logical conclusion might altogether dissolve the very presumption of historical continuity between past and present texts, thereby reducing literary legitimacy to a function of immediate social context (Adams 1971: 555ff.). While this sentiment drove an increasingly large wedge between the critics and practitioners of the arts (with the critics retaining the sense of historical continuity and artists obscuring it), the sciences largely refused to take that next step into what we now call 'post-modernism'.

Method of Modernist Metahistory: Time-discounting

The modernist rewriting of the past as the anticipation of the present is crucial for explaining a feature of Western culture that has often been held responsible for the rise of science. Inspired by Francis Bacon, Western inquirers supposedly did not fear any backlash effect from their experimental attempts to bend nature to their collective will. Psychologically speaking, such organized fearlessness is most easily maintained under two conditions:

(1) that the inquirers themselves do not need to live with the consequences of their actions (perhaps because these occur in remote times and places, as in the case of environmental hazards);

(2) that history is rewritten to such an extent that by the time these consequences occur, they will seem to have been what were desired all along.

These two conditions can be summed up as the systematic conversion of a psychic liability into a virtue. It is the phenomenon that economists call *time-discounting*, or simply *discounting*, whereby people prefer the satisfaction of immediate desires to distant ones, even if the distant ones are likely to be greater, and the satisfaction of the immediate desires may jeopardize the satisfaction of the distant ones (Ainslie 1992). A simple example is the overweight person who gorges himself on food now, though it undermines his long-term interest in losing weight.

There are two schools of thought on the rationality of discounting. Welfare economics is largely founded on the patent irrationality of discounting. Following Arthur Pigou, welfare economists have argued that the function of the state is to compensate for the human tendency to exaggerate the longevity of one's immediate interests which leads us to discount the future too heavily (Price 1993). Compensation is provided by progressive taxation, social security insurance and other schemes to redistribute income, as well as incentives to invest rather than consume. Indeed, John Rawls (1971) raised this counter-discounting sensibility to a philosophical verity: that is the best way to motivate a sense of social justice is by persuading agents that, at some point in their lives, they may end up in *anyone's* current situation, hence, they should not treat their present position, however favourable, as indefinitely extendable into the future. When forced to decide from behind such a 'veil of ignorance', power relationships metamorphose into epistemic ones: When I accept a certain income redistribution scheme, I am not so much manipulating other people's lives as rationally planning for a prospect that may well become my own (e.g. the expectation of my eventual retirement). Consequently, one is placed in a frame of mind whereby the strong proprietary distinction among selves dissolves into an undifferentiated sense of social belonging (Parfit 1984). But for all its admirable qualities, had the welfare mentality successfully pervaded all spheres of activity in the West, it is unlikely that the 'fearlessness' attributed to the Scientific Revolution would have ever been manifested.

This last observation suggests that we may not do justice to discounters by reducing their worldview to the mistaken induction that one's immediate desires will continue in perpetuity. Rather, discounters should perhaps be given credit for the meta-level realization that it is precisely this propensity for preferring the temporally near to the temporally far that will continue in perpetuity. In that case, by the time people reach the point at which they should regret their earlier decisions, the consequences of those decisions will have shaped what they now desire, thereby turning regret into acceptance, if not outright endorsement, as in the case of the phenomena associated with that complement of 'sour grapes', the phenomenon of *sweet lemons* (Elster 1984). Thus, the overweight person who continues eating and thereby fails to curb his obesity in the long term may eventually come to devalue the significance of achieving a 'normal' weight.

In other words, discounting is a rational strategy if one believes that the past is just as constructible as the future, or as the French political sociologist Raymond Aron so elegantly put it: 'The past is never definitively fixed except when it has no future.' Such a sensibility gives the overdeterminationist history of science its plausibility because, somewhat paradoxically, the idea that reality has a deep invariant structure to which our access may nevertheless improve over time presupposes that, if need be, we can discount long-standing beliefs in favour of a recent discovery, whose significance is sufficiently deep that we can imagine that it would have been recognized as such by our intellectual ancestors, had they stumbled upon it. Indeed, this argument was used to great effect by Galileo in response to the Jesuits who wondered why, if his astronomical findings were true, had they not been registered by the Biblical patriarchs or Church fathers. Answer: They never had the opportunity to look through a telescope—but had they, they would have presumably been drawn to Galileo's conclusion.

Although many cultures have had elaborate literary recording practices, one increasingly prominent tendency in Europe set it apart as a culture ripe for the collective discounting of historical memory illustrated by both Galileo's response to the Jesuits and the retrospective colonization of the past noted in the previous section. That tendency—much to the consternation of the Roman Catholic Church—was the relatively uncontrolled production and distribution of printed materials (Febvre and Martin 1971). Most commentators on the significance of the printing press for the Scientific Revolution have focused on the sheer availability of books as making it unnecessary for new authors to summarize received wisdom before setting out their own contributions (Eisenstein 1979). One could simply presume that readers had access to earlier works and then devote one's energies almost exclusively to one's own novel findings or conceptions. In this way, the pace of intellectual progress quickened. However, at least as important was the fact that books were disseminated from many different sources that often contradicted one another, especially in their characterization of the past and its legacy for the present. This simultaneous proliferation of texts erased any clear temporal source of knowledge. In this way, the continual rewriting of the past to serve present needs became a matter of implicit practice, if not official doctrine. For, while it is common nowadays to associate scepticism about the search for historical origins with the work of Foucault, Derrida and other post-modern thinkers, such scepticism was already present in the early modern period—say, in the popularity of self-consciously mythical 'philosophical history' during the Enlightenment. Indeed, as the elusiveness of the spatiotemporal coordinates of the West's 'Scientific Revolution' continues to illustrate, even today relatively few branches of history are underpinned by an incontrovertible chronology.

As Hegel first realized, when pondering exactly who or what could have a 'history', the establishment of canonical historical trajectories has required substantial state intervention, typically involving the mass indoctrination of students on state-sanctioned school textbooks. Not surprisingly, political history has been most susceptible to this treatment, in which the familiar litany of dates of particular leaders, treaties and battles serves to underscore the legitimacy of the current regime. Indeed, it is often forgotten that the advent of professional historical scholarship, especially its reliance on period documents as the main evidentiary base for historical claims, coincided with the rise of what is known as *Whig History*, that form of overdeterminationism whereby the past lends its weight in support of present-day entitlement claims. The connection, as the young Lord Acton observed, is that heavy documentation interpreted in one's favour creates a burden of proof against any attempt to challenge the legitimacy of one's rule (Butterfield 1955).

However, in the West, this practice has to be set against the backdrop of an increasingly democratic political process, one which encourages contesting parties to revise continually the relative significance assigned to the dates in the textbook chronologies, as well as the national goals they are meant to subserve. The most famous programmatic statement of the sociology of knowledge, Karl Mannheim's (1936) *Ideology and Utopia,* is fairly read in this context, one in which the multiple schemes of historical significance projected by the spectrum of political parties in Weimar Germany together constituted ripe opportunities for 'shifting the goalpost' with each successive government, as would be expected of a society that was given to time-discounting. And lest one think that this phenomenon is confined to weak democracies struggling through volatile times, it is worth recalling how often such 'strong' and 'stable' democratic regimes as the United States and the United Kingdom have altered the definition of such basic indicators of welfare as 'unemployment' and 'inflation' in ways that have made political expediency seem to coincide with some sense of increased technical sophistication. Pigou and Rawls notwithstanding, this is discounting raised to the level of statecraft.

However, as the last sentence suggests, it would be a mistake to regard this tendency as confined to politics. If anything, the modes of time-discounting in science are *more* pronounced. A good way to understand these modes is in contrast to the sense of history championed in recent years by Fernand Braudel and Immanuel Wallerstein (1991). Here history is defined as the superimposition of temporal perspectives of varying durations, ranging from the point-like events that characterize moments in political history, through the periodic rhythms of business cycles, to the more enduring features of the natural environment that provide the background conditions for a people's long-term survival. These perspectives are distinguished primarily in terms of the extent to which past conditions continue largely unabated into the future. In contrast, the temporal perspectives enumerated below, while equally seen as superimposed, are based on the extent to which future conditions can reconstitute those in the past. In other words, the presence of time-discounting reflects the capacity for *reflexive historical awareness.* To see exactly what I mean, consider the following three senses in which the future can reconstitute the past, which together represent an increasingly reflexive sense of historical awareness. I shall then illustrate these with familiar contemporary orientations to the history of science.

(1) *What you do now changes what you can do later but not what you originally wanted.* This includes (at the individual level) self-

defeating wants and (at the collective level) so-called 'counterfinality effects', the most notorious of which is perhaps Garrett Hardin's 'tragedy of the commons', whereby the individually rational actions of farmers eventuate in a collectively disastrous consequence, the depletion of grazing land. This perspective, used to justify arguments for environmental protection, presupposes that neither the past nor the future can be changed. Rather, past wants remain constant and are simply thwarted in the future.

(2) *What you do now changes what you will want.* This includes what social psychologists call 'adaptive preference formations', whereby one learns to like an outcome only once it has occurred, as in the complementary phenomena of sour grapes and sweet lemons noted earlier. Here, future wants are reconstituted in the name of 'learning from experience', which is little more than an epistemological way of saying that past wants remain the same but future ones change.

(3) *What you do now changes what you will have wanted.* Here history is rewritten in the comprehensive fashion depicted by George Orwell in *1984,* such that one no longer notices that a change has occurred. In other words, past wants are reconstituted to make it seem as though the future naturally continues the attempt to satisfy them. Any evidence of disruption is removed from the historical record, and hence the West's much-vaunted sense of 'fearlessness' is manufactured.

Now for the corresponding examples from the history of science:

(1) *The historian's perspective:* A good example is Kuhn's (1970) account of scientific puzzle-solving as eventuating in anomalies, the accumulation of which then precipitates a crisis in the paradigm. From this standpoint, a scientific revolution is always a long-term unintended consequence of the pursuit of normal science, not something that comes about by design. 'Rationality' in this context means nothing more grandiose than the research strategies that have proven most immediately useful to the scientific community, regardless of their long-term prospects.

(2) *The philosopher's perspective:* Here history appears as a collective learning process, in which scientists are presented as regularly having to choose from among two or more theories. Thus, they must explicitly decide whether or not to continue their current research trajectory on the basis of its 'track record'. One thing that typically becomes evident from this track record is that long-

standing conceptual problems with one's research programme can be offset as being of 'merely philosophical' interest, once enough empirical success has been registered (Laudan 1977).

(3) *The scientist's perspective:* This is the 'positivist' textbook history of science that is reconstituted with each paradigm shift. It makes a science's current problem set seem to be the one towards which all prior research had been implicitly heading. Even apparent diversions and errors are cast as being guided by an 'invisible hand' or 'cunning of reason', the full sense of which is typically realized by today's scientists, after having eluded their precursors.

Taken in succession, these three temporal perspectives can be read as constituting a 'slippery slope' of ever-increasing rationalization for whatever happens. Each stage is a bit more reflexive in that the future is discounted a bit more heavily in anticipation that such a rationalization will occur. The Western philosopher of science who has been most mindful of this slippery slope is Paul Feyerabend. Feyerabend (1975) became notorious for proposing a 'methodology' expressly designed to reduce the prospect of scientists making only the most conservative revisions of their theories when faced with discrepant empirical results. To do so, according to Feyerabend, would be tantamount to an adaptive preference formation, as scientists would be wishfully presuming that the discrepancy they have discovered exhausts everything that is wrong with their theory—a rather improbable presumption, especially when made in the name of 'methodological conservatism', a principle that is most naturally taken to mean that the anomalous finding is only the *beginning* of one's epistemic problems and that consequently one should presume that more trouble lies ahead. Nevertheless, by minimizing the significance of such anomalies, a research programme can easily appear to be 'progressing' by ever finer increments that gradually converge on a comprehensive understanding of the domain to which its theories putatively refer. The Feyerabendian antidote is to render the same evidence base compatible with as many different theories as possible, the coexistence of which would serve to remind the scientific community that even its historically privileged theories may be ultimately shown to have been radically off-base. This counter-discounting strategy may be compared with the one Rawls implements in order to justify the redistribution of income under the guise of welfare. Specifically, both aim to instil an openness to alternative futures by defying our inductivist instincts to act as if our current situation will continue unabated in the future.

Islam as Metahistorical Hybrid

Islam presents an interesting challenge to both under- and over-determinationist metahistoriographies of science. However, the challenge is easily lost amidst the blizzard of discourse surrounding *multiculturalism*, which, at least in its Western incarnations, tends to homogenize the sense of 'difference' that non-Western cultures supposedly provide to the West. Take a canonical multiculturalist work, such as Edward Said's (1978) *Orientalism*, which dwells on the portrayal of Asiatic peoples in European literature and scholarship over the last 250 years. By treating the exoticization of Asia as itself a uniform phenomenon ripe for deconstruction, Said unwittingly continues the tendency to homogenize difference. Of course, after Said's work, the exoticism has gone into reverse: The phenomenon of orientalism is now taken to reveal more about how Europeans think than about how Asians think—but the lumping of 'Them' versus 'Us' remains (Ahmad 1992). A casualty of this approach is that Westerners rarely appreciate the difference between a culture whose roots are virtually independent of Europe's and a culture that is largely an alternative development of the common heritage it shares with modern Europe. In terms of Asia, China epitomizes the former, Islam the latter. Here it is instructive to examine a group of cosmopolitan Muslim intellectuals collectively known as the 'Islamic revivalists', for whom Islam's unique historical status posed a specific practical challenge (Al-Azmeh 1993, Rahmena 1994). Writing in the late nineteenth century, they upheld Islam as the revealed word of Allah, but admitted that the Muslim world had fallen into decline, due largely to the spiritual failure of its own leaders, which made it vulnerable to the West's opportunism. However, the Islamic revivalists found themselves in an awkward position with regard to Muslim backwardness. On the one hand, they wanted to claim for Islam many of the achievements, especially in science, that the West took to be definitive of its own sense of progress. On the other, while conceding backwardness, they still wanted to distance any future development of Islamic science from the more patently materialistic and destructive features of Western science. Thus, we may think of these reformers as having agreed on the following five theses:

(1) Islam and Christianity have essentially the same origin and represent alternative ways of continuing that common heritage.

(2) Initially, Muslims deliberately refused to follow the West's path of allowing science to become so autonomous from the rest of society that it frequently threatened its moral constitution. In the words of Sayyed Hossein Nasr (1968), a recent Islamic epistemologist, the West has pursued 'knowledge' at the expense of 'wisdom'.
(3) Nevertheless, Islam's own historical trajectory must equally be rejected, as it has led to an ossified doctrinal consensus that has failed to accommodate to changes that have taken place in the world over the last five centuries.
(4) The revival of Islam must do more than project a utopia that ignores the Muslim world's long period of decline. Rather, it must learn from that history and selectively appropriate elements of Western history that could aid Islam's revival.
(5) The challenge of modern Islam, then, is to recover—at least rhetorically—that original fork on the road where Christianity and Islam parted company and to pursue a form of knowledge infused with the teaching of the *Qu'ran* that is at the same time recognizable as epistemically progressive.

Looking over this list, what is striking is the extent to which the mutual understanding of Islam and the West, until very recently, has rested on common historical misunderstandings. Two sorts of misunderstanding are especially apparent from the onset of Islamic revivalism. The first is the failure to see that the social conditions for promoting science in Islam and the West were still not so far apart, even perhaps as late as the time the revivalists were writing. The second is the failure to see that the West has benefited from what the economic historian Alexander Gerschenkron (1962) has called the *relative advantage of backwardness* vis-à-vis Islam. Let us consider each, in turn.

In What Sense did the West Surpass Islam in Science in 1900?

This may seem to be a strange question to ask, one that gives rise to the suspicion that I have capitulated to 'political correctness' in its worst sense. However, I mean only to signal an issue raised earlier, when discussing the relative strengths and weaknesses of the two metahistoriographies of science—namely that to explain the origin of a concept or technique is not to explain its diffusion. On the one hand, several philosophically inspired distinctions threaten to turn this observation into a platitude. Logicians routinely invoke the 'genetic fallacy' to ward off those who would judge the validity of an argument on the basis of its origins. Philosophers of science have long distinguished contexts of 'discovery' and 'justification' to similar effect. On the other hand, even relatively sophisticated historical accounts of science, such as Thomas Kuhn's (1970), continue to focus their narratives entirely on front-line researchers (as opposed to those who apply or teach the knowledge), who are themselves presented as completing the vision set out by some heroic predecessor, as in the case of the 200-year 'Newtonian paradigm' in physics. At the start of this essay, I noted that scientists are especially prone to see things in Kuhn's way. However, in his critique of Needham, Sivin (1982) makes the obvious, but all-too-neglected, point that the Scientific Revolution did not occur overnight in the West. He points out some very substantial time-lags between the emergence and adoption of revolutionary scientific ideas. (This point would not have surprised the man who introduced the expression 'Scientific Revolution' into English shortly after the Second World War, Herbert Butterfield, since his benchmark of a revolution was the message of Jesus Christ, the institutionalization of which was achieved only with great difficulty over the course of several centuries.) One of Sivin's own vivid examples is that the modern science of anatomy and physiology, which was founded by Andreas Vesalius and his sixteenth-century colleagues, did not figure in Western medical practice for another three centuries. In other words, explaining the spread of science across the face of Europe itself turns out just as problematic as explaining its spread across the rest of the world.

Nevertheless, despite their ideological differences, the three great Islamic revivalists Ahmad Khan (1817–98), Jamaluddin Afghani (1839–97), and Muhammad Abduh (1849–1905) thought they could explain why science (supposedly) spread so much more quickly in the West than across Islam. They were in agreement that Islam's failure to adopt the Western practice of separating Church and State was the source of the natural sciences' failure to gain a proper footing in the major Muslim universities, be it Aligarh near Delhi (Khan), the University of Istanbul (Afghani), or the El-Azhar in Cairo (Abduh). What the reformers had in mind, of course, was not the removal of religious training from the curriculum, but rather the elimination of theological oversight on scientific subjects. Yet, while Western observers encouraged these reforms, it was not because the soul of Western higher education was fully in the hands of secular authorities in the

West, or even that the elimination of theological oversight had demonstrably improved Western institutions. Indeed, the country where the laity's victory over the clerics was most loudly advertised—France under the Third Republic—was also the one whose institutions of higher learning were in most disrepute, mainly because 'secularization' had opened the door to faculty-blurring criteria of intellectual distinction and public notoriety, a tendency that arguably continues to the present day (Zeldin 1967). Moreover, one should not conclude from the fact that ecclesiastical authorities were not directly involved in the governance of European nation-states that they had lost all control over the reins of power in the West. Even in the scientifically most advanced country of the time, Germany, the number of university students matriculating in all the natural sciences combined did not equal the number in theology until 1898 (Inkster 1991: 97). Even more striking is the case of Germany's world-famous interdisciplinary research institutes, the Kaiser Wilhelm Gesellschaften (now called the Max Planck Institutes), which pooled the resources of the academy, the state, and industry to develop the first of what are now regarded as 'big science' projects. The force behind their development was the country's leading theologian, Adolf von Harnack, who, in the first decade of this century, identified science and the military as the twin bases of Germany's spiritual superiority. The secret of the West's ongoing success, then, was not the marginalization of Christian theology, but its mobilization in support of the natural sciences. Lest we forget, William Whewell, the man who coined 'scientist' in English as the name of a profession (the training for which deserved a place in the Oxbridge curriculum) was himself confirmed as an Anglican priest.

Over the last century, it has also been common to bemoan the fact that, in the Muslim countries, the diffusion of scientific and technical innovations was largely restricted to the military institutes and not integrated into the general university curriculum. In particular, the decline of the Ottoman Empire has been attributed to such a blockage in the transmission of scientific ideas (Lewis 1982). However, this pattern is not so very different from the West's own experience. In Britain, France, Germany, and the United States, the first institutions of higher education to make natural science and technology a part of their ordinary curriculum were likewise the military academies, which had begun to evolve into fully-fledged polytechnics by the end of the eighteenth century, as ambitious rulers followed Napoleon's lead in regarding civil engineering projects as a productive way to deploy their professional soldiers in peacetime. In any case, this was over a century after the so-called Scientific Revolution had taken place and nearly a full century before these technical subjects were generally seen as fit for genteel civilian consumption (Barnett 1967). How long, then, did it take for science to spread across the West itself?

The answer varies from country to country, of course, but it generally corresponds to the introduction of compulsory mass education towards the end of the nineteenth century and the insinuation of scientific studies, often quite basic, in vocational training at roughly the same time. Indeed, the most significant difference between the educational patterns in European and Muslim countries was to be found not in the tertiary sector—where the Muslim figures were often more impressive than the European ones—but at the primary and secondary school level, where the Europeans were much more effective at what we would now describe as 'laying down the infrastructure for an information society' by ensuring basic levels of literacy and numeracy. Thus, 100 years ago, Egypt's El-Azhar enrolled over 10,000 students in a country of 5.5 million, whereas the combined enrolment of *all* British universities in the same period was 20,000 students out of a population of 39 million (Crabbs 1984, Halsey 1992). The university curriculum in both countries was highly biased towards theological and humanistic studies, and the academic elite were generally suspicious of scientific and technical subjects. However, upon examining educational differences below the elite level, it becomes immediately clear why Britain was the resident colonial power in Egypt, and not the other way around. Unfortunately, because the history of science is still conceptualized—in both East and West—as the circulation of elites, the overriding significance of a policy as humble as the provision of basic technical training for the masses can be easily neglected.

When Did the West's Original Liabilities Turn Into Virtues?

It is worth recalling the various senses in which the West has enjoyed the 'relative advantage of backwardness' vis-à-vis Islam. From the mid-eighth to the mid-seventeenth centuries, the Muslim realm was both farther-flung and more tightly knit than the feudal kingdoms and city-states scattered across the face of Christendom. Indeed, a commonplace among medieval historians is that 'Europe' would never have been such an attractive idea to rally around, had the continent's inhabitants not shared the threat of an expanding Islam. In Henri Pirenne's words, 'No Charlemagne without Muhammad'. A more controversial but probably no less true claim is that the Mediterranean regions of Europe, most often cast as innovators in science and technology during the Renaissance, were acting more in the 'Japanese' spirit of reverse engineering with the intent of producing technically improved versions of Muslim (and, for that matter, Chinese) innovations. However, the interconnectedness of Muslim rule, which enabled relatively good patterns of communication among scholars in Spain, Egypt, and Persia, also made it easier to subject intellectual life to direct political control. In some cases, this had a clearly salutary effect. Consider the famed Muslim drive for synthesis, which made Arabic the first genuinely cosmopolitan language of thought, one that integrated the heritages of Graeco-Roman, Judaeo-Christian, and Indo-Iranian cultures under a single conceptual rubric long before Latin came to play that role. That drive was nurtured by the pedagogical needs of students training for civil and imperial service. The downside of political control became apparent once religious and secular authority was synchronized to such an extent that the Muslim world became a *de facto* theocracy. This did not come about very quickly or by design, but nevertheless was the result of taking seriously the idea that a common faith was the ultimate source of legitimation for so many people being encompassed under a common regime.

Although Islam's sense of the sacred, like Christianity's, is located in the believer's encounter with the Word of God as transcribed in His Holy Book, professional interpreters of God's Word, the *ulama*, soon emerged. While this development bears some resemblance to the priestly class of the Roman Catholic Church, there were two important differences. First, Christendom never enjoyed the degree of political integration needed for the enforcement of a religious orthodoxy, however much St Augustine would have wished otherwise. Heretical religious leaders inspired by a bewildering array of imaginative readings of the Bible can be found throughout the history of Christianity. In most cases the Church was powerless to do more than threaten them. Moreover, as the Church's own material stakes grew, it was forced to depend on the goodwill of secular leaders for protection. Consequently, institutions developed—most notably the confessional—which made it relatively easy for kings and princes to be absolved from the morally dubious aspects of their lives. A second difference, one resulting from the Church's accommodation of the religious and the secular, was that the preservation of Christianity became increasingly identified with monastic and scholastic specialists who resided in abbeys and universities out of the line of political fire. By contrast, the Muslim *ulama* held increasingly influential secular positions alongside their religious functions which enabled them to mediate, survive and even flourish in the midst of dynastic infighting.

When Jamaluddin Afghani sought a Western model for breathing new life into Muslim civilization, he turned to the great Protestant reformer Martin Luther, whose promise of spiritual rebirth was a response precisely to the Church's uneven influence on worldly affairs. Moreover, the regimes proposed by such reformers as Luther and John Calvin resembled what had already existed in the Muslim world for the past 250 years: activities were permitted only if they could be suffused with religious significance. In practice, this turned out to be a very stiff imposition, as it proscribed not only activities that overtly contradicted religious doctrine but also ones that did not clearly contribute to the maintenance of the faith. Not surprisingly, both Lutheran and Calvinist states were much more hostile than the Catholic Church to the propagation of Copernican astronomy, given that it added little to Ptolemy's precision in charting the heavens and was accompanied by an explanatory framework—the sun-centred universe—that defied both Scripture and commonsense, and hence threatened to create another level of mediating authority. For its part, the Catholic Church issued a characteristically divided judgment on the work of the Polish cathedral canon, Nicolaus Copernicus. While the Jesuits were persecuting Galileo for Copernicanism in Rome, they were also promulgating Copernican doctrine as part of their missionary work in China. Interestingly, the Protestant sects that failed to establish political sovereignty tended to be still more liberal about the new scientific ideas, most notably in Britain and France. The point here is not to credit or blame certain religions for their attitudes towards science, but rather to highlight the indirect epistemic advantage afforded by the political incapacity to impose a doctrinal consensus. By the end of the eighteenth century, Western political theorists, following the lead of Baron de Montesquieu and James Madison, would be commending this 'incapacity' as a design feature of constitutional government, under the rubric of 'checks and balances'.

Even as characteristically a sign of Western strength as capitalism should be regarded as resulting from the relative advantage of backwardness. Starting with Robert Merton's (1970) adaptation of Max Weber's *The Protestant Ethic and the Spirit of Capitalism,* it has been commonly assumed that the link between the rise of modern science and capitalism lies in some direct connection between science and technology, with the latter interpreted as the primary mechanism of wealth creation. Francis Fukuyama (1992) is only the most recent prominent exponent of this sentiment. However, the history of Western science speaks somewhat differently to this issue. Until the late nineteenth century, virtually all economically relevant technological innovations had been introduced by people whose formal scientific training was spotty at best. This point is often obscured by the eagerness with which academic scientists of the day sought to *explain* why these innovations worked as well as they did, once they had already been deemed successful at a practical level. Not only did such explanations provide still wider societal legitimation for the technologies in question (as that they were deemed reliable according to known scientific principles), but they also legitimated the academic scientists themselves, whose relevance to the liberal arts curriculum was still in question. However, to locate the real sense in which the fates of science and capitalism have been coupled, one would do better to study the *autonomization* of, respectively, education and the economy that marked the devolution of the Roman Empire and the inadequacy of its feudal replacements. Immanuel Wallerstein (1991) has stressed this point with regard to the rise of capitalism, but it is only now beginning to be forcefully made with regard to the rise of science in the West (Huff 1993).

According to Wallerstein, the increasing inability of imperial bureaucracies to redistribute farming surpluses and the local inefficiencies of manorial agricultural production opened the door to the organized forms of piracy and forced exchanges that eventually metamorphosed into 'trade routes' and 'market towns'. This is the context in which the West's modern preoccupation with limitless growth should be understood—as a kind of insurance against risk. The insatiable desire to accumulate wealth and transact goods—sentiments that were deemed intemperate by the classical Greek philosophers—were born of the need to pre-empt an anticipated loss, be it the result of decadence from within, barbarism from without or simply the Biblical decline of 'Man'. Whereas Aristotle genuinely believed in the prospect of a steady state economy in which people grew, made, bought and sold only that which was necessary for their subsistence, during the Middle Ages this image was increasingly replaced with one in which economic well-being was bound to deteriorate if not actively maintained, and perhaps if not overcompensated (Arndt 1978). From this need to concentrate so intently on the maintenance of economic well-being came capitalism.

By contrast, built as it was on arid land, Muslim civilization was held together by state-supported irrigation projects, whose construction and maintenance required the coordination of vast networks of technical workers. Because advances in science were thus readily seen as advances in statecraft, a considerable investment was made in developing the *madrasas*, the first residential colleges. However, the fates of these colleges were tied to the fortunes of their political patrons. Islamic law did not guarantee the colleges the corporate autonomy that would enable their inquiries to continue in perpetuity. That idea—the university—was a Western innovation, one born of the disorganization of feudal Europe, where rulers were often more than happy to grant autonomy to self-constituting groups that agreed to abide by minimal standards of political loyalty (Dorn 1991: 110–21). It was still a long way from establishing institutions of pure inquiry to legitimating the natural sciences as we know them today. However, as Needham realized, a decisive point was that under Roman law, the category of *universitas* covered not only (or even primarily) institutions of higher learning but also craft and trade guilds. It is worth observing that while this equality of legal status enabled a kind of commerce between 'scholars' and 'artisans' in the cities that had not existed in the great civilizations of the East, this intermingling of classes was bemoaned at the time as symptomatic of social disorder. Indeed, if most intellectuals of the High Middle Ages had had their way, they would have preferred the strict separation of 'head' and 'hands' that characterized classical Athens and Rome. In other words, what we, following Needham, now consider to have been the economic preconditions of the Scientific Revolution were regarded in their day as signs of *decline*.

An important reason why the founding of the universities did not eventuate in the Scientific Revolution until another half millennium had passed was that the revolutionary ideas had to live down their heretical cachet. The most universalist of universal claims about science—that even God is bound by the laws of nature—first reached Europe as a Muslim idea that was seen as threatening to the omnipotence of God and, by extension, the absolute authority of his Roman Catholic mediators. This fear materialized as the 'Condemnation of 1277', in which the Bishop of Paris expelled sympathizers of the great Muslim philosopher Averroes from the university. Had the decree been issued just a few years earlier, Thomas Aquinas might have been himself expelled for his interest in synthesizing Averroes' commentaries on Aristotle with Church doctrine. In any case, the Averroists fled to Northern Italy, where they spent the next three centuries training most of the scientific leaders of the Renaissance, most notably Galileo. Pierre Duhem (1954) famously identified 1277 as the year of the birth of modern science, since once the Averroists had been formally divested of their theological responsibilities, they were able to pursue natural inquiry in secular exile.

Have We Reached the End of Science, Eurocentrism, or Both?

Pierre Duhem put forth his celebrated thesis about the institutional origins of modern science shortly before the start of the First World War. He was among the first Western thinkers to suggest that Europe may have originated the institutional, but not necessarily the intellectual, side of the Scientific Revolution. By contrast, in the various schemes of historical progress put forth in the nineteenth century, the superiority of the natural sciences was more often attributed to the superiority of the West than vice versa. One would be hard-pressed to find a Hegel, Comte, or Spencer argue that the natural sciences could just as easily flourish in the East as in the West. In other words, Westerners generally treated the natural sciences as a peculiar accomplishment of their own culture—an accomplishment, even when held in highest esteem, that was regarded as uniquely Western. So, when did all this unabashed ethnocentrism come to an end, ushering in the era of so-called scientific universalism? The answer lies in the first voluntary and successful appropriation of the natural sciences by a non-Western country, Japan, which over the period 1870–1900 became one of the six leading military and industrial powers in the world. The force of this achievement was driven home by Russia's 1905 defeat at the hands of Japan, widely advertised at the time as the first non-Western victory over a Western nation. Had the Japanese beaten the Westerners at their own game? It would be more correct to say that the rules of the game changed once Japan took to the field and won. Western merchants and diplomats, who had for years tried to sell all manner of things to the Far East, were surprised at how the Japanese eagerly sought training in European chemistry and physics but showed little interest in acquiring such classical status markers of 'civilization' as knowledge of the fine arts, philosophy, and literature. Here, too, the relative advantage of backwardness played an important role (Fuller 1996). That the Japanese could match Western scientific and technical performance without having to master the classics of Western culture was the sort of surprise that defined *for Westerners* the sense in which the natural sciences were 'universal'.

Of course, Japan was not an isolated case. The heyday of European imperialism had unwittingly produced other 'standpoints'. The emergence of colonial centres of scientific research, sometimes staffed primarily by assimilated non-Westerners, led humanists in the metropolitan powers to reconceptualize the 'uniqueness' of Western science as a matter of *contingency*, as if it were only by accident that the natural sciences had emerged in Europe rather than, say, China, India, or Egypt. By posing the question of science as one of historical accident, rather than of historical necessity, 'history of science' as a recognizable field of study came into its own. The force of this question was to suggest *both* that the natural sciences were within any culture's reach *and* that Europe's domination of the globe was by no means guaranteed in perpetuity. In any event, it shifted scholarly interest from science as the reflection of more general European attitudes to science as a relatively autonomous enterprise that was easily exported (through imperialism) or imported (by Japan). Here it is worth recalling that imperial expansion was often promoted as a way of addressing (or at least diverting attention from) signs of economic and cultural decline that were already occurring at home. Little surprise, then, that France was the country in which history of science was first recognized as a field of inquiry, since it was the major European power most publicly preoccupied with the prospect of succumbing to the type of 'degeneration' which threatened to turn it into the new 'sick man of Europe', a title which by that time had come to be associated with the Ottoman Empire (Pyenson 1993).

In the twentieth century, the distinction between over- and

underdeterminationist metahistoriographies has grown clearer as the source of the contingency associated with the rise of science in the West shifted from its initial overdeterminationist interpretation to its current underdeterminationist one. Thus, instead of stressing that the West need not have been the leader in science, historians nowadays are more likely to claim that there need not have been a 'science' in which the West has so clearly excelled since the seventeenth century. But whether one adopts an under- or overdeterminationist framework, the decline of science in the West can be just as easily posed as its rise. And inasmuch as the emergence of science has been taken as virtually synonymous with the global ascendancy of the West, alternative accounts of the decline of science in the West may encourage a non-Eurocentrist orientation that treats the history of science and the history of the West as only contingently interrelated. With that end in mind, consider the following chart:

	Overdetermined	Underdetermined
Retrospective	China fell behind	China was never in the race
Prospective	West may lose the lead (East becomes the leader in science)	West may drop out of the race (Science declines as the West declines)

If, on the overdeterminationist view, the Chinese lost the 'race' to the Scientific Revolution, then it is entirely possible that the West will relinquish its lead in the future to, say, Japan or the Far East more generally. But if, as underdeterminationism implies, the Chinese were never in the science race in the first place, then correspondingly the West may drop out in the future. The question, of course, remains as to whether this would herald the 'end of science' altogether.

But how might such a thing happen? While any answer to this question is bound to be speculative, one good indicator is that the social functions jointly satisfied by the pursuit of scientific knowledge would come to be individually satisfied in separate practices (Fuller 1994). The university was crucial for the advancement of science in the West mainly because it provided a common space for the production and distribution of knowledge (through the university's research and teaching functions, respectively) that permitted an intellectual 'synergy' to develop between the demands of innovation and tradition. In the thirteenth century, academics began to take a professional interest in why craft innovations worked as well as they did. By the end of the nineteenth century they found themselves in the business of producing such innovations, often on a large scale, in the case of chemistry research. But given the recent 'market-driven' challenges to the institutional integrity of the university, it is easy to imagine, on the one hand, the privatization of innovation as intellectual property and, on the other, the formalization of tradition through vocational certification. The former function would be increasingly performed by entrepreneurs and their industrial backers, while the latter would be left to academics, whose courses of study and examinations would determine who can enter which field of employment. As the production and distribution of scientific knowledge grow steadily apart in this fashion, we could witness a quick reversal of the last half millennium of Western history.

An important sign of this reversal would be that the meanings of words associated with 'science', and 'knowledge' more generally, would start losing their clarity (Fuller 1996). Indeed, today's oxymorons may metamorphose into tomorrow's common parlance. Thus, a prominent clue that the end of science might be near is the popularity of Derek de Solla Price's (1986) expression 'Big Science', a bit of Orwellian Newspeak that since its debut in 1961 has come to mean 'great science'. Big Science literally refers to expensive research projects, first in physics and chemistry and more recently in molecular biology, that are affordable only by the more affluent nations but on which all nations increasingly set their sights (Crowther 1968: 317–20). This global fixation on Big Science has obscured the fact that in the past great science was what Kuhn would call 'revolutionary science', cheap in its conduct but dear in its consequences—yet not so dear as to discourage the revolutionary impulse altogether. The desire to complete the 'big picture' put forth by a Newton, Darwin or Einstein has required the mass retraining of scientists in new techniques and methods, new ways of seeing the world, and sometimes the development of new instruments of investigation. This reorientation, while invariably resisted by the scientific orthodoxy, was at least financially tolerable. But as science has come to be so thoroughly involved in the economic and political maintenance of the societies housing its pursuit, any truly revolutionary project in science today would pose as great a threat to societal stability as a political revolution normally would. Thus, as Price himself first observed, while the diminishing half-life of scientific journal literature is indicative of a rapidly advancing research frontier, nevertheless all of this activity occurs within a relatively narrow conceptual spectrum, reflecting the need to make the most of science's fixed capital resources.

This turn of events suggests that science in what I have called its *post-epistemic mode* (Fuller 1994, Fuller 1996) may be modelled on an idea that has long been thought discredited, namely that of *oriental despotism* (Wittfogel 1957). This is the so-called 'Asiatic mode of production', Marx's dystopic portrayal of Eastern regimes whereby the dominant class does not own the means of production but instead controls the state and economy through its ability to extract rents from villagers and landholders, which are then used to finance large-scale public works projects. In the emerging 'information-based', 'knowledge-intensive', or simply 'credential' society, the new rents are the training costs that students must incur in order to possess the credentials required for being employable, if not actually employed (Fuller 1997). Academics would shift completely from being producers of innovative, potentially disruptive knowledge to being the main gatekeepers of the processes of societal reproduction. In return, the fees paid by students would feed into the proprietary research that academics are increasingly encouraged to pursue in cooperation with industry. Most of what transpires here is not ultimately intended for classroom consumption. Indeed, depending on the legal protection afforded to research as intellectual property, its fruits may entirely lose their status as 'public goods', which would in turn raise the spectre of 'fields of knowledge' being quite literally mapped, ploughed and mined in a system of social–epistemic relations that one commentator has not unfairly dubbed *information feudalism* (Drahos 1995).

References

Adams, H. (1971), ed., *Critical Theory since Plato.* New York.

Ahmad, Aijaz (1992), *In Theory: Classes, Nations, Literatures.* Verso, London.

Ainslie, G. (1992), *Picoeconomics.* Cambridge, UK.

Al-Azmeh, A. (1993), *Islams and Modernities.* London.

Arndt, H. (1978), *Economic Development: The History of an Idea.* Chicago.

Asante, M. (1988), *Afrocentricity.* Trenton, NJ.

Barnett, C. (1967), 'The Education of Military Elites', in Laqueur and Mosse (eds), pp. 15–36.

Bernal, M. (1987), *Black Athena: The Afroasiatic Roots of Classical Civilization.* New Brunswick, NJ.

Butterfield, H. (1955), *Man on His Past: The Study of the History of Historical Scholarship.* Cambridge, UK.

Cohen, H.F. (1994), *The Scientific Revolution: A Historiographical Inquiry.* Chicago.

Crabbs, J. (1984), *The Writing of History in 19th Century Egypt.* Detroit.

Crowther, J.G. (1968), *Science in Modern Society.* New York.

Diop, C.A. (1991), *Civilization or Barbarism?* Brooklyn, New York.

Dorn, H. (1988), *The Geography of Science.* Baltimore.

Drahos, P. (1995), 'Information Feudalism in the Information Society'. *The Information Society,* 11: 209–22.

Duhem, P. (1954/1913), *The Aim and Structure of Physical Theory.* Princeton.

Elster, J. 1984, *Sour Grapes.* Cambridge, UK.

Febvre, L. and H. J. Martin, (1971/1958), *L'apparition du livre.* Paris.

Feyerabend, P. (1975), *Against Method.* London.

Fox Keller, E. (1985), *Reflections on Science and Gender.* New Haven: Yale University Press.

Fukuyama, F. (1992), *The End of History and the Last Man.* New York.

Fuller, S. (1993), *Philosophy, Rhetoric and the End of Knowledge: The Coming of Science and Technology Studies.* Madison: University of Wisconsin Press.

——— (1994), 'Rethinking the University from a Social Constructivist Standpoint'. *Science Studies* 7 (1): 4–16.

——— (1996), 'Social Epistemology and the Recovery of the Normative in the Post-Epistemic Era'. *Journal of Mind and Behavior* 17:93–8.

——— (1997), 'Life in the Knowledge Society: A Case of Some Really Artificial Intelligence'. *Theory, Culture and Society* 14:143–55.

Gellner, E. (1988), *Plough, Sword, Book.* Chicago: University of Chicago Press.

Gerschenkron, A. (1962), *Economic Backwardness in Historical Perspective.* Cambridge MA: Harvard University Press.

Halsey, A.H. (1992), *The Decline of Donnish Dominion.* 2nd edn, Oxford: Clarendon Press.

Huff, T. (1993), *The Rise of Early Modern Science: Islam, China and the West.* Cambridge, UK.

Inkster, I. (1991), *Science and Technology in Industry.* London.

Kuhn, T.S., (1970/1962), *The Structure of Scientific Revolutions.* 2nd edn, Chicago.

Laqueur, W. and G. Mosse, eds (1967), *Education and Social Structure in the 20th Century.* New York.

Laudan, L. (1977), *Progress and Its Problems.* Berkeley.

Leplin. J. ed. (1984), *Scientific Realism.* Berkeley.

Lewis, B. (1982), *The Muslim Discovery of Europe.* New York.

Mannheim, K. (1936/1929), *Ideology and Utopia.* London.

Merton, R. (1970/1938), *Science and Technology in Seventeenth Century England.* New York.

Nasser, S.H. (1968), *Science and Civilization in Islam.* Cambridge, Mass.

Needham, J. (1971), *The Grand Titration: Science and Society in East and West.* London.

Parfit, D. (1984), *Reasons and Persons.* Oxford.

Price, C. (1993), *Time, Discounting and Value.* Oxford.

Price, D. (1986/1961), *Little Science, Big Science ... and Beyond.* 2nd edn, New York.

Pyenson, L. (1993), 'Prerogatives of European Intellect: Historians of Science and the Promotion of Western Civilization', *History of Science* 31 (1993): 289–315.

Rahmena, A., ed. (1994), *Pioneers of Islamic Revival.* London.

Rawls (1971), *A Theory of Justice.* Cambridge, Mass.

Rose, H. (1994), *Love, Power and Knowledge.* Cambridge, UK.

Said, E. (1978), *Orientalism.* New York.

Shapin, S. and S. Schaffer (1985), *Leviathan and the Air-Pump.* Princeton.

Sivin, N. (1982), 'Why the Scientific Revolution Did Not Take Place in China—or Didn't It?' *Chinese Science* 5: 45–66.

Van Sertima, I. ed. (1986), *Blacks in Science: Ancient and Modern.* New Brunswick, NJ.

Wallerstein, I. (1991), *Unthinking Social Science: The Limits of Nineteenth Century Paradigms.* Oxford.

Weatherford, J. (1988), *Indian Givers.* New York.

Winner, L. (1977), *Autonomous Technology.* Cambridge, MA.

Wittfogel, K. (1957), *Oriental Despotism: A Comparative Study of Total Power.* New Haven.

Wolpert, L. (1992), *The Unnatural Nature of Science.* London.

Zeldin, T. (1967), 'Higher Education in France: 1848-1940', in Laqueur and Mosse (eds), pp. 53–80.

6

Patrick Petitjean

Needham, Anglo-French Civilities and Ecumenical Science

For scholars who are not specialists on 'Science and China', but who are interested in the transfer of scientific knowledge between different cultural areas, Needham's 'The Roles of Europe and China in the Evolution of Ecumenical Science' is of immense value.[1] This essay mainly focuses upon Needham's specificity within what Gary Werskey has called the 'visible college'—a group of leading British scientists who dedicated themselves to science and socialism—and more particularly on Needham's institutional commitment to UNESCO, including the project 'Scientific and Cultural History of Mankind' (SCHM).[2] This essay attempts to understand the manifold relationships and ties between the members of the 'visible college' and the French scientists with whom Needham and his friends shared many commitments.

Both these points are not detailed in Werskey's book. His project was to write a collective biography of the small group of British scientists who, in the 1930s, embraced socialism. Naturally, Werskey was inclined to overestimate what these five scientists (J.D Bernal, J.B.S. Haldane, L. Hogben, H. Levy and J. Needham) shared together, and to underestimate their differences and individualities, their

[1] *Journal of Asian History*, 1967, 1, 1.

[2] Gary Werskey, *The Visible College. A Collective Biography of British Scientists and Socialists in the 1930s.* London, Free Association Books, 1988.

relations with other colleagues (for instance P.M.S Blackett, E.H.S. Burhop, M. Wilkins, J. Huxley, J.N Pirie or J.G Crowther) and their participation in international institutions or networks. Needham's worldview deserves a separate analysis, and cannot be seen as a mere variant of Bernalism.

The *Visible College* is principally oriented towards the participation of this social group within British society. But what about the international networks with which Bernal worked since the 1930s, and which became his main commitment in the 1950s (for instance the World Federation for Scientific Workers, WFSW)? What about Needham's role at the Sino-British Scientific Commission during the war? And what of his struggle for an international scientific service? What about professional collaborations/associations, the visits to foreign laboratories, the International Council of Scientific Unions (ICSU) meetings, congresses, etc.? Laboratories at Cambridge in the 1920s and in 1940–2, or the French College de France in 1936–9 were intellectual melting pots, not merely in terms of effective international scientific practice, but also in providing the space for the emergence of new conceptions about the international function of science. UNESCO, with Huxley and Needham, played the same role. Such associations helped Needham and others change their views about science and about society.

Three sets of questions are discussed in this essay, though its frame is too restricted to pretend to be a systematic and exhaustive Anglo-French collective biography.

What are the various levels of international collaboration between scientists, particularly for international cooperation and exchange? For the scientists under scrutiny, how did institutional and professional associations combine with more ideological affiliations and common participation into political or social organizations or international networks? What are the relevant aspects of social determination, political will, professional and intellectual interests, and historical contingencies? What is the source of the coherence, if any, of their various commitments? Did their involvement in various networks produce new intellectual or political agendas? What about anti-Eurocentrism (in science and, more generally, in civilization), which was so much a part of UNESCO's atmosphere, and which has been the landmark of the encounter of Needham and Febvre?

No collective biography of the French scientists and socialists in the 1930s has ever been written. Such a work would have to analyse the global nature of the social, political and professional commitments of these scientists, and Werskey's book would have enabled a fruitful comparative Anglo-French study. Some books do exist, but either they analyse intellectuals (among whom some are scientists) as a whole and their political commitments,[3] or they are mostly concerned with the struggle undertaken by some physicists for the institutionalization of science.[4] Nevertheless, what were the various commitments of prominent scientists from both sides of the Channel? Are traditional explanations, such as social determinism (the growing influence of science under capitalism in the 1930s drove some scientists, as a social group, to become politically active in order to make their presence felt within the power structure) or political activism (the organized propaganda of communists among scientists to mobilize fellow travellers for the USSR and the communist parties) sufficient? Our main interest is to follow the networks in which Needham participated with Bernal, Rapkine or F. Joliot-Curie. We can thus evaluate the exact importance of cross-influences, to throw light on possible common relations between these scientists.

Among members of the 'visible college', Needham was the only one deeply involved in the building of an international scientific organization (UNESCO) and in international professional activities, as opposed to Bernal, whose international commitments were of a more political nature. Needham's Anglican devotion is well known, as well as the manner in which he always managed to reconcile spiritualism and Marxism. It should be added that he sometimes appeared to be closer to Julian Huxley's 'scientific humanism' than to Bernal's scientism. Huxley was the one who invited Needham to UNESCO, and they went on to cooperate at UNESCO for many years after their official responsibilities had come to an end. Their proximity lasted far longer than the alliance between the socialist and liberal scientists, described by Werskey. Needham was also the only one among the members of the 'visible college' to participate in the Society

[3] David Caute, *The Fellow-Travelers. Intellectual Friends of Communism.* New Haven and London, Yale University Press, 1988. Christophe Charle, *La Republique des universitaires,* 1870–1940. Paris, Seuil, 1994. Pascal Ory, *La Belle illusion. Culture et politique sous le signe du Front Populaire, 1935–1938.* Paris, Plon, 1994.

[4] Spencer Wyart, *Scientists in Power.* Harvard University Press, 1979.

for the Freedom of Science a typically liberal society.[5] Was Needham a liberal among Marxists, or a Marxist among liberals?

Why was Needham the only member of the 'visible college' who was not satisfied with a general rejection of European colonialism, but effectively opposed Eurocentrism in science and the history of science? How were his institutional activities and his intellectual trajectory and the coherence of his commitments related? What was the attitude of Needham towards the Radical Science Movement of the 1970s, consequent on his changing perception of science.[6] Even if during the early years the BSSRS[7] was thought to be some kind of a continuation of the 'visible college', it appeared very soon that the radicalization of the 1970s was based on the criticism of science itself, and not of the use/abuse models of science (science as value-neutral) which underlay Bernalism. These movements also radically rejected socio-economism, the main ingredient of Bernalism. The radical criticism of the 1970s found echoes in some of Needham's papers, even though science in China was spared such criticism. Needham, at least, partially understood what these radical critics were trying to say, and cooperated with these movements.

Rather than analyse all the issues raised here, I will concentrate upon the various levels of Anglo-French cooperation and some of their consequences.

Anglo-French Linkages

The essay follows individuals through their scientific and political, personal and institutional, official or trade union commitments. On the British and French sides, these individuals were mainly physicists

[5] William McGucken, 'On Freedom and Planning in Science. The Society for the Freedom of Science, 1940-1946', in *Minerva,* 16, 1 (1978), pp. 42–72.

[6] See, for instance, Hilary and Steven Rose, 'Radical Science Movement into the 80's', in *Science for the People,* no. 47, Autumn 1980, pp. 10–13. Hilary and Steven Rose (eds), *The Radicalization of Science. Ideology of/in the Natural Sciences,* The MacMillan Press Limited, London and Basingstoke, 1976. Hilary and Steven Rose, 'The Two Bernals: Revolutionary and Revisionist in Science?', in *Fundamenta Scientiae,* vol. 2, nos. 3–4, 1981, pp. 267–86. Gary Werskey, 'Making Socialists of Scientists: Whose Side is History on?', in *Radical Science Journal,* 1975, nos. 2–3, pp. 13–48.

[7] BSSRS is the acronym for British Society for Social Responsibility in Science, founded in 1969. Joseph Needham addressed the BSSRS tenth anniversary meeting

and biologists, with socialist commitments. J. Needham, J.D. Bernal, L. Rapkine, P. Langevin and L. Febvre are our main characters, with J. Huxley, J.G. Crowther, F. Joliot-Curie, P. Rivet and P. Auger in the second line, and there are many others backstage. Combining characters and chronology, nine scenes are introduced.

Needham and Rapkine (1925)

Joseph Needham is at the centre of this story. He is our first character, jointly with Louis Rapkine who was his main French partner. They first met in the summer of 1925 at Roscoff,[8] where Dorothy and Joseph spent their vacation and met Louis Rapkine,[9] Rene Wurmser, Marcel Prenant, etc.

Needham was initiated to Marx through Rapkine, discovering new perspectives.[10]

> Our discussions considerably opened the perspective of people who grew up in the somehow insular, or at least cramped atmosphere, typical of England in general and Cambridge in particular. He gave us some indication of what could be life for people who, by accident, came into the world in East-European countries where living standards were much lower than in the West.[11]

Like Needham, Rapkine never joined the Communist Party (hereafter CP), and always claimed a 'spiritual' (or ethical) dimension for his

---

in 1979 about his role in the movements in the 1930s. See also ch. 5 (written by Needham) in Roses (1976) and Joseph Needham, 'An Eastern Perspective on Western Anti-Science', in Joseph Needham, *The Moulds of Understanding,* New York, Saint-Martin's Press, 1976, pp. 295–304.

[8] See Henry Holorenshaw, 'The Making of an Honorary Taoist', in Mikulas Teich and Robert Young, *Changing Perspectives in the History of Science: Essays in Honour of Joseph Needham,* London, 1973, p. 8. Roscoff is a marine biological station in Brittany.

[9] Dozens of letters from Rapkine to Needham, since 5 September 1925, are preserved at the Pasteur Institute in Paris (Rapkine archives).

[10] Joseph Needham, 'Science and Politics', in Ben and Viviane Karp, *Louis Rapkine, 1904–48,* North Bennington, Vermont, The Orpheus Press, 1988, pp. 91–4. Born in a working-class family, which emigrated from Bylorussia to Canada, Rapkine went to France in the 1920s with a Rockefeller grant and became a French national in 1939. He kept in close touch with Needham until he died in December 1948. He visited Needham's laboratory in Cambridge for the first time in January 1926.

[11] Joseph Needham, speech delivered at the memorial meeting for Louis Rapkine, Association des Travailleurs Scientifiques, Paris, 13 October 1949. (Needham archives, Cambridge Library, J184).

commitments, referring to Spinoza as a Jewish philosopher.[12] During the general strike in Britain of 1926, when Needham substituted for train drivers on strike, Rapkine[13] expressed his horror at the prospect of a bloody revolution. If the rotten social system should be changed, he refused the path of bloodshed: 'To kill one's fellow-being is never justified.' According to Needham, Rapkine had always been sympathetic to the British 'science and society' movements.[14] Bernal's, Needham's and Rapkine's roads crossed many times till 1948.

After Rapkine's death, two memorial meetings were organized in London[15] (9 March, 1949, by SVS) and Paris[16] (13 October, 1949, by ATS[17]), bringing together the main characters of our story, and official representatives of scientific institutions like the French CNRS. A Rapkine Fund was instituted after the London memorial meeting.

Altering the setting from professional to personal, from social to intellectual, and forging new connections with other scientists, the relationship between Needham and Rapkine was probably the most inventive in our story.

Langevin (1931)

In the next narrative—international institutions—the new characters are Paul Langevin and J.B.S. Haldane who attended a meeting in Madrid in May 1933 to discuss 'High Culture and the Emergence of an Universal Moral'. Marie Curie, Miguel de Unamuno, and Paul Valery were among the other participants. The function of science was discussed, and in responding to Haldane, who denounced the

[12] Needham in Karp (1988), pp. 91–4.

[13] Needham to Rapkine, 5 May 1926 and Rapkine to Needham, 17 May 1926. Rapkine archives, Pasteur Institute.

[14] Joseph Needham in Karp (1988), pp. 91–4.

[15] Needham archives, Rapkine Fund, J180/J182. SVS is the Society for Visiting Scientists.

[16] Needham archives, Rapkine Fund, J184. ATS is the Association des Travailleurs Scientifiques.

[17] In the area of physiology, we come across Brachet (from Belgium), who worked with Needham, then cooperated with him in UNESCO, and joined the Belgian CP in 1944. Two other French physiologists regularly crossed Needham's path: Wurmser (especially in helping scientists escaping from Nazi tyranny) and Henri Laugier (UNESCO).

negative effects of science, Langevin argued that science also remedies the harm it does.[18]

The Madrid meeting was organized by the 'Intellectual Cooperation Organization',[19] the precursor of UNESCO between the wars, attached to the League of Nations. But for Gilbert Murray, the British participation in the ICO was meagre. The ICO was French dominated, both in terms of the personalities involved, as well as in terms of the general conception of intellectual, cultural and international scientific cooperation. The IIIC was the main channel for the activities of the organization: regular meetings, exchange of books, directories (on education, literature, museums), etc., so as to develop a 'society of minds', as Paul Valery called it, independent of all governments. Such a society would promote a strong movement for peace. Science was nothing other than an intellectual activity. Needham, in his general survey of scientific international relations,[20] made, in a few lines, a very severe criticism of the ICO action.

The institute was terribly elitist, and mainly consisted of Western intellectuals. Western civilization was the model. This had been criticized in the 1920s by Bannerjee,[21] its Indian member. The commission's conception of culture was supposed to be universal, but was in fact purely European, and Bannerjee proposed, without success,

[18] Manuscript of Langevin's talk in Langevin archives, ESPCI-CHR (179/22), Paris. As early as 1933, two years after his mission to China, Langevin proposed views that appear familiar in the 'Scientific and Cultural History of Mankind' project: cultures cannot escape mutually influencing each other; the planet is becoming smaller on account of the new means of communication; China is an 'admirable country: there is between China and us, and more generally between the Far-East, including India and us a possibility of mutual enrichment'.

[19] ICO was mainly composed of a commission within the League of Nations, the International Institute for Intellectual Cooperation (IIIC), and the International Bureau for Education. The commission included known intellectuals, like Marie Curie and Albert Einstein.

[20] Joseph Needham, *Science and International Relations,* Blackwell Scientific Publications, Oxford, 1949, p. 14, which is a revised version of the 1948 Boyle Conference. See also UNESCO/Prep. Com./Nat. Sci. Com./12 July 1946, 'Taches et fonctions de la section des sciences', p. 24. Some authors are more indulgent: see Brigitte Schroeder-Gudehus, *Les Scientifiques et la paix,* Presses Universitaires de Montreal, 1975. According to her, the IIIC did not concur with the International Council of Scientific Unions, with which an agreement was signed only in 1937.

[21] Schroeder-Gudehus, 1975, ch. v.

that the Commission should pay much more attention to non-European people. The only significant action undertaken by the ICO was the mission to China in 1931 (October to December) in order to assess the nation's problems. The team included Langevin, C.H. Becker (Berlin), M. Falski (Poland) and R.H. Tawney (London).[22] Langevin was specifically in charge of the universities and their laboratories. This mission was followed for some years by the exchange of university professors.[23]

Langevin returned with enthusiastic accounts of the Chinese civilization and its people. He was worried about the risk of North America's culturally colonizing China, and insisted on the necessity of a new version of civilization, a Chinese path that synthesized traditional civilization and modern science, a path that rejected the American and European ways. Chinese pupils, he felt, must take the time to assimilate their own culture and to return to the sources of knowledge, without limiting themselves to the useful applications of modern science. For Langevin, China has to learn how to use 'the powerful instrument for material and spiritual liberation that scientific methods constitute'.[24]

The presence of Langevin in our story naturally brings in Lucien Febvre, with whom he cooperated within the CIS, the 'Semaines de Synthese' and the *Revue de Synthese* (1930). Febvre was in charge of the social sciences and Langevin of the exact sciences. The unity of the natural and social sciences, in tandem with the notion that science provided the basis for a general intellectual synthesis, were the two main guidelines for the CIS. Henri Berr and his colleagues proposed editing a general history of humankind, many years before the

[22] See the mission's report, 1932, IIIC publications.

[23] *L'Annee de la Cooperation Internationale*, 1932 onwards. See also Pham-ti Tu, *La Cooperation intellectuelle sous la SDN*, Droz, Geneve, 1962.

[24] See the report. See Bernadette Bensaude-Vincent, *Langevin. Science et vigilance*, Paris/Berlin, 1987. His ideas about the Chinese road to modern science and civilization were not so common in 1931, the year when colonialism was at its zenith in France with the great colonial exhibition. See also 'European Scientific Thought', manuscript of a conference held in Beijing (1931). Langevin archives, ESPCI-CRH (178), Paris. Regarding the evolution of civilizations, Langevin was still influenced by the discussions at the 'Centre International de Synthese' (CIS), founded by Henri Berr (1925), with Lucien Levy-Bruhl (*The Primitive Thought*).

commencement of the UNESCO project. This history valued science, but only as a form of higher intellectual activity, the so-called Greek miracle that founded a new (European) civilization.

Significantly, Langevin also cooperated with the *Encyclopedia Française*, that had been managed by Febvre since 1933. Two other characters were involved in this project: Paul Rivet and Jean Perrin. Their model was Diderot's *Encyclopedia*, where science was the principal unifying factor. The objective of the project was to emancipate the masses through a programme of cultural and scientific indoctrination, and to awaken a critical sense among them. Langevin and Febvre shared the same cult for living science and humankind.[25]

With the founding of UNESCO, institutional international cooperation took another direction. Needham's choice symbolized a triple move: from the French to the British group, from intellectuals to scientists, from Eurocentrism to worldviews. Two branches of knowledge came closer, Langevin moved from science to philosophy, and Febvre moved the other way. Ten years later, in another institutional frame, in another political context, Needham and Febvre similarly approached each other, giving birth to a new encyclopaedic project, the SCHM.

Against War and Fascism: Political Connections (1934–1936)

John Desmond Bernal, Paul Rivet and Frederic Joliot-Curie now join our company. To fight the rise of fascism, the 'Comite de Vigilance des Intellectuels Antifascistes' (hereafter CVIA) was founded in April 1934, with an 'appeal to workers'. But it rapidly commenced functioning as the 'popular front' of French intellectuals, with Langevin (representing the communists, though he became a party member only in 1944) and Rivet (the socialists). The CVIA immediately inspired a great deal of interest among British left-wing scientists, who had just come to power in the Association of Scientific Workers (hereafter AScW) and launched the Cambridge Scientists Against War Group (hereafter CSAWG).[26] Bernal's first trip to Paris to meet Langevin[27] and F. Joliot-Curie took place in 1934. Referring to the

25 Bensaude-Vincent (1987), ch. x.

26 Werskey (1988), pp. 234–6.

27 Bernal, preface to Pierre Biquard, *Paul Langevin, Scientifique, Educateur, Citoyen,* Paris, Seghers, 1969, p. 5.

CVIA, back in London, he participated in the constitution of the 'Society for Intellectual Liberty', oriented towards writers[28] (such as Virginia Woolf) though this group was less politically committed than the CVIA. In November 1934, in their letter to *Nature*, C.B.O. Mohr and N. Wooster referred to the CVIA as an exemplar of an association opposing war and analysing science's relations with war.[29] In 1936, a delegation was sent to Paris to establish a common cause against fascism and war.

CSAWG gained both audience and credibility in testing the evacuation plans in the case of bombings, and the efficacy of gas masks that were provided by the army.[30] This scientific expertise resulted in CSAWG becoming an official partner for governmental agencies. The CSAWG played a role similar to that of Langevin during the war of 1914–18: the population had to be protected as far as possible from the effects of the weapons of mass destruction.[31] Further, CSAWG prepared left-wing scientists to participate in the national war effort. As for Langevin, J. Perrin and F. Joliot-Curie, there were no ideological or political barriers preventing them from joining the national efforts.

In 1936, Langevin was sent to London by the 'Comité mondial de lutte contre la guerre et le fascisme' to develop ties of solidarity with Spain, and to convince British authorities to intervene on the side of the Republicans. The 'Society for Intellectual Liberty' conveyed the content of his mission.[32] The 'Comité mondial ...' organized (Paris, September 1936) an international conference showing solidarity with Spain, and Bernal was chosen as co-president.[33] Langevin had introduced Bernal to this committee.[34]

[28] Biquard (1969), p. 5. See the speech delivered by Bernal at the memorial meeting for Langevin (London, 23 and 25 May 1947, organized by the AScW, the SVS and the WFSW), pp. 16-19. Langevin archives, ESPCI-CRH (112/16), Paris.

[29] This letter is quoted by E.H.S. Burhop, 'Scientists and Public Affairs', in Maurice Goldsmith and Alan MacKay, *The Science of Science*, London, Souvenir Press, 1964, p. 34. Mohr and Wooster were active members of AScW.

[30] Werskey (1988), pp. 234–6.

[31] Speech delivered by Bernal at the memorial meeting for Langevin (London, 23 and 25 May 1947, organized by the AScW, the SVS and the WFSW), pp. 16–19. Langevin archives, ESPCI-CRH (112/16), Paris.

[32] Langevin archives, ESPCI-CRH (30), Paris.

[33] Langevin archives, ESPCI-CRH (32), Paris.

[34] Bernal, preface to Michel Rouze, *Frederic Joliot-Curie*, Paris, Les Editeurs Français Reunis, 1950.

In addition to other associations—among which was the Royal Society—the AScW also helped receive European scientists fleeing to Britain. In Paris, Rapkine founded the 'Comite d'accueil des savants Etrangers' in August 1934, and got in touch in 1937 with the British Academic Assistance.[35] He met Bernal several times in 1938–9 on this count.[36]

This period had two main characteristics—political connections were deeply intermeshed with professional ties, and rooted in professional associations or institutions. This rise gave birth to 'popular front' movements,[37] like the AScW, the Division for Social and International Relations in Science (within the British Association for the Advancement of Science (hereafter DSIRS-BAAS) or, in a way, *Nature*, in Britain or CVIA and 'Jeune Science' in France.

'Palais de la Decouverte' (1937): Building Scientific Institutions

The physicist Pierre Auger and J.G. Crowther now make their first appearance in our story, in which Bernal and Langevin are in the forefront. French and British scientists simultaneously acted for the defence of science, its popularization and organization. These three dimensions were present in the creation of the 'Palais de la Decouverte', which was in the same tradition as the French Encyclopaedia, or quite in tune with Langevin's activities in the popular education movements. IIIC, CIS and the Encyclopedia were the main partners. Dedicated to the popularization of science and built for the International Exhibition of 1937, the Palais was actually the temple of science and modern rationalism. On this occasion, a large scientific conference, with many disciplinary congresses, was organized. Langevin, J. Perrin and F. Perrin, Wurmser, Auger and F. Joliot-Curie were the official committee. Needham attended the biology congress, Bernal and Crowther were among the invited British scientists.[38] A 'month for intellectual cooperation' was organized, in addition to which

[35] Needham, speech delivered at the Rapkine memorial meeting, ATS, Paris, 13 October 1949. Needham archives (J184).

[36] Louis Rapkine to Sarah Rapkine (Rapkine archives, Paris).

[37] Werskey (1988) pp. 237–48. F. Joliot-Curie joined SFIO (Socialist Party) in 1934, and left it in 1936 because of the Party's refusal to intervene in Spain. He joined the Communist Party in 1942. See Pierre Biquard, *Joliot-Curie*, Seghers, Paris, 1961.

[38] Needham archives (H32-H34). See also Bensaude-Vincent (1987), ch. x.

there was a meeting of the 'Society of Minds', a conference of national commissions for intellectual cooperation, etc.

The popularization of science was the main task of the Rationalist Union, a movement in which Langevin and Bernal participated.[39] Langevin was to deliver three lectures in 1939 for the Union, the first one on 'the relations between science and general activities'. He began by saying that he had just received 'the beautiful and voluminous book from my friend Bernal. (...) I started reading it, and it looked so important to me that I desired bringing it to your attention. I found its introduction so interesting and so linked with what I intended to speak of, that I took it as my guide'.[40] And Langevin proceeded with his lecture by translating Bernal's introduction.

French scientists actively participated in the political struggle for the organization of scientific research.[41] Their British colleagues, who were confronted with shortage of funding due to the crisis of 1929 and general governmental disinterest, admired the results of this struggle,[42] especially Irene Joliot-Curie's participation (for some weeks, then substituted by J. Perrin) in the Popular Front Government. The presence of a Nobel prize-winner in the government appeared an unachievable dream for British scientists.

Bernal often claimed having the same ideas as Langevin:[43] 'As far as social implications of our work were concerned, Langevin and I followed identical roads, and I always remembered his words: if we

[39] Founded in 1930 by Henri Roger, Ernest Kahane and Henri Laugier. Langevin had been the president. Bernal was a member of the Honorary Committee. On Bernal's death, the Union paid its tribute in a special issue of *Cahiers rationalistes,* December 1971, no. 287, which included various papers of Bernal and a biography by Alan MacKay.

[40] Conference manuscript. Langevin archives, ESPCI-CRH (179/15), Paris. Bernal's *The Social Function of Science*, 1939, has not been translated into French, which was a barrier to the influence Bernal and his friends claimed to have had in France. Passages from the translation (by Langevin himself) of the introduction have been published in *Cahiers rationalistes* no. 75, 1939, pp. 114–34. Langevin translated another Bernal paper, 'La Science et le sort des hommes', in *La Pensee,* no. 5, October–December 1945, pp. 129–32.

[41] Jean-François Picard, *La Republique des savants. La recherche francaise et le* CNRS, Paris, Flammarion, 1990.

[42] J.G. Crowther, *Fifty Years with Science,* London, Barrie and Jenkins, 1970, pp. 183–7.

[43] J.D. Bernal, preface to Pierre Biquard (1969), p. 6.

are not doing our scientific work, somebody else will do it, but if we neglect political action, soon science will disappear.' Langevin has been long known in Britain, since he worked in the Cavendish Laboratory at the end of the nineteenth century. Since his scientific accomplishments were renowned, his British colleagues were attentive to his social and political commitments. Langevin's influence upon Bernal 'was decisive. It was a considerable help to see such a notorious and famous scientist to express so definite social ideas (...). Langevin's ideas became progressively a gathering point for British and French intellectuals fighting against Fascism.'[44] Langevin's influence also extended the national organization of science, and the manner in which he conceived the relationship between scientific and social or political commitments.

A national homage to Langevin and J. Perrin was organized at the Sorbonne and the Pantheon on 15 November 1948. Bernal was the only foreign scientist to speak there. Besides, another colloquium was held at the 'College de France', on the organization of scientific research.[45] Bernal presided over the first session, and F. Joliot-Curie the second one, symbolizing the Anglo-French relations for the organization of science.

Scientific Mobilization for the War Effort, and National Unity (1939–1945)

The stage now moves to London, and more precisely to a 'dining club', the 'Tots and Quots',[46] where since the 1930s, Solly Zuckerman gathered twenty young scientists having diverse ideological commitments, 'a stimulating society for exchanging scientific ideas and news, and deepening scientific friendship'.[47] Among the guests were Bernal, Crowther, Waddington, and others. The mobilization of scientists for the war effort was primary on the agenda on both sides of the Channel in 1939. The CNRS was constituted for that purpose.[48] The 'Tots and

[44] Ibid., p. 7.

[45] Langevin archives, ESPCI-CRH (112 for the official session, and 112/28 for the colloquium), Paris.

[46] Solly Zuckerman, *Scientists and War. The Impact of Science on Military and Civil Affairs*, London, Hamish Hamilton, 1966. Crowther (1970).

[47] Crowther (1970), p. 211.

[48] CNRS was founded in October 1939 and assumed all tasks for scientific

Quots' concentrated upon war and science and, when a French delegation (with Langevin and Auger among others) arrived in London in February 1940 in order to strengthen scientific cooperation during the war, it was naturally invited to the Club. The lack of preparation of French scientists in war left a disastrous impression. The French scientific attaché gained a permanent invitation to the Club. As 'the anxiety among the younger scientists was even greater',[49] Crowther was sent to Paris to constitute a British–French scientific cooperation committee. At the turn of April 1940, Crowther met the Joliot-Curies, Langevin, F. Perrin, Laugier, and Auger in Paris. The Anglo-French Society of Sciences was founded on Crowther's return to London. Dirac and F. Joliot-Curie were co-presidents, Crowther and Auger general secretaries, and Langevin, Laugier, Rapkine, F. Perrin, Wurmser, I. Joliot-Curie, Bernal, Haldane, Blackett and Zuckerman were among the committee members.[50] Physicists dominated the association on both sides and there were a few physiologists introduced by Rapkine, but neither Needham nor Huxley appeared to have participated, and Haldane joined later. Bernal suggested that this society was the result of the will of 'young scientists who began to think that science should play its part in the resistance against the military strength of Nazis',[51] and thus to nourish the official but dying relationships that were dominated by conservatives. Rapkine was sent to London in May 1940 to organize the reception of French scientists, but the French defeat in June brought the activity of the Anglo-French Society to an end.

Through professional connections Joliot-Curie's collaborators, Halban and Kowarski, could leave France for London on the 18 June, with the French heavy water stocks. Laugier and other scientists sought refuge in London. But after the Anglo-French naval battle of Mers-el-Kebir, Rapkine and Laugier left London for the USA. With the help of Laugier for some months, and with funding from the Rockefeller Foundation, Rapkine helped save 40 of the 52 French

mobilization, as defined in July 1938 and which were earlier held by the 'Haut Comite de Coordination' which was presided over by J. Perrin. See Picard (1990), pp. 72–84.

[49] Ibid., pp. 213–15.

[50] Ibid., p. 218.

[51] Rouze (1950), preface

scientists who were considered to be in danger.[52] 'The great figure that emerged in that time was Louis Rapkine.'[53] A few months before D'Day, the British–French stage revived with the same characters.

Rapkine was back in London in September 1943. According to Crowther,[54] he was the one who called for Rapkine, to help activate the new 'Society for Visiting Scientists', dedicated to supporting exiled scientists or visitors, among them the French Londoners. But Rapkine soon took over as director of the French scientific bureau in London, in charge of the rehabilitation of French science. Following the Liberation of France, he organized scientific exchanges between laboratories and sent many British scientists to Paris to deliver lectures (Dirac was the first to visit Paris and Crowther the second, in December 1945) to raise the academic level of French scientists coming out of isolation.[55]

During the summer of 1944, F. Joliot-Curie often commuted between Paris and London, and Crowther tried to revive the Anglo-French Society, which held its first conference in London in January 1945.[56] Rapkine, Auger, the Joliot-Curies, F. Perrin, Bernal, Haldane, Blackett, Crowther and others attended. But, like many such Anglo-French societies launched with enthusiasm at the end of the war, this Society had no future. Rapkine finally went back to Paris at the end of 1945 to create his own laboratory at the Pasteur Institute. After having obtained a gift of $350,000 from the Rockefeller Foundation for the reconstruction of CNRS, Rapkine concentrated upon his professional activities.[57] F. Joliot-Curie was also deeply involved in the reconstruction of French science, the CNRS in 1945 and the newly born CEA in January 1946.[58]

Science at war, and the national reconstruction effort, had the

[52] Obituary of Louis Rapkine in *Nature,* no.191, by Crowther.

[53] Crowther (1970), p. 218.

[54] Ibid., pp. 241–2.

[55] Ibid., pp. 264–70.

[56] Ibid., pp. 261–2.

[57] Rapkine to Biquard, 30 October 1946, in which he expresses his desire to confine himself to scientific activities. Biquard archives (8/1), ESPCI-CRH, Paris.

[58] Biquard (1961), p. 85. The orientation was the same as in 1939: scientific research is the basis for both knowledge and industrial development. CEA: Commissariat à l'Energie Atomique.

left-wing scientists cooperating with the state scientific institutions, where they played a major role. Bernal with Mountbatten, Needham in China, Crowther at the British Council, Rapkine with the French bureaus abroad, F. Joliot-Curie, Febvre and Laugier in the CNRS. What was left of the socialist perspective?

Putting an 'S' in UNECO (1943–1945)[59]

Julian Huxley was the last to join this company and he stayed in the front line with Needham and Febvre till the end of our story.

Between 1943 and 1945, Needham published three memoranda[60] on international scientific cooperation, which he was able to implement within UNESCO's science division which he headed between 1946 and 1948. CAME was trying to build a post-war organization dedicated to the reconstruction of education in the former occupied countries,[61] and this effort was extended to the cultural domain. But science was considered part of education and culture, with no specific approach. The applications of science were left to another UN agency. Needham opposed this conception, and argued for the inclusion of science in the future programmes of the organization.

Needham refused to separate the applications of science from pure science, and refused to establish a separate science agency that would lack credibility to negotiate with governments and would weaken the scientific unions themselves.[62] In letters written towards the end of February 1945 from the USA, and before leaving on 19 March for China, Needham proposed for the first time to include an 'S' in

[59] On Science and UNESCO, the most recent reference paper is Aant Elzinga's, 'Unesco and the Politics of International Cooperation in the Realm of Science', in Patrick Petitjean (ed.), *Les Sciences coloniales, figures et institutions,* ORSTOM Editions, Paris, 1996. This is the second volume of the proceedings of the ORSTOM Conference, 'Twentieth Century Sciences: Beyond the Metropolis' (Coordination: Roland Waast).

[60] Memorandum I, Chungking, July 1944: 'On an International Science Cooperation Service', Memorandum II, London, December 1944, 'Measures for the Organization of International Cooperation in Science in the Post-War Period', addressed to the British Parliamentary and Scientific Committee, Memorandum III, Washington, 15 March 1945, revised Chungking, 28 April 1945, 'The Place of Science and International Scientific Cooperation in Post-War World Organization'.

[61] CAME: Conference of Allied Ministers of Education (British dominated).

[62] Needham (1949), p. 21.

UNECO.[63] Among the addressees were H. Dale, F. Joliot-Curie, Kefauver, etc. He then wrote the first short draft of his third memorandum, on 15 March. CAME created a science commission, whose general secretary was Crowther. Various CAME commissions or plenary meetings rejected the inclusion of 'S' during the spring of 1945,[64] as well as at the San Francisco conference. In April and May 1945, Needham sent the final version of his third memorandum to hundreds of scientists and politicians through correspondents like Crowther and F. Joliot-Curie from China. On his way from China to Moscow, in June 1945, for the 220th anniversary of the Moscow Science Academy, he met the American delegation at Teheran Airport. This meeting was, according to Needham, essential to convince his American colleagues.[65] F. Joliot-Curie was also at Moscow with an important delegation of French academicians, while Huxley and Bernal were with the British ones[66]. The bombing of Hiroshima was decisive for the inclusion of 'S'. Rapkine, in London, received the first and the third memoranda, but F. Joliot-Curie was the one chosen to dispatch it in France.[67]

Needham's strategy appeared clearly in a letter to Joliot-Curie.[68] Having met him in London, in early 1945, Needham realized that the UNECO project was close to realization. He was consulted regarding the ability of such an organization to act as the international scientific service that he had proposed in his memoranda. Needham answered positively on two counts: first that science appears in the name of the organization, and that the applications of science should be considered with pure science. He sent the authorities a complete list of scientific activities to be included in the future organization. But Needham felt

[63] Needham archives (D8, D9, D10). First version is in D10. Kefauver had the responsibility for the UNECO project.

[64] Needham archives (D11, D16).

[65] Needham (1949), p. 21.

[66] Maurice Goldsmith, *Frederic Joliot-Curie,* Laurence and Wishart, London, 1976, pp. 174–80. Goldsmith argues that the WFSW had been conceived during the Moscow festivities.

[67] Needham to F. Joliot-Curie, 19 May 1945. Needham archives (D25).

[68] Needham to F. Joliot-Curie (letter in French), 1 March 1945, from Washington. Needham archives (D9). The following quotations are taken from this letter.

that many obstacles were present in Britain, and asked Joliot-Curie for help.

Lately, I tried to influence politicians and scientists in order to obtain a full and fair participation of science in this organization. I rely on you to do whatever you can in the same purpose(...) I am writing on the very same day to Sir Henry Dale, Sir Richard Gregory, Pat Blackett and Darlington with the same demands. But I have one more thing to say. An organization for international scientific co-operation, which would give us this supra-national loyalty we are all looking for, is here now in statu nascendi. But to this initiative from Washington, don't we prefer other initiatives to be added? I mainly thought of USSR. I think that active co-operation with Russians is one of the main problems we have to face. We hope the Washington leadership should be shared with 'leadership' from Moscow. For this point, you appear to me to have a special position. Your relations with Russians must be excellent. You also represent the Western science of the continent. I am wondering whether it would not be possible for you and me to travel to Moscow, invited by Russians, to explain the historical opportunity for an important 'leadership'. We will have to explain to Russians all the advantages they will draw from their participation in UNESCO. I already gave M. Paniushkin, their ambassador, a copy of my first memorandum. But a personal contact is now necessary. Without it, UNESCO would become as ineffective and limited as the UNRRA.

As evident, British opinion is divided. The elders, Sir Henry Dale, A.V. Hill, and the like, whatever their role in international scientific co-operation, were skeptical about any organization that they do not actually favour my proposal of an international service. On the opposite side, Blackett, Darlington, and all the group of scientists under 50, who besides are beginning to dominate the Council of the Royal Society, completely agree with such a service. (...) Meanwhile our Louis Rapkine will always inform you of what is happening in London.

It is true that Dale did not seem to value Needham's active role in promoting international scientific cooperation. He disregarded the telegram that Academia Sinica sent to CAME to support Needham's proposals. He encouraged Needham to stick to his scientific activities in China, rather than trying to be the maestro of international science.[69] After receiving the third memorandum, Ernest Baker also kept Needham at a distance. He agreed to Needham's services, but only

[69] Sir Henry Dale to J.G. Crowther, 24 May 1945. Needham archives (D17).

outside UNECO.[70] Huxley supported Needham's proposal and even suggested that the British government appoint him as scientific adviser.[71] What Werskey called the scientific popular front, the alliance between radicals and liberals, was effective, through Needham and Huxley, in the foundation of UNESCO. During the London conference, *Nature* published an issue on UNECO (still the official title), with papers from E.F. Armstrong, Huxley, Crowther, Bernal and Needham.[72] A little earlier, Gregory, the editor, indicated his support for Needham after the first memorandum appeared.[73] *Nature* also published an editorial and a summary of this memorandum.[74]

Notwithstanding Needham's speculations, F. Joliot-Curie had not been very involved in UNESCO. Though member of the French delegation to the founding conference at London, and then vice-president of the French National Commission, he did not seem to have been an active participant. He considered UNESCO to be too dependent on the USA. For him, WSW represented the international unity of science far better than through intergovernmental cooperation within UNESCO.[75]

The French delegation to the London conference was headed by Leon Blum, former prime minister of the Popular Front Government and leader of the SFIO. Laugier and Febvre as members, were part of the delegation. Laugier opposed the American–British project, and proposed to keep the former IIIC as an autonomous body of intellectuals, to reform it and to obtain more funding (which was

[70] Sir Ernest Baker to Crowther, 10 August 1945. UNESCO archives, CAME correspondence IV.

[71] Huxley to the Foreign Office, 14 August 1945. Letter reproduced in *Le Courrier de l'Unesco,* 36e annee, no.10, October 1985, pp. 23–6. Huxley explicitly supported Needham's idea for the 'S', for science had to include education and culture. 'I would venture to suggest that Dr Joseph Needham, FRS, now head of the British Mission to China, could be secured for a few months for this purpose' (scientific adviser to the Government to prepare the conference).

[72] *Nature,* no. 3967, 10 November 1945, vol.156, pp. 553–61. Needham's paper is a summary of his third memorandum, with the title 'The Place of Science and International Scientific Cooperation in Post-War World Organization'. The editorial in *Nature,* 6 October 1945, supported Needham's proposals.

[73] Needham archives (D4).

[74] *Nature,* no. 3917, 25 November 1944.

[75] Goldsmith (1976).

seen as the main problem). The balance of power after the war was not in favour of French proposals, and the American–British project was kept, with minor modifications.

Being in China, Needham was unable to participate in this conference, though he tried hard to obtain an invitation.[76] But his ideas regarding the function of science in such an international organization triumphed. Not surprisingly, he was called by Huxley (who took charge as president of the Preparatory Commission on 1 March 1946, and was elected as the director-general at the first meeting) when he was still staying in China, to organize the science division in UNESCO. This entailed a number of telegrams from Dorothy Needham and Huxley in February and March to get in touch with him.[77] Needham came back to Britain at the end of April 1946.

The UNESCO office moved to Paris in September 1946, just before the first general conference held in November that year. Febvre, a friend of Leon Blum, actively participated, unlike F. Joliot-Curie, in the building of UNESCO from the London conference onwards. Febvre was in charge of reviewing the Huxley preparatory report in October 1946 for the French National Commission.[78] He particularly scrutinized the 1947 programme for social sciences. Febvre rejected the proposed title (humanities in the plural and in French, the meaning being different in English): the word gave priority to Greco-Latin sources for culture, and privileged Western civilization over others. It is 'a risk to raise a suspicion', or 'to force the perpetuation of Western hegemony in culture'.[79] The question of the relation between cultures already led Febvre and the French delegation to suggest that the London conference deliberate upon the feasibility of a 'vast

[76] Needham to Crowther, 5 August 1945. Needham archives (D19). Needham had just learnt of the decision to hold the UNESCO conference on 1 November. He wanted to attend it to provide science an important place in UNESCO. He was sure that Dale and Hill would do everything to prevent his participation in the conference. He wrote to Crowther to mobilize Huxley, Blackett, Darlington, Bernal and some friendly diplomats for help. However, he did not succeed.

[77] Needham archives (D25).

[78] Langevin archives, ESPCI-CRH (60/4), Paris: report by Lucien Febvre on Social Sciences submitted to the national commission for UNESCO, October 1946. Hereafter called Febvre (1946).

[79] Ibid., p. 10.

comparative enquiry about civilizations: the problem is to understand how, and from which common ingredients, a human culture could be built in the future, a human culture that will be devoid of show and disguise.'[80]

UNESCO's orientation was a subject of controversy during its early years: to act directly, or indirectly for peace. Febvre shared the view that the organization

> help the development of science and the progress of civilization, and to put their services at the disposal of the masses through a good educational system, this is, at the same time, to move away from man the most dangerous of all plagues, war. Now, the ideal climate for an acting science, for an efficient civilization, for a civilization unable to mutilate the human being, is liberty.[81]

Febvre thus supported UNESCO's priorities, refusing to limit it to direct propaganda for peace.

The Science Division (1946–1948)

Needham's organizational programme for international science took root within the scientific unions (ICSU), his model for peacetimes, and in various organizations built during the war. From the scientific unions, he retained the idea of direct participation of scientists, as scientists. From the war committees, he took the idea of large funding, travelling facilities, the network of organizations, but incorporated the idea of the autonomy of scientists from the government.

For Needham, the main utility of such an organization was not for Western scientists. The goal was to strengthen the exchange of scientists as you moved away from the centres of science. International science is very different when seen from the 'bright zone' and from the peripheries. A *laissez-faire* policy could solve some problems, but this was exclusively from an Euro-American point of view. An international organization for science[82] was a necessity before all for

[80] Ibid.

[81] Report by Febvre, in Notes et etudes documentaires, no.1080, 26 February 1949, pp. 9–13, La Documentation francaise (documents related to the reception of Torres-Bodet on 22 January 1949 by the national commission for UNESCO after the Beirut conference). Hereafter called Febvre (1949).

[82] UNESCO/Prep. Com./Nat. Sci. Com./12 July 1946, 'Science and UNESCO. International Scientific Cooperation. Tasks and Functions of the Secretariat's Division of Natural Sciences', p. 8.

less advanced countries, and it concerned a large part of Europe that had faced the destruction of war.

'Together with what could be called the UNESCO's "peripheral principle", through which one tries above all to help who needs the most to be helped, that is the scientists beyond the "bright zone", one has to mention a corollary: no important intervention inside the "bright zone".'[83] According to Needham, it would be absurd to allocate funds to promote exchanges that may transpire even without these funds, and Western scientists, who feared the bureaucracy, will be more at ease. The key word for the UNESCO Science Division was 'to facilitate'. Detailed proposals came in the wake of this general orientation.

According to Needham, the Science Division was then committed to organizing scientific exchanges where and when they encountered difficulties. The division was not supposed to be concerned with exchanges between European scientists, but was completely dedicated to organizing exchange programmes between European countries and the Third World. This strong affirmative action was undertaken by Needham and his collaborators, and its impact has been important.[84] During this period UNESCO was a melting pot of scientists from many cultures, producing a very creative and stimulating atmosphere. The SCHM project embodied this very atmosphere and values. But in the early 1950s (the turning point being the resignation of Torres-Bodet and his staff in November 1952), UNESCO definitely fell into the hands of an administration without autonomy from governments.

At the end of his mission Needham drew up a balance sheet of what he had tried to do, and some bitterness about his colleague appeared in his remarks:

I am frankly rather tired of the people who sit in their laboratories and never give a thought for their colleagues at the other end of the world who are working in difficult conditions and even desperate need. If they were to travel about the world and visit the places which are really remote, those are the conditions they will find.[85]

[83] Ibid., pp. 8–9.

[84] Within the Science Division, according to Needham, one may encounter five Chinese, four Indians, three French, three British, two Argentinians, two Brazilians, two Belgians, two North Americans and 12 from other countries. See Needham (1949), p. 24.

[85] Needham (1949), p. 29.

This was an appeal for solidarity with colleagues working far away, and to put an end to 'parochialism' among European scientists.

In fact, war committees rapidly proved that they were not good models for science organization in peacetime, when funding was far behind what was expected. The Science Division within UNESCO did not develop as an international organization for scientists, but as an intergovernmental agency for science. International idealism did not survive after Needham's departure from UNESCO in mid-1948.

The 'Scientific and Cultural History of Mankind' Project (1946–1951)

At the end of the 1940s, and before the cold war, our company splits into two branches, with different kinds of associations and political dynamics. J. Perrin and Rapkine had died, Needham, Huxley, Rivet, Auger and Febvre chose the institutional path, UNESCO, while F. Joliot-Curie, Bernal and Crowther chose the communist-influenced WFSW. The SCHM project was probably the most instructive achievement of UNESCO during its early years. In dozens of meetings, papers, letters, dating to the period 1945–51, one can discover contemporary debates about science and society, science and civilization, progress and peace, and above all the trial of Eurocentrism.

Needham, Febvre, Huxley and Rivet, were the main actors involved in the development of the SCHM project. The idea of publishing a history of humankind essential to the reconciliation between peoples, was one of the big projects that gave UNESCO its identity during the early years, when there still existed the hope of having a world without war (this hope vanished in the 1950s during the cold war) and when UNESCO's life was dominated by intellectuals rather than by governmental representatives. The authorship of such a project has been claimed, more or less strongly by Huxley, Needham, Febvre and Rivet, or has been attributed to them by some of their friends. The official UNESCO story attributes the proposal of this history to Huxley's inaugural lecture as president of the Preparatory Commission in March 1946.[86] However, Huxley himself credited Needham with the idea.[87] Febvre explained in 1949[88] that the idea came independently

[86] Julian Huxley: *UNESCO: Its Purpose and Philosophy.* Washington, Public Affairs Press, 1947 (first published in 1946 by the Preparatory Commission).

[87] Julian Huxley, *Memories II,* London, George Allen & Unwin Ltd, 1973, p. 54.

[88] Febvre (1949).

to the minds of an historian (himself), an ethnologist (Rivet) and a biologist (Julian Huxley). But Moraze in 1956 credited Febvre for the proposal.[89] It is true that there were numerous encyclopaedic projects prior to this, and UNESCO's was the convergence of many former projects.

Febvre felt that the proposal regarding an enquiry into the history of civilizations presented by the French delegation at the London conference in November 1945 was the starting point of this

> very important project: which is the history of peaceful relations that, from their origins, human groups and societies maintained with each other. To show that, from the most ancient times, humans lived peacefully with other humans, that they communicated with each other by exchanging and borrowing their particular wealth, whether be they tools, technical procedures, domesticated animals, improved plant varieties; thus a whole network of peaceful relationships never ceased to embrace a world which we only want to see coming apart; finally that there are no small people, no poor or indigent civilizations which, once upon a time, had their moment of invention, and which somehow contributed to the edification of our large arrogant civilizations that, in fact, only live from what they have borrowed.[90]

Febvre's guns were targeted at Eurocentrism. He insisted on the priority of the study of civilizations during the meeting of the French National Commission in October 1946 in preparation for the first UNESCO general conference:

> Westerners can no longer consider their civilization as capable of conquering as such and without resistance, all human beings constituting the world—even if such a conquest had ever been possible and legitimate. Entire portions of this civilization have been unsettled, discredited by Westerners themselves. On account of this unsettlement, one can assist in the world in awakening numerous ethnic groups that inherited original civilization which they did not wish to renounce. An immense task of confrontation, or even opposition to these civilizations and of what the West calls the civilization, and which is in fact its own civilization, has begun. UNESCO has to direct this task, to organize it on scientific and disinterested lines.[91]

[89] Charles Moraze, Obituary of Lucien Febvre, in *Cahiers d'Histoire Mondiale,* 1956, vol. III, pp. 553–7.

[90] Febvre (1949).

[91] Febvre (1946), p. 18.

The French delegation at the Paris conference proposed then to initiate such an enquiry. But in fact it took one more year to see the project adopted at the Mexico conference in December 1947. A decision was taken to launch preparatory studies, to publish books for scientists and for ordinary readers, 'allowing a better comprehension of scientific and cultural dimensions of the history of humankind and showing the mutual dependency of peoples and cultures, and their respective contribution to humankind's common patrimony'.[92]

Nothing happened before the summer of 1948, when Huxley organized some internal meetings and called for an expert group to implement the decision.[93] Auger had succeeded Needham as Head of the Science Division. On 25 and 27 October, the UNESCO directorate (among whom were Huxley and Auger) met Needham, Febvre, Taha Hussein, Salles and Burckhardt. Despite Needham, the group, following Febvre and Auger, gave priority to a synthetic work of scholars, from which other volumes may be prepared for the public. Febvre and Huxley rejected the idea of drawing up an encyclopaedic balance sheet of scientific contributions and civilizations, preferring instead a work whose unitary aspect would be given by a point of view of human fraternity and not by a political doctrine. This would give a sense and direction to the project. The first draft of the plan was written for the Beirut conference, with the intention of highlighting all that concerned exchanges between nations, their common elements and the interactions produced by scientific and cultural factors. The project would emphasize the tendency of the modern world towards unification and integration.

Unfortunately, the Beirut general conference (November 1948) took a wrong turn when both sub-commissions on the 'exact and natural sciences' and 'cultural matters' met. Despite the presence of Rivet, Febvre and Needham, USA and Norway stalled matters and a

[92] Meeting of the SCHM committee, 26 July 1948. UNESCO archives, SCHM8 (2.31[2]). At the following meeting, on 29 September 1948, Huxley stressed the same idea: the book had to insist 'upon the part played in history by discoveries, inventions, new ideas; upon the production of new ideas by new facts, and of new facts by new ideas; upon the consequences of exchanges between peoples in that field.' UNESCO archives, SCHM8 (2.31[2]).

[93] The session records, the experts' final report and the draft of the document for Beirut are in UNESCO archives, SCHM7 (2.31[1]).

compromise resolution proposed by the Egyptian was adopted. The experts' project appeared to be too political, too ambitious and too expensive. The situation was more favourable during the general conference itself, and the project was finally approved. Needham, who was sent to Beirut by WFSW and not by his government, campaigned for this project, but was let down by his British colleagues.[94]

In spite of Febvre's and Rivet's enthusiasm, the Beirut conference (where Torres-Bodet substituted Huxley), did not significantly change the situation.[95] Furthermore, a British–French rivalry emerged: UNESCO's staff criticized Febvre and Rivet for appropriating the collective project, and for refusing Huxley's cooperation.[96] Auger tried to sidestep Needham. Finally, the ICPHS was consulted through Febvre, and ICSU through Needham ('The interest of the Febvre/Needham collaboration is that they had, from the start, decided to agree with one another').[97] And Miguel Ozorio de Almeida was asked to present a report on the project for the UNESCO executive committee.[98]

In his report to the permanent committee of ICHS, Febvre drew attention to the special objective of SCHM:[99] 'It pretends to influence

[94] Needham to David X, 1 December 1948. Needham archives (D74). Not only did the British delegation vote against his report, but one of the delegates, Leigh Aston, publicly dissociated himself from Needham.

[95] On 1 December 1948, Rivet enthusiastically declared 'History is badly constructed for it is incomplete, and nationalistic in tendency. Textbooks record every war and make no mention of people who did not fight but worked to promote what we call culture. The narrow point of view adopted by these books, which leaves out of account contributions by foreign peoples to the progress of humanity, is erecting formidable barriers to international understanding.' Needham archives (D161).

[96] Thomas to Mayoux (philosophy and civilizations division), 14 February 1949, and Mayoux to Thomas, 25 February 1949. UNESCO archives (SCHM8, 2.31[2]).

[97] Thomas to Mayoux (philosophy and civilizations division), 14 February 1949 and Mayoux to Thomas, 25 February 1949. UNESCO archives (SCHM8, 2.31[2]).

[98] Almeida was a Brazilian physiologist, a friend of Laugier, Lapicque and Wurmser, and had been pushed into this position by Auger. Published in August 1949, his report was severely criticized by UNESCO's staff and Huxley, and soon forgotten. See his report in *Cahiers d'Histoire mondiale*. vol. 1, 1953–4, pp. 962–85.

[99] Febvre report to ICPHS, May 1949. In *Cahiers d'Histoire mondiale*. vol. 1, 1953–4, pp. 954–61.

mentalities to eradicate the war virus'.[100] What was needed was not only school books but a collective scientific history to train schoolmasters, to synthesize global knowledge, in short a world-wide history. Febvre then presented his general plan of six volumes: one on general considerations, two on analytical developments, two on synthetic approaches, and one for the conclusions. Volume I, dealt with the important questions raised by anthropology and human biology, and then by other human sciences. Volumes II and III, on exchanges and 'all which have been subject to circulation': technical knowledge, systems of ideas, beliefs, material objects, animals, etc. 'From that, will emerge the image of a moving humanity since its origins, traveling permanently through a perpetual series of transcontinental migrations.'[101] Within UNESCO, Febvre found a way to open a new channel for his intellectual agenda, with the benefit of cooperating with Needham. Volumes IV and V proposed a synthetic summary within a geographical frame: what each part of the globe had received from, and given in exchange to, other parts of the globe, beginning with Asia. 'From this picture, would emerge the idea that separations in the world are mere illusions, and that the earth never ceases to diversify, to enrich, to mutually fertilize with streams of peaceful exchanges.'[102] Volume VI, was to be a conclusion and synthesis, essentially recapitulating the main stages of historical development. Febvre had no difficulties in obtaining ICPHS support for his aims and general plan.

Huxley sent UNESCO long comments on Ozorio's report, where he clearly explained his own conception of the SCHM.[103] Huxley wished to write a natural history of civilization, with a strictly scientific approach, and refused the ideological or political aims that, he thought, Lucien Febvre and Miguel Ozorio were proposing. The development of civilization must be understood as the continuation 'on the human and social level of the general evolutionary process which had previously led to the attainment of that level by evolving life'.[104] With

[100] Ibid., p. 954.

[101] Ibid., p. 960.

[102] Ibid.

[103] Memorandum by Julian S. Huxley on the report of Professor Ozorio de Almeida on the SCHM, July 1949 (UNESCO archives, SCHM8, 2.31 [2]).

[104] Ibid., p. 3.

such an approach, Huxley did not need an underlying philosophy, but a real theory. 'Without some such unifying theory, any such book will end up to become colourless, or just a catalogue of facts.'[105] According to Huxley,

> in a given period of a given civilization, a dominant system of ideas influences scientific and technical development, and not only the inverse. One has to draw from the theory of evolution a system of ideas that will produce a scientific approach to history, which should not be displaced from 'the hard facts of evolutionary reality'.[106]

After the Paris conference, a new group of experts was set up.[107] Among them were Needham, Febvre and Rivet (for ICPHS). From 12 December to 17 December 1949, the committee held ten sessions to finalize the aims and the general plan of the SCHM. Diverse issues were discussed, and where Febvre and Needham appeared to show a great deal of agreement and to animate three main debates:

What gives history its unity and homogeneity? As opposed to Florkin and Shryock, who wished to stick to a purely encyclopaedic approach, Febvre, with most of the committee, proposed a more didactic approach: 'the only aim is to do good history to preserve peace', but the first volume has to include a 'political mental reservation: the peaceful mental reservation that has been UNESCO's fundamental idea' for SCHM.[108] The various questions have to be studied in order 'to eliminate the reasons of misunderstandings between peoples', in particular anthropology and ethnology which, in their origin, favoured to a certain extent racist prejudices, and which have then to be thought differently.

Is a chronological approach, or a problematized one better? Febvre,

105 Ibid.

106 Ibid., p. 5. This evolutionary humanism appeared to be opposed to social biologism, but was very different from Febvre's proposals. Needham was not prepared to go as far as Huxley, who in those days kept away from Eugenism but sometimes seemed to share with Teilhard de Chardin the idea that evolution is a rise towards an ultra-humanity.

107 The records of the drafting committee are in UNESCO archives, SCHM23, 2.633(1). Other experts were Florkin (Liege) for ICSU, Ciasca (Genoa), Shryock (Johns Hopkins), Piaget (Geneva). The UNESCO administration attended the meetings.

108 Ibid., 2nd session.

with the support of Needham, opposed Rivet who wanted to start from a chronology of the main discoveries, to identify the main stages crossed by humankind, to show that 'no discontinuity arrested this march towards progress', and to establish 'successive balances which put so clear rhythms into progress'.[109] Rivet's proposals were of a positivist (the stages of progress of humankind) nature and founded on a Eurocentric history of civilization, and whose outcome was the Western scientific civilization. Needham, followed by Febvre, refused such a chronology, which was not relevant to the history of other civilizations. For them, exchanges should constitute the core of the SCHM. Finally, there was general agreement that 'the chief emphasis should be on the exchanges or borrowings between peoples and countries, of the factors which have contributed to human civilization, including inventions, discoveries, ideas and culture'.[110]

What is the part played by science and techniques? Febvre's stages of human development perspective accorded science and techniques a central place. He was followed by Needham, but with a very different justification: science and techniques are the vectors, by nature, of more exchanges and deeper comprehension between peoples than are the fine arts, religions and philosophies which divide more than they unify, and which are more difficult to transfer. Subsequently, the emphasis had to be put on science, and religion had to be relegated to the end of SCHM. Shryock supported Needham. Febvre, with Piaget, challenged the hegemony of science and asked for an equal share. Despite Needham's pressure, exerted till the fifth session, religion and philosophy gained a full place in the SCHM.

The final plan did not differ much from Febvre's proposals of 1948.[111] Back in London, Needham wrote to Huxley to inform him about the 'astonishing progress' made by the committee and about

[109] Ibid., 2nd session.

[110] Ibid., 3rd session.

[111] The final plan was outlined by Needham in his letter to Huxley: Ia: Brief introduction of general fundamentals (from Febvre)—Ib: The contribution of the successive ages of man to the growth of civilization (from Rivet)—II and III: The mutual indebtedness of all cultures (from Febvre)—IV and V: The special characteristics of the main historical cultures and civilizations (from Needham)—VI: Synthesis and conclusions (from Febvre). Seven experts shared the responsibility for the volume. Needham got no. II, Rivet no. Ib and Febvre no. V.

Rivet and Febvre's magnificent performance. He was happy that nothing had been left of Ozorio's hazy plan. Confirming that Huxley had been pushed out of the project he should have been responsible for, he wrote: 'I trust you will not imagine that I look upon your absence from the progress of this work with anything than shame and disgust. I have a good deal that I should like to tell you about what I have heard concerning your relation with it, and the extreme disinclination of UK officials to agree to your continued association with it.'[112]

Cortesao wrote the first draft of the expert report and submitted it to the seven participants.[113] Only Needham proposed a significant amendment which summarized his conceptions.[114] The evolution of humankind extends biological evolution. Successive stages evolve towards higher and higher levels of organization, with the progress of knowledge about, and the control of, nature.

When national commissions were consulted during the spring of

[112] Needham to Huxley, 17 December 1949. Needham archives (D164).

[113] Close to Torres-Bodet, Cortesao resigned with him.

[114] Needham to Cortesao, 14 January 1950, Needham archives (D165). Stating that 'hypocritical things were said at the meeting about perfectly objective history', he considered that nevertheless 'we have to adhere to the final sentence about the uniting and divisive'. He proposed an amendment that was three times longer than the original draft. It said that: 'the committee understood, from the information before it, that the general conference and the director-general had had in mind the preparation of a work on the history of mankind which should contribute powerfully towards the mutual understanding of all people at present, on the basis of their achievements and cooperation in the past. The committee therefore elaborated a plan of contents which was presented in this paper. After an introduction, talking about fundamental knowledge about Man and the world in which he finds himself, there would be a second part describing the series of chronologically successive stages in the progress of humanity in social organisation and control over, and understanding of, Nature. The third part will be concerned with exchanges and transmissions in all branches of human knowledge, practice and experience; demonstrating the mutual indebtedness of all peoples, and bringing out the fact that there are no people or cultures which have not contributed elements of essential value to the total human patrimony. The fourth part will outline the various patterns of the great cultures and civilisations, their particular world-outlooks which were characteristic of them, and which, though not transmitted in former times, are now fusing into the world-picture of universal man. The fifth concluding part would be of a synthetic character. In so far as the attainment of perfect historical objectivity might be considered to be impossible, the committee felt that emphasis might well be placed on the factors which have united mankind throughout history, rather than on those which have divided the various peoples.'

1950, the critics were substantial, and support limited. The British commission, in spite of the presence of Needham and Cortesao, did not accept the project.[115] Only France—Febvre presented the report—supported the experts' conclusions, and pointed the source of difficulties as 'due to the obstinacy with which so many representatives of so-called "European" or "Western" civilization regard the latter—their own—as the only true civilization'.[116] Febvre, and Needham, were determined adversaries of Eurocentrism.

The general conference at Florence in the spring of 1950, decided however to speed up the effective realization of SCHM, and to call for an editorial committee.[117] Huxley was back on stage,[118] together with Paulo de Barredo Carneiro, Turner, Moraze (Febvre's pupil)—the three leaders of the committee—and a dozen members, who included none of the previous experts. The committee was an international one and met for the first time in December 1950.[119] The previous plan was rejected as 'both overcautious and unsuitable for the dawning of a world consciousness'. Chronology was back, and this became the beginning of a new story.

In 1951, Febvre was pushed out of the project, but received, in December, as consolation, the directorship of a journal, the *Cahiers d'Histoire mondiale*.[120] However, Torres-Bodet tried to have him back,

[115] 13 March 1950, London: records in Needham archives (D167).

[116] Report 5Ç/PRG/2, UNESCO archives (SCHM7, 2.225).

[117] Needham archives (D170). Needham tried to propose Bernal's name for the committee, recognizing that Bernal 'has an encyclopaedic knowledge, but', he said, 'I am afraid it might not be politically wise'. UNESCO archives, SCHM7, 2.31(1). Carneiro was a Brazilian biologist, exiled in Paris since 1937, who founded the 'Maison d'Auguste Comte' in Paris.

[118] Professor Charles Moraze, whom I met in March 1997 in Paris, insisted upon the part played by the Rockefeller Foundation and by US representatives in UNESCO, to replace those members believed to be 'red', and to install Ralph Turner as the leader of SCHM project. He remembered resisting the move with other colleagues (among whom was Huxley) and succeeded in getting Carneiro elected instead. We might infer then that, Huxley, as a liberal, was more acceptable to the US than other UK left-wing scientists and historians (like Needham). Huxley, as former Director-General of UNESCO, was already associated with the project.

[119] UNESCO archives, synthesis report: SCHM8, 2.31(2)—and report of the executive council: SCHM7, 2.31(1).

[120] The *Cahiers d'Histoire mondiale* also published Carneiro's reports about the state of work, and other UNESCO documents about SCHM.

with Rivet, as correspondents of the editorial committee, but they declined.[121] When Febvre died in 1956, Moraze, remembered Febvre's participation in the first assembly of UNESCO, with Blum at his side, his role in the French encyclopaedia, in the CIS and the *Revue de Synthese* with Berr, in the organization of scientific research with F. Joliot-Curie and J. Perrin.[122] He remembered his three-year-long struggle for SCHM:

> This had been his most beautiful victory. In Florence, the proposed plan was Febvre's. Inside ICPHS, the proposed plan was Febvre's. Later, UNESCO had to move from the plan to its implementation. It might be that non-European representatives were wary by this cumbersome character, too European in their eyes.[123] In brief the effective launching of our world history appeared to be subordinated to this 'democratic' measure: Febvre's self-effacement. Who can imagine how hard this self sacrifice has been for him?[124]

The World Federation of Scientific Workers— WFSW (1946–1951)

At the second international association of scientists after the war, we meet Crowther, Bernal and F. Joliot-Curie among our characters. The origin appears to have been largely Anglo-French cooperation, beginning with the first contacts in the 1930s to the 1945–6 meetings: The debate (organized by SVS) was on the social implications of atomic weapons, with F. Joliot-Curie, Auger, Bernal and Blackett in September 1945,[125] the AScW conference in February 1946 on 'Science and the Welfare of Mankind',[126] with Huxley, Bernal, Blackett, Dorothy

[121] Rivet to Torres-Bodet, 14 March 1957, and Torres-Bodet to Rivet 30 April 1951. UNESCO archives, SCHM42, 2.92(1). Rivet wrote: 'As I was a promoter of the project, alongwith my great friend Professor Febvre, UNESCO felt that it should go for a totally new committee. I do not want to get into the details of its objectives but they were certainly different from what Febvre and I have outlined in the UNESCO general assemblies at Mexico and Beirut. Now I had only one option, either agree with the new ideas or criticise them. I opted, on the contrary, to avoid this alternative between a conformist and a critic.'

[122] Moraze, Febvre obituary, in *Cahiers d'Histoire mondiale* (1956), vol. III, pp. 553–7.

[123] See footnote no.118. Moraze probably referred to US interventions

[124] Ibid., p. 556.

[125] J.G. Crowther (1970), pp. 246–7.

[126] Ibid., pp. 271–6.

Needham (Joseph was still in China), and a message from F. Joliot-Curie.

The WFSW was founded in London, in July 1946.[127] F. Joliot-Curie, the first president (C.F. Powell replaced him in 1957), traced the genealogy of this progressive organization to the official Anglo-French Society. Bernal, the ideological father, was deputy-president and Crowther general secretary (Biquard replaced him in 1955). The British AScW, then presided over by Blackett, and the French ATS were the two pillars of the new organization, that sought to mobilize scientists' unions or professional groups with the exception of official scientific unions.[128] Needham was present at the founding meeting, on behalf of the UNESCO Science Division.

The problems with UNESCO appeared as early as 1947, and North American political pressures delayed an agreement between UNESCO and WFSW for months, to the point that Needham found it 'inadvisable to conclude an agreement' in April 1947. He then used his ability as a lobbyist, and Auger was his best supporter: 'the moral is that if WFSW is to play the part we all believe it should play, and which we all want, a great deal more effort must be put into winning the active support of Executive members beforehand.'[129] The agreement was accepted in July 1947, and Crowther was sent to the Mexico assembly as an observer. At Beirut, Needham himself represented WFSW.

The first general assembly took place in Prague, in September 1946, and the difficulties were numerous: difficulties with UNESCO, difficulties in mobilizing a new generation of young scientists, difficulties of finding a space between UNESCO and ICSU, complicated debates about the freedom and planning of science.[130] In 1948, AScW accounted for more than half of the WFSW members.

The cold war deepened the crisis between liberal and socialist scientists, and put an end to popular front associations. The World Congress of Intellectuals for Peace (Wroclaw, August 1948), where Huxley (in his individual capacity and not as the head of UNESCO),

[127] Ibid.

[128] The 'Association des Travailleurs Scientifiques'(ATS) was a group which had been created during the Resistance as a branch of the 'Front National Universitaire.'

[129] Needham to Crowther, 20 April 1947. Needham archives (D27).

[130] Introductory report by J.G. Crowther (Needham archives, K329), 10 September 1948.

Bernal and I. Joliot-Curie participated, symbolized this rupture.[131] Russian delegates refused all compromises with non-communist intellectuals and forced a resolution that converged with the requirements of Soviet diplomacy.[132] Huxley did not sign this resolution.

In 1950, WFSW's agreement with UNESCO came to an end. AScW was about to leave the federation.[133] According to Bernal, WFSW was criticized for its inaction and extremist views.[134] F. Joliot-Curie accepted many concessions to prevent the departure of AScW. But after the second assembly (Paris and Prague, April 1951), the approval of the Soviet Union altered the equilibrium inside WFSW.[135] There was a big difference between the Eastern Unions (numerous, state-supported and rich) and the Western Unions (not numerous, state-opposed and poor). WFSW was no more in the position to play an important part among Western scientists.

Conclusion

Common Features

Almost all our characters were members of the academic elite: Cambridge men, Fellows of the Royal Society, Paris University professors, Fellows of the 'Academie des Sciences', even Nobel laureates. They pursued their scientific activities, and did their best to obtain scientific institutional recognition by their peers. In the 1930s and 1940s, their radical political commitment did not contradict their institutional strategy. This situation changed with the cold war in the 1950s.

Most of these scientists turned to socialist activism, and were not merely ideological sympathizers, during the 1930s. Furthermore, some

[131] Goldsmith (1976), pp. 184–7.

[132] Cf. Huxley (1973).

[133] Ibid., pp. 175–80.

[134] Meeting, December 1950, in Paris, between Bernal, Crowther, Biquard and F. Joliot-Curie. Joliot-Curie Archives (F4), Institut Curie, Paris.

[135] In 1948, WFSW claimed 24,000 members at Prague, mainly from Britain and France. In 1951, the total membership went up to 44,000, with 29,000 Chinese scientists. Only 12,000 members were from non-communist countries. Then, the Soviet scientists joined the WFSW, causing a shift in the balance. See Needham archives (K327–K335).

of them chose to become official members of socialist and communist parties, which was quite uncommon until then, even among politically committed French intellectuals. The symbolic rôle of an intellectual was gauged by his critical autonomy, and they participated mainly in peripheral left-wing organizations, like the 'Ligue des Droits de l'Homme' or the CVIA. Direct participation in political parties (and in elections) was an exception till the end of the 1930s.

Far from being isolated from their colleagues, these scientists were emblematic of a large section of the scientific community. They were involved, in Britain as well as in France, in what Werskey calls 'scientific popular fronts', like the AScW, the DSIRS-BAAS, the French Rationalist Union and CVIA. In such organizations, socialist scientists were associated with reformist colleagues, some of them adhering to liberalism or scientific humanism. However, these popular fronts rapidly dispersed after the war, since they had fulfilled their socio-professional agenda. Further attempts to launch other popular fronts in the 1950s, like the WFSW, presented other agendas, that contradicted this status of a large front among scientists.

All our characters were committed to the defence of science, its development, funding and organization, and to the employment of scientists. In France, they headed the main scientific institutions and even participated in the government. In both countries, they were active members of various agencies and ad hoc committees. The defence of science was also an intellectual battle for rationality, against Nazism and other so-called anti-science movements.

Most of them participated in activities dedicated to the popularization of science, through public conferences, books, columns in newspapers, radio broadcasts. They spoke, and wrote for thousands and thousands of people, and this reflected their social commitment. They also participated in various movements addressing the problems of education, or new ways of teaching.

The historical development of science was interpreted within the general frame of scientism, and was reinforced by it. The 1930s were symptomatic of a resurgence of triumphant positivism. Our progressive scientists headed this movement. The use/abuse model of science was unchallenged. Most of our characters felt that the abuse of science might be suppressed by a socialist organization of society, which, others felt, should be complemented by predominance of pure science over material applications and the social responsibility of scientists. Even if there were differences, and if, for instance, Bernal's position changed, they believed that science had become the main social force, and that it was pushing in the right direction. Bernal's and F. Joliot-Curie's scientism was a form of strong socio-economism, but Needham's scientism was based on the conception of science, in itself, as the main source of ethics and moral values. This conception was close to that of Langevin. For Needham, ethics was a biological necessity in the long term: 'I always thought that the sociological evolution was the continuation of biological evolution and you should wait centuries and centuries to see the result, and ethics are inside the process itself' (the process of scientific development). For Langevin, science was at the same time the daughter and the mother of democracy. It was the chief factor in transforming mentalities, and subsequently the only means to ban war. The spiritual contributions of science were more important than the material aspects. And above all, 'science has taken a considerable advance upon justice'.[136]

In the 1930s, an important criticism of capitalism came from progressive scientists. According to this criticism, the problems with capitalism resided in its inability to let science realize all its potentialities, for the benefit of humankind. Science was progress, and to fight for science was to fight for socialism. This identification between a fully developed science and socialism was not uncommon in the 1930s among the scientific community. With strong roots in the scientific community, our characters' political activism was not strong in social commitment. They could continue to participate in elite institutions and networks, and this was even desired by the communist parties. In other words, the political commitments of elite scientists were compatible with scientific careers.

Most of them officially claimed adherence to Marxist philosophy, but one can ask if it was the actual philosophy underlying their scientific work or their representation of science. M. Prenant and J.B.S. Haldane tried to combine biology and dialectical materialism in the 1930s and 1940s,[137] but showed more reserve after the Lysenko

[136] Bensaude-Vincent (1987), pp. 112–17.

[137] See, for instance, J.B.S. Haldane, *La Science en marche*, Paris, PUF, 1952, a collection of short articles published in newspapers: 'Dialectical materialism is a method of thinking and action, and not a dogma. It is difficult to see how it could influence decision-making in such controversies' (Lysenko versus Vavilov), p. 220.

case. Needham saw dialectical materialism in the 'levels of organization' of the living world.[138] But for F. Joliot-Curie himself, it was more an ideological proclamation than a practical tool for his scientific work. Langevin's and J. Perrin's Marxism have been deconstructed elsewhere, and scientific idealism and humanism are more consistent with their actual philosophy.[139] Bernal's ideas in relation with Marxism are probably more controversial.[140]

For them, Marxism appears to be reducible to a variety of socio-economic determinism underlying a vision of scientific development. For instance, when translating Bernal's introduction to *The Social Function of Science*, Langevin chose 'Science as a factor for moral and social evolution' as the title of his talk. According to B. Bensaude-Vincent, 'Langevin remains unconcerned with Bernal's explicit attempt to consider past and present science as an institution depending on existing social order and impregnated with the spirit of dominant classes. After, like before, he goes on saying that science determines the organization of society'.[141] Langevin's ardent rationalism is the key to his interest in Marxism, identifying scientific rationalism and dialectical materialism. By developing science, capitalism showed the possibility of ushering in universal welfare, but prevailing social structures oppose the realization of this welfare. Capitalism created the conditions for its substitution by another social system in which science could develop all its beneficial potentialities. Capitalism's social structures (looking for profit) as well as its ideological ones (the apology of individualism) became antagonistic to the plain development of science and techniques.[142] Fortunately, USSR's demonstration of science under socialism provided a political alternative for left-wing scientists.

[138] Needham wrote a preface for Marcel Prenant's book *Biology and Marxism*, London, 1938.

[139] See Bensaude-Vincent (1987), and Mary-Jo Nye 'Science and Socialism: The Case of Jean Perrin in the Third Republic', in *French Historical Studies*, 9, 1975, 141–69.

[140] Hilary Rose and Stephen Rose, 'The Two Bernals, Revolutionary and Reformist in Science?', in *Fundamenta Scientiae*, vol. 2, ní3–4, 1981, pp. 267–86.

[141] Bensaude-Vincent (1988), p. 205.

[142] Rose and Rose (1980), Such an analysis was shared by the official communist parties, but also by many other left movements.

Eurocentrism and Functionalism

Only Needham seemed to have made a clear distinction between Western and modern or universal science. As early as 1937, he proposed to study the Chinese contribution to modern science, offering dominant ideas which considered Greek science as a 'miracle', and the essential watershed in the history of civilizations. After his visit to China in 1931, Langevin seemed to be on Needham's side. At the Second International Congress of History of Science (London, 1931), the Soviet delegation criticized internalism and highlighted the socio-economic conditions shaping scientific development. They retained the cumulative model of the growth of scientific knowledge, according in the same movement a superior status to the science produced in Europe. Bernal's affirmation that 'Science is a collective enterprise, it belongs to no country or race' was probably the most common thinking among scientists, whether Marxist or liberal. But unquestioning science was a means of legitimating the imposition of the science which really exists under capitalism, and not an utopic ideal science.

In fact, Eurocentrism is consubstantial with Bernal's socio-economism, as the organization of science-based industrial societies is the model for socialism when liberated from profit.[143] There is little in *The Social Function of Science* about science in non-Western countries, and most remarks of this nature are relegated into the distant past.[144] Bernal is clearly not exempt from common European prejudices

[143] Remember the adoption of Taylorism by the USSR.

[144] Bernal (1939). Quotations are from the 1967 edition (MIT Press), pp. 207–10. Until the end of the nineteenth century, 'the pursuit of science was almost confined to Western Europeans either in their American colonies or at home'. In ancient civilizations, 'science was suddenly and abruptly introduced'. No science existed in India before the twentieth century. Indian scientists were seeking national affirmation, but had to learn science from the British, and to be under their domination, which Bernal finds contradictory. The Indian scientists show 'a mixture of submissiveness and arrogance that inevitably affects the quality of the scientific work. Indian science is noted at the same time for originality of its conceptions and experimental processes, and for extreme unreliability and lack of critical faculty in carrying out the work itself.' Consequently, Bernal affirms that the development of science will be possible only after national independence, and that, meanwhile, 'the best workers for Indian science today are not the scientists but the political agitators who are struggling towards this end' (p. 208).

about non-European scientists. USSR's successes in Central Asia were considered to have occurred in a scientific vacuum.[145] Modern science could only be diffused from the West. Colonialism was not rejected because of its nature (to impose industrial and scientific capitalism and civilization), but because of its modalities and consequences. Bernal denounced in a very conventional way the racism against Indian scientists, their exclusion from scientific institutions.[146] But science being beneficial by nature, on account of its social dynamics, as well as its material applications, would oppose colonialism as it opposes capitalism. It is not surprising then that some of our characters, like J. Perrin, played an active role in the development of colonial science. Auguste Chevalier, the chief of French colonial scientists in the inter-war period, joined the 'fellow-travelers' in 1944.[147] The analysis of how colonialism destroyed indigenous knowledge systems in occupied countries was not common in those years.[148]

It is not surprising, since it was in consonance with their scientism, that most of our characters, when talking of, or when involved in,

[145] 'The history of the Soviet Republic of Central Asia shows how rapidly science can be built up and how eagerly it is seized on by a population starting from a medieval standard of culture'. J.D. Bernal, 'A Permanent International Scientific Commission', in *Nature*, no. 3967, 10 November 1945, pp. 557–8.

[146] Jerome Ravetz, 'Bernal's Marxist Vision of History', in *Isis*, vol. 72, 1981, pp. 393–402. His economism accounts for his poor understanding of the origin of modern science (p. 397). And he does not see science as an instrument of colonial oppression: 'the genocidal horrors practised by European imperialism on its conquered lands (what we call now "Third World"), involving the technologies of warfare, of primary-products of exploitation and of various addictive poisons, did not interest Bernal greatly' (p. 398).

[147] A colonial botanist, Chevalier joined the patronage committee of *La Pensee*, the journal of communist intellectuals and their fellow travellers, in 1944. He published an article, admiring the agronomy of USSR, in *La Pensee*, October–December, 1944, pp. 38-43 ('La Science agronomique en France et en URSS'): 'We will have learnt a lot from USSR, which has made tremendous progress during the last 25 years, mainly in applied sciences, particularly the agricultural science institutes'.... 'Science is in the forefront .... In no other country science has played such an important role and contributed so much to general progress in so short a period. Is it not worthwhile, when we have so much distress in our country today, to follow the USSR in its application of research and of scientific methods and thus contribute more to human welfare?'

[148] Among exceptions: Pierre Boiteau, 'Biologie et colonialisme', in *La Nouvelle Critique*, November 1952, pp. 76–88.

international scientific organizations, adopted what is usually called a 'functionalist' approach.[149] The specific intellectual and social status given to science implied that the international scientific community should be a model for international relations; that scientists had to play a specific and important part in international relations; and that practising science and developing transnational scientific organizations were means of developing supranational loyalties, and would help weaken nation-states, ensuring the possibility of greater peace. The aim was to develop functional (scientific, technical) tasks to be delegated to supranational institutions, resulting in a progressive erosion of national sovereignties. Scientists, being 'naturally world-minded', had to be at the forefront of the process of passing beyond nation-states. For Needham, a world functionally integrated through scientific cooperation would represent a superior level of organization of humankind.[150] 'There is no field which has a stronger tradition of international community work than that of natural sciences, and it can indeed be taken as a model for other fields.'[151] When proposing his international service for scientific cooperation, Needham requested that governments recognizing a space for scientists at the national level, do the same at the international level, because scientists were, by nature, internationally minded and were not nationally minded, and because science transcended national borders.[152]

Like Needham, neither Bernal, nor Joliot-Curie escaped from functionalism.[153] The famous remark of Bernal: 'In its endeavour science is communism' is a typically functionalist conclusion:

Already we have in the practice of science the prototype for all human common action. The tasks which the scientists have undertaken—the

[149] See, for instance, Brigitte Schroeder-Gudehus(1975), where she analyses Needham's thesis.

[150] Quoted by Schroeder-Gudehus (1975).

[151] Needham (1949), p. 23. And again: 'In general, it might be said that there is a universal desire among scientists to see better international scientific contacts in the coming years' (in UNESCO/Prep. Com./Nat. Sci. Com./12, p. 7, July 1946).

[152] Joseph Needham and Dorothy Needham, *Science Outpost.* Papers of the Sino-British Science Co-operation Committee, 1942–6, London, The Pilot Press, 1948 ('International Science Co-operation in War and Peace', p. 270).

[153] See, for instance, 'Only science allows to create within humankind the consciousness of its unity as a working community', in J.D. Bernal 'La Science et le sort des hommes', *La Pensee*, no. 5, Oct.–Dec. 1945.

understanding and control of nature and of man himself—is merely the conscious expression of the task of human society. The methods by which this task is attempted, however imperfectly they are realized, are the methods by which humanity is most likely to secure its own future. In its endeavour, science is communism.[154]

If science is the model, the practices of scientists are the exemplar, and Bernal lyrically describes scientific activities, with the same words Polanyi would have used.

A Peculiar Historical Conjuncture in the 1930s

For scientists too, the 1930s represented an important political fracture:

- The 'moral crisis of science' which followed the First World War, and which inaugurated the abuse of science for war, had mostly faded away, and confidence restored. The growing aspiration for scientific organization and funding was the main preoccupation of the 1930s.
- The economic crisis, with massive unemployment and misery, the inability of capitalism to satisfy minimal social needs, while science appeared to conceal many possibilities to do so.
- The image of science in the USSR had approached its zenith in terms of social and political recognition, as well as in terms of benefits accruing to the people.
- The rise of fascism, and the Nazi abuse of science were interpreted as an attack against rationality and science itself. Fascism presented a double danger for science: it destroyed science in occupied countries and in others, its ideas 'reinforce everywhere the forces of obscurantism and injure the spirit of science (...). The development of science in Fascist countries is a clear indication of the fundamental incompatibility between either the theory or the application of science and the tendencies of capitalist economic and political development'.[155] Since fascism was an extension of capitalism, it strengthened the identification between the defence of science and the struggle for socialism.

The British 'visible college' and the French equivalent were then the product of two processes: a socio-professional dynamics (the social

[154] Bernal (1939), p. 415.

[155] Ibid., pp. 220–1.

rise of scientists and their demands for a greater place in society and for participation in power) and an historical political fracture (Nazism, building of the USSR, social crisis) which pushed them into political activism for socialist ends. These two processes combined in the 1930s, giving it an historical singularity. With some differences and many similarities, there existed both in Britain and in France, facilitating common associations. At this particular historical conjuncture, the classical French notion of a politically committed intellectual was applicable to many of their British colleagues. In the name of science, universalism, reason, they fought for socialism. More generally, in the 1930s, British intellectuals reacted to political events collectively as a social group, and possibly for the first time. But, according to Stephan Collini, the war brought out the differences between French and British intellectuals.[156] Due to the position occupied by Britain at the front-line for the battle for freedom and liberty, left-wing (dissident) intellectuals reintegrated the national community in Britain, as opposed to intellectuals in France and continental Europe.

Bernalism gave some kind of theoretical legitimacy to political commitment which did not contradict either scientific elitism or the hierarchies concerned. This might explain why scientists adhered to socialism and the USSR far longer than other intellectuals.

One cannot understand these movements, their deep roots in their communities, without taking into account both dynamics. And this also explains why all attempts to launch such movements after the war failed, in historical situations, under different political conditions, but also when science and scientists had gained a very different place in society. A radical science movement could no longer be rooted at the top of scientific institutions, scientism had become conservative, and USSR lost its prestige in the 1950s.

Bernalism and WFSW

The high status accorded to science was an inheritance of Comte's and Spencer's conceptions at the end of the nineteenth century, which then reflected the ideology of the liberal and colonial bourgeoisie: progress, order and imperialism.

[156] Stefan Collini, 'Intellectuals in Britain and France in the Twentieth Century: Confusions, Contracts—and Convergences', in Jeremy Jennings (ed), *Intellectuals in Twentieth Century France: Mandarins and Samuraïs*, London, Macmillan, 1993.

To derive from science all advantages it can offer, there must exist an intimate linkage, perfectly materialized for each stage, between science and social progress: one has to foresee the needs, to study and modify the structures of society, to solve production and distribution problems, and finally to control the results of its application, so that they do not take unexpected or undesirable directions.[157]

For Bernal, 'Science will come to be recognized as the chief factor in fundamental social change'.[158] What is left then of class struggle or mass initiative? When science is given such a part, the differences between socialism and capitalism fade away. The scientific revolution is the new social revolution. 'Comrades, don't bother, we are constructing socialism through science for you', as Bernalism had been analysed by *Radical Science Journal* in the 1970s. And Bernal insisted 'Science implies a unified and co-ordinated and above all, conscious control of the whole social life'.[159] Such social control in a science-dominated society does not portray a pleasant future. Scientism had been the main ideology for the construction of the Soviet state under Stalin, and there was no room left for contradictions and differences in such a scientifically organized society.

Science was social order, and this could explain the fascination of the 'USSR scientific society' for Bernal and some others.[160] All retained, from their visits to the USSR memories of the organization of scientific research, planning, rapid scientific development, which Haldane particularly exalted.[161] The 220th anniversary of the Moscow Academy of Sciences in June 1945 had been a big social event for all scientists, whether progressive or conservative.

[157] J.D. Bernal 'La Science et le sort des hommes', *La Pensee,* no. 5, Oct.–Dec. 1945.

[158] Bernal (1939), p. 414.

[159] Ibid., p. 409.

[160] Most of our characters went to the USSR in the 1930s, and Crowther was there on a dozen occasions. In 1931, after the congress, he took about 20 scientists with him (among whom were Bernal, Haldane, Pirie and Huxley) and about 40 physicists. See Crowther (1970), pp. 84–5. See also David Caute (1988). Langevin travelled to Moscow in 1928 and 1931. Needham himself co-edited with J.S. Davies an apologetic volume entitled *Science in Soviet Russia*, London, Watts and Co., 1942,

[161] Haldane (1952), p.200. Writing about the success of the Sedov cruise, he explained: 'Such a work totally refutes the charge that the Soviet science was only interested in the discovery of immediately useful facts. On the contrary, these discoveries would take place in the end, leading to the welfare of everyone.' After

At the end of the war, in a way, Bernalism had triumphed with capitalism. His analysis of the social function of science did not contradict capitalism. On the contrary, further developments of capitalism were based on the innovative processes initiated by science and technology. But then what of socialism?

During the war, the left wing merged with official organizations. In the immediate post-war period, one may wonder whether left-wing participation in official meetings signified its general political influence, the personal weight of personalities like F. Joliot-Curie and Bernal, notwithstanding their political colour, or the ideological alignment of dominant bourgeois ideas about science, as a consequence of scientism. Before the cold war, it was easy to move from WFSW to official institutions, as a consequence of the left-wing participation in the national war effort. Together, scientists contributed to the defeat of Nazism. After 1948, such relationships exploded.

The cold war, the Lysenko affair, led to a general ideological withdrawal of left-wing scientists towards a pretence of the neutrality of science, when not oriented to the defence of the freedom of science. Science planning was reduced to a mere science policy.[162] If Langevin refused in 1939 to put ideology in science as Bernal did, he would have shared much with Bernal's apology of scientific neutrality of the 1950s. If WFSW managed to appropriate the prestige of participation in the war effort, and could even count Nobel prize-winners in its fold, after Lysenko and the events of 1956 it did not function anymore and this did not prevent the marginalization of WFSW. The way WFSW toed the Soviet line, and its absolute weakness in defending its own former leaders—like Malek in 1968—completed the process of its being discredited.

Anglo-French Associations and Joseph Needham

Within this specific context of the 1930s, new associations developed which contributed to bringing different groups of British and French scientists together, far beyond strictly professional relations (unions, international congresses, scientific exchange programmes). Professional relations evolved, and were sometimes reinforced by other associations.

Portuguese and Spanish navigators, and after Cook, Livingstone and Scott, 'the torch was in the hands of young socialist workers in USSR'.

[162] Rose and Rose (1981).

The main characteristic of these associations in the 1930s was the ability to jump from one level to another: connections were often professional and political (in the larger meaning of the word), with mixed organizations like the Anglo-French Society, unions, anti-fascist movements, and so on. International organizations played an important role towards the end of the 1940s. This was the case of UNESCO and WFSW, which represented political and professional commitments. This mixture should not be regretted: such associations generated a creative dynamic, and revealed the interplay between science and politics that is usually hidden. This interplay was one of the issues discussed during the preparatory meetings of the SCHM. But when Febvre and Needham agreed to defend some kind of a political (didactic) outlook, they differed in their motivations: one was going from science to politics, and the other from politics to science.

After the war, two different dynamics emerged that ultimately appeared to be contradictory. The first one channelled through UNESCO, and especially the SCHM project, which represented a radical rupture with the pre-war Eurocentrism (and Francocentrism for French intellectuals). But UNESCO's fruitful period ended with the cold war, and with its interstate and administrative evolution. The second one is WFSW, where the scientists were deeply rooted (even more than in UNESCO) in official institutions. It is interesting to notice that the WFSW occupied the place refused by UNESCO at the end of the 1940s (when it chose to give priority to general educational, cultural and scientific activities, with indirect effects on pro-peace efforts—and refused to be confined to peace propaganda): WFSW occupied mainly the ground of peace campaigns (and professional demands as well as the promotion of science). But even on that ground, WFSW's direct influence vanished rapidly to the profit of wider movements like Pugwash.

British–French associations appeared to have played an important part, going from one association to the other, but the networks included scientists from other countries, like Belgium (Brachet, Florkin), Brazil (Ozorio de Almeida, Carneiro) and probably many others. The story still has to be written.

Back to Needham

The significance of Needham today goes beyond the ideological proximity to the scientism of the 1930s and an outdated Bernalism. He differed from ordinary scientism on two original counts: his opposition to Eurocentrism, and his claims for ethics in science. His evolution bears an inverse relation to Bernal's. And it allowed him to be friendly with the radical science movements of the 1970s, and to agree with Febvre.

His reference to Marxism, mainly to historical materialism, allowed him to introduce the idea of the historical contingency of modern science. He explained that the universality of science came from its ability to amalgamate the various contributions, even with a cumulative model. Universality is some kind of a historical construction. In doing so, Needham not only rehabilitated the civilizations of the past, but also, for the present and the future, affirmed their ability to contribute to a universal science. This was the positive side of the scientific optimism of the 1930s, pursued also by UNESCO. As Needham said in the 1970s:

> In the days when Bernal was writing, there was never any doubt that the natural sciences were essentially beneficial to mankind. The principal gravamen of the attack of these scientists on capitalism was that it prevented science from exerting its beneficial effect. Now, the situation has utterly changed. There is widespread fear of science and no longer any conviction that its activities are always for the benefit of mankind, or even would be under socialism.[163]

Inversely, in the 1970s, WFSW was considered poorly on account of its defence of a value-neutral science by the radical science movements.

Within this specific domain, Needham is still an exemplar for global activism, for science and post-Stalinist socialism. His various commitments influenced each other and gave his thought a contemporary relevance.

[163] Karp (1988), pp. 91–4. Similar views are expressed by Needham in a book in French, where he answered questions by Didier Gazagnadou (Joseph Needham [1993]), *Un Taïste d'hommeur—autobiographie. De l'embryologie a la civilisation Chinoise,* Paris, Editions du Felin/UNESCO, p. 108). Speaking of the contemporary rejection of science, Needham added: 'I realise that the very thought of atom bomb, the issues of ozone holes and the greenhouse effect, and genetic manipulation, etc. are disturbing facts and compel us to be critical of science. However, I still believe that capitalism and science are irreconcilable, as this system is based on profit. It will continue to invest in nuclear weapons and the production of war gas' (p. 108).

7

Shiv Visvanathan

The Strange Quest of Joseph Needham

John Haldane was once asked by a cleric what he had learnt of God on the basis of his studies of nature. Haldane replied, 'an infinite fondness for beetles'. My story goes one step further. Haldane goes to his classless heaven, God they say had a fondness for eccentric Marxists, especially Gramsci and Rosa Luxemburg. Haldane asks God what impressed him about Marxism. And God replies, 'the infinite variety of Marxisms', a diversity that years of Stalinism had not blurred. The epoch of Joseph Needham reflects the infinite variety of Marxisms.

Consider just the scientists of the period. John Desmond Bernal, Marxist, crystallographer, admirer of Joyce, author of *The Social Function of Science* and *The World, the Flesh and the Devil.* There was Haldane himself, one of the few scientists who experimented on himself, a British eccentric who spent his last years in Calcutta and Bhubaneswar.

The Indologist Francis Zimmerman has produced the fascinating paper 'Why Haldane Went to India?' (Zimmerman 1996). Zimmerman traces the remarkable attempt to link and embed population biology in local modes of thought. In this, Haldane was following his obsession with diversity, not just at the level of biology but of modes of thought. We witness next the writings of Lancelot Hogben and J.G. Crowther who attempt to popularize science without creating a populist one. *The Social Relations of Science* and *Science and the Citizen* are testimonies to that effort. It is to this who's who of English Marxists that Needham belonged. Needham was just as maverick in style, whether it was in his Marxism or in his embryology. What marked them all was the dislike for what he called 'fixed boundary lines'. Max Weber is supposed to have once said: 'I want to see how much I can stand.' Needham in a similar way could be epitaphed as claiming that he wanted to see how much he could combine.

In tracing the strange quest of Joseph Needham, we have to emphasize his struggle against dualism. Needham's was not so much an obsession with class or political economy. In fact, it is at that level that he is at his most banal. Needham's work was an obsession with the ethics of knowledge, its civilizational scope, it politics, its career, its hidden violences. Needham's contributions along the three domains, China, embryology, Marxism, can be read in this context. In each there is a search for plurality struggling with unity. With China, it is an obsession with civilization. Unlike most Marxists who located their problematique at the level of class, political economy, the mode of production, Needham's Marxism was also an obsession with the other as a civilization. His biology, particularly his embryology was an escape from reductionism, an attempt to show that Marxism could provide a form of organicism, an axiomatics to evade the violence of reductionism, either as mechanism or vitalism. He articulated this through his idea of levels of organization, through the concept of the system. In Needham's essays there is an attempt, or the hint of an attempt, to differentiate between system and whole. The whole is not only more than the sum of the parts; it is emergent, evolutionary, numinous. It is mysterious in that it can never be grasped completely. Needham's Marxism was also an ethics, a theory of knowledge, and he tried to show that Marxism was not only French and German, but also English. His least-known book, written under the pseudonym of Henry Holorenshaw, was a study of the genealogies of Marxism in the Levellers and Diggers. For him, Marxism was a pluralist sensibility and he located it in and as a religious ethic. Joseph Needham was a religious man, a Christian who sought to define religion in a new way. Influenced by Rodolf Otto (Otto 1958) and Collingwood (1946), Needham's creativity sought to balance incommensurability with universality. It is this search for a plural Marxism that sought roots in Tao, in Whitehead, in local struggles, different strands that we shall seek to unravel. In all this, there is an attempt to articulate what I shall call cognitive justice, an effort to extend the notions of equality and fraternity beyond the domain of economy to knowledge.

Needham's work has to be understood within the general approaches to the historiography of science.

The approaches to the historiography of science can be understood in terms of two oppositions (see Figure 1). The first is the distinction between the externalist and internalist critiques of science. Briefly put, the externalist approach to the understanding of science focuses on the social organization of science. It analyses the role of professional societies, universities, industrial research laboratories and multinationals. Included within such a perspective would be questions like the relation of science to war or the difference between science in capitalist and communist systems. The internalist approach concentrates on the nature of science itself as a mode of cognition. Such an approach would examine the attitude of modern science to nature, its mathematization or the vivisectionist violence of the scientific experiment.

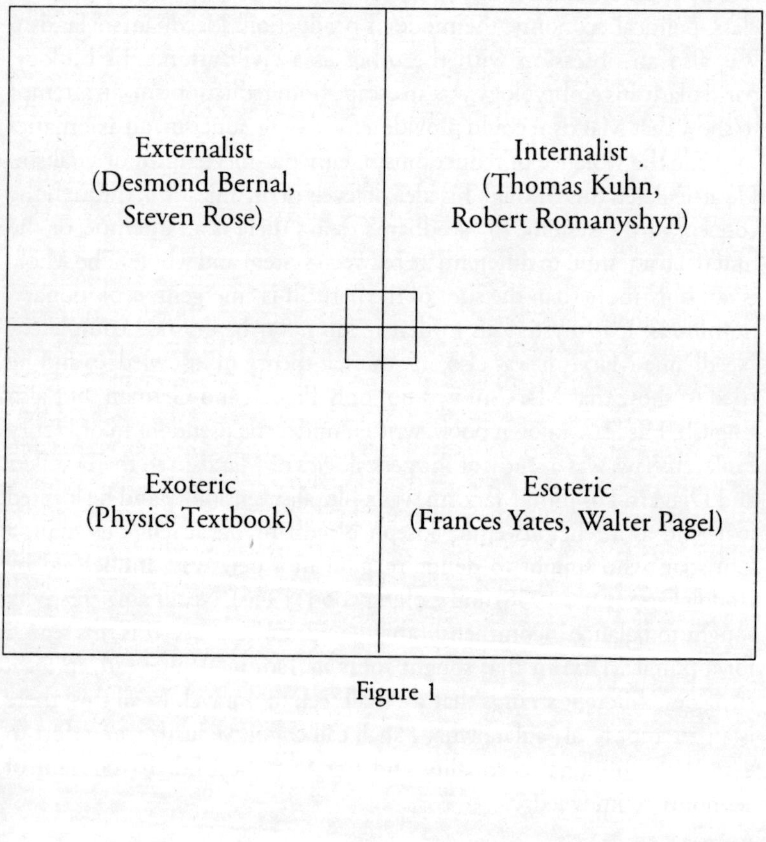

Figure 1

Complementing this is the opposition between exoteric and esoteric approaches. The esoteric concern is with the gnostic, occult or alchemical roots or readings of modern science. Questions of esoteric linkage would include the impact of alchemy on Newtonian science, the kabbalah on Freudian analysis or the *shakta* tradition on J.C. Bose's botany (Nandy 1995). The exoteric approach would restrict itself to a more overt linear view of science. Such a view is present in any conventional history of science where Archimedes, Galileo, Newton, Einstein form a series.

This list of oppositions gives us a fourfold grid over which any scientific event can be mapped out, both as to its statements and its silences. It is within such an overall grid that we shall approach Needham's work.

I want to emphasize that my essay is a form of methodological naivete, a triptych of naivetes. I am neither a China hand, nor an embryologist, nor a Marxist, but I believe it is only by sinning thrice that we can return to Needham's questions as an open work.

I borrow the idea of *The Open Work* from Umberto Eco (Eco 1989). Eco emphasizes, in fact celebrates, the importance of ambiguity. I think such a notion is important. Needham's work is monumental, over twelve volumes on China, half a dozen books of essays, tomes on embryology and histories of embryology. It is like confronting a monument. Looking at the sheer enormity, the graniteness of it, one is awed. As one moves between familiarity and distance, one realizes what it is. Derek de Solla Price (Solla Price 1973: 17) dubbed it but never followed it up beyond the act of labelling. It was a *museum imaginaire,* a notion developed by Andre Malraux in his *Voices of Silence* (Malraux 1974).

Malraux talks of the strangeness of a museum. Even in the West, the museum is a recent entity, while in the non-West it was virtually unknown. A museum is an act of estrangement. It removes works of art from their context, disembedding them of their qualities of sanctity, adornment, possession. Earlier, a Gothic statue was a component of a cathedral. Similarly, a classic picture was tied up with the setting of its period; whereas the modern art gallery not only isolates a work of art from its context but combines it with rival or even hostile works. 'The gallery became an anthology of pictures and the subsequent reproduction of these pictures has created what might be called the "fictitious" arts by systematically falsifying the scale of objects, by presenting oriental seals the same size as decorative reliefs on pillars, amulets like statues' (Malraux 1974: 14). This act of juggling juxtaposing creates a certain playfulness, a virtual reality which subverts the original purpose of the work. Needham's *Science and Civilization in China* is a portable anthology on the subject, a text of exhibits on agriculture, geomancy, astronomy and so on. There is even here an act of disembedding. As Nathan Sivin, one of Needham's finest collaborators and critics, has pointed out, Needham's work removes the Chinese sciences from their contexts. Bennett adds: 'When a template designed by Aristotle is placed over Chinese science, some of the disciplines fit nicely—mathematics, astronomy, medicine for example—while others are distorted or have simply no place at all.' Under Needham's organizational scheme, the latter are classified as the pseudo sciences, as for example geomancy. Bennett notes that Sivin seeks to compare not individual isolated innovation but complete lines of development seen in relation to their backgrounds. While Sivin's work is formidable, it does not provide the playfulness of Needham's volumes. The latter is like a huge pack of cards which we can shuffle and reshuffle. China moves from being a distant other to mirror, and double, from exotic strangeness to siblingship. In the very nature of this, subversiveness feeds on itself and what is lost, or becomes least memorable, is Needham's original question. 'Why did China not develop the sciences in the way the West did?' We move more to the question as to whether a different array of sciences was possible. Needham's work as a *museum imaginaire* makes his question less linear and more polysemic than he intended it. The original 'Weberian' question was a linear patriarchal one. The Needhamian question was less linear and weakly patriarchal. Yet Needham sticks to the idea of a universal science. Science is one, he claims. We can see it most clearly in his critique of Seyyed Hossein Nasr. Needham claims that 'there is only one unitary science of Nature approached more or less closely, built up more or less successfully and continuously, by various groups of mankind from time to time'. This means 'one can expect to trace an absolute continuity between the first beginnings of astronomy and medicine in Ancient Babylonia through the advancing natural knowledge of medieval China, India, Islam, and the classical world, to the break-through of late renaissance Europe, when, as has been said, the most effective method of discovery itself was discovered'. Needham explains that while there are many who share the view, there is another point of view which he associates with Oswald Spengler. To Spengler, the sciences produced by different civilizations are like separate worlds, 'with their own frames of reference, and not subsumable into a single history and a simple ever growing structure' (Needham 1976: XXIII). Needham dubs this the Spenglarian question. Probably the finest exponent of this was Seyyed Hossein Nasr, the historian of Islamic science.

Needham argues that Nasr wishes away the failure of Arabic civilization to produce a modern science. Far from regretting it, he makes a positive virtue of it, rejecting belief in any integral, social evolutionary development of science.

Nasr states,

> The history of sciences is often regarded today as the progressive accumulation of techniques and the refinement of qualitative methods in the study of nature. Such a point of view considers the present conception of science to be the only valued one; it therefore judges the sciences of other civilizations in the light of modern science, and evaluates them with respect to their development with the passage of time.

Such a perspective, says Nasr, fails to present 'Islamic science from an Islamic point of view' (Nasr quoted in Needham 1976: XXXVIII).

If Needham sees Western science as the great instauration, Nasr sees Islamic science as the creative refusal, an affirmation of a different world, a refusal to profane nature. He claims,

> Neither in Islam, nor in India, nor the far East was the substance and the stuff of nature so depleted of a sacramental and spiritual character, nor was the intellectual dimension of these traditions so enfeebled, as to enable a purely secular science of nature ... to develop outside the matrix of traditional intellectual orthodoxy' (Needham 1976: XII).

It was a world where *scientia* was subordinated to *sapientia.* Needham constructs it as a debate between the sacred and the secular and of the question of the oneness of science. He quotes Santillana to claim there is no going back. 'Copernicus and Kepler believed in cosmic vision as much as any Muslim ever did, but when they had to face the moment of truth they chose the road which was not that of *sapientia* ...' Needham argues that Nasr's statement is a threat to the equality of science. His politics demands (1) the unity of science, (2) the equal competence of all races for science, and (3) that whether Greek, Jew, white or black, they all do science. Within the kind of battles Needham was fighting against biological fascism, racism, Darwinism this is a viable answer. It is a theory of cognitive equality based upon the equal scientific competence of all races.

But the idea of one science requires the standardization of knowledge. Is equality only to create a fraternity of sameness or can we have fraternity of difference within a modern world? Can we have an alternative science or sciences? For Needham, such a view is an anathema, given the years of Lysenkoism and racist science. At that level one needs his secular notion of equality. One celebrates it. Yet, at another, given particularly the crisis of violence and ecology, one needs a different view of science. One is tempted to cite that granddaddy of papers in this debate, Lynn White's, 'The Historical Roots of the Ecological Crisis' (White 1968). White argues that for all its secularist rituals, science still remains Christian. It still develops its methodologies from the Christian assumptions of nature, that nature was made for man's use. Christian notions of time and of nature mark science as much as they determine the original structure of the detective novel. Is there a link between Christianity and the current structures of violence in science? Is it an internalist, esoteric problem or merely one demanding an externalist solution?

Needham approaches this issue in his paper 'History and Human Values' (Needham 1976b), realizing that this new critique of science is not a fascist science or a revivalist one but a grounded critique from scholars like Roszak, White and the political movements of the 1960s and 1970s.

He begins with his dogma: 'There is only one modern science and it is identical under every meridian. There is only one logically controlled experimentation, only one application of mathematical hypothesis and their testing by statistical methods.' Modern science is one and immaculate. What creates the ethical challenge is the application of science. Fundamental to the Needhamian approach are still the two basic axioms of conventional science policy:

(1) The myth of the immaculate conception of science.

(2) The myth of the fall.

The first holds science to be neutral and, if well used, can solve the problems of poverty and disease. The malaise lies in the politics of use. Needham is sympathetic to the Roszakian world of counterculture, but he claims that the problems of genetics, nuclear weapons, health are questions of ethics. He makes fascinating suggestions in this context. He agrees with Lynn White that the problem of ecology is one of ethics and the ethics of Christianity is generally illiterate or idiotic about a sacramental notion of nature and 'ill equipped to deal with the problems posed by the physio-chemical and the biological sciences'. The basis of such a humanistic ethic, he feels lies in 'the sages of Confucianism and Mohism, the philosophers of Taoism and legalism'.

The Confucian ethic was an optimistic one. It believed that man was fundamentally good. The doctrine of the fall was alien to it. Second, it was not caught in the dualistic trap of mind and matter. China, says Needham, was a profoundly non-Cartesian society. It provided the most powerful basis for a non-obscurantist organicism. In fact, Needham suggests that the cultural revolution was a Chinese attempt to overcome Cartesian dualisms of mental–manual, soul–matter and the democratic opposition of red–expert.

The cultural revolution encouraged artisans, craftsmen to take the initiative and persuaded managers to spend time at the bench. But the organicism of China was not only one of method, but also of community. The idea of the Confucian ethic questioned atomism not only of matter but of the individual. Needham could be criticized in terms of the later reassessments and the genocidal violence of the cultural revolution but I wish to redirect him in a more playful way.

Needham suggests that the mindsets of Tao or Buddhism might be more useful to a nuclear physicist than Christianity. There is a touch of the Sapir–Whorf hypothesis, of the linguistic structure of American Indian languages being more amenable to the formulations of quantum mechanics. If Whorf's ideas are at the level of the cognitive, Needham's stick to the ethical. Yet at one point in *The Grand Titration* he remarks that 'if environmental conditions were reversed as between Europe-Africa—all else would have been reversed too—all the great names of the heroic age of science, Galileo, Malphigi, Vesalius, Harvey, Boyle would have been Chinese'. For Needham, a Chinese Galileo would have been scientist first, and Chinese second. Yet, it is tempting to ask if a Chinese Galileo would not have been different. Could we visualize a different biology or different physics?

One playful response to this came from the Japanese novelist Junichiro Tanizaki. In his *In Praise of Shadows,* Tanizaki talks about the lost world of the Japanese house, where electric lamps have substituted for lanterns and flush toilets of sparkling porcelain replaced the wooden toilet. He muses how different everything would have been if we in the orient had developed our own science. 'Suppose we had developed our own physics and chemistry, would not the techniques and industries based on them have taken a different form, would not our myriads of everyday gadgets, our medicines, the products of our industrial art—would they not have suited our natural temper better than they do?' (Tanizaki 1984: 7).

Tanizaki's preoccupation is at the level of aesthetics and comfort. Tanizaki claims that ink would not have been blue but black and that paper would not have been the Western kind. But the objections are more than aesthetic. The work of C.V. Seshadri, J.P. S. Uberoi and Dharampal have shown that the critique of science is not purely about aesthetics or ethics. There is a theory of knowledge implicit in it. And this becomes fundamental in the ecological and feminist critique of science. Here the internalist model of science, its objectivity, its notion of experiment are itself brought into question. Would Needham have approved of a more feminine science? How would some of his collaborators, who were women, respond to the feminist work of Harroway, Rapp, Butler and Hilary Rose?

The feminist critique is different from the Marxist critique. The Marxist critique was an externalist one. Some feminists used gender as an equivalent of class, substituting for class relations the model of patriarchy. But the last decade has seen a critique of a more fundamental kind, where science is seen as patriarchal, Baconian, vivisectionist, monocultural, hegemonic, museumizing and genocidal. This, I must emphasize, is not to be dismissed as anti-science. It is a philosophical and democratic attempt to create an alternative science. It is here I believe that the Needhamian science becomes an open work. What begins as a Weberian patriarchal question of a linear universal science moves to a weaker patriarchal, more pluralist idea of science, challenging the reductionism of Marxism, biology and history. But there are hints, ambiguities which go further. Can feminism add new content to the universal idea of science? Hints of it exist in Bray's work. Francisca Bray, who wrote the volume on agriculture in *Science and Civilization,* published an essay on the Green Revolution in the *Scientific American,* an incisive critique of the monocultural directions of modern agriculture. More specifically, could Needham have gone the way of Hilary Rose?

Hilary Rose, along with her Marxist biologist husband Steven Rose, has produced a series of books on a Marxist approach to science. One thinks in particular of *Science in Society* (1969) and *The Radicalization of Science* (1976). What began as a confident and positive approach became a source of discomfort. The nature of the transition is what interests me and it is captured in Hilary Rose's later book *Love, Power and Knowledge* (1994), subtitled 'Towards a Feminist Transformation of the Sciences'. It was, as Rose remarks, a struggle with a variety of selves. The displacements begin with the fact that the labour of women is left out of Marxist economy. The next issue was the question of gender in science, the equivalent of the Weberian question, why are women absent in science to the question of the representation of women in science. The word representation has two meanings. It refers in a democratic demographic sense to the active presence of women in laboratories. It refers also to the way in which women become objects of science, of how the category woman is constructed by science. There is a move from externalist to internalist questions. There is then the third question of whether woman's labour, the woman's ideas of her body can alter the notion of rationality in science. It is a move from woman's voice to a feminist transformation of the sciences, especially in the way reproduction, demography, hysteria are constructed. The interesting question is: Would Needham have made an equivalent transition to the possibility of an alternative science? One doubts this in the personal sense, but one would like to suggest that the Needhamian project is moving in that direction.

In fact, I would like to suggest that the official question that sparked the project is not the only fruitful one. The archive of asides, footnotes and observations on the critical issues of science in Needham demand an aggregation. One needs not just a recovery of the Chinese voice in science. The eminent sociologist Graham made this point when he said that we have to wait for the development of Chinese interpretations of this period. But I want to follow a different argument.

The very multiplicity of causes that were required for science seems to imply that it was a form of 'moral luck', a luck that it consolidated over time. Bernard Williams, in an essay with that very title, suggests a simple example (Williams 1981: 23). He says think of a painter. Let us for the purposes of discussion call him Gauguin. Gauguin abandons family and career to go to Tahiti and paint. There was a risk element in this, a sense of agency. Williams writes that one justifies Gauguin's decision in terms of the painter's popularity. But what if his ship had sunk on the way to Tahiti or if he had been a bad painter like Winston Churchill? Would we have approved of the painter's decision to abandon his family? Would we have been amoral about this on grounds that his creativity needed it? Along with Williams, I am not denying causation. All I wish to emphasize is the serendipity, the luck that a science happened, 'chose' the mechanical metaphor, linked itself to the bourgeois, abandoned the idea of *anima mundi* and became what it was. If science were less an agency of domination and control, would we look with the same approval on it? Second, I would like to suggest that the moral luck of science seems to be running out. The sheer facticity of genetic engineering, nuclear weapons, the rise of iatrogeny, the ecological devastation of reductionist science, all seem to suggest this. In this frame, the question remains: has the moral luck of the scientific method run out? There is a lot in Needham that, despite official caveats, seems to suggest this. The Weberian question then changes into a different quest. One then asks: Is there in other cultures an alternative ethic for science or a way of constructing an alternative science? Would this alter the conceptions of the future UNESCO to which Needham provided the original S?

II

Joseph Needham was as profoundly maverick as Gramsci or Luxemburg. Gramsci in his *Prison Notebooks* asked that Marxism be translated into Sardinian. Needham attempted an equivalent project with fascinating consequences. For Needham, Marxism was not something French or German but also English. In his attempts to create a *swadeshi* Marxism, Needham located a role model for himself in the Levellers and Diggers.

Needham believed the England of the Puritan Revolution was at as crucial a juncture as Russia after the Bolshevik revolution. Both were periods of change and transition, creating epistemic breaks at the level of knowledge and politics. The seventeenth century saw the true transition from the middle ages to the modern world, the beginning of modern science and the end of religious autocracy. They were differences. In one, science was being born and religion was still powerful. In the other, religion seemed a declining force. Both were creative movements where the grammar of the relationship had to be reworked. Needham's book on the revolution is his least-known book.

There is a hint that he saw himself as the John Reed of the Puritan Revolution. The Puritan Revolution was the English equivalent of *Ten Days that Shook the World.*

The figure that fascinated him is Gerrard Winstanley. Needham claimed that Winstanley was along with Milton and Bunyan, 'among the greatest of the 17th century Englishmen'. He sees in Winstanley's attempt an effort not only to redeem the commons but to see the world as one. In his *Law of Freedom,* Winstanley claimed, 'The Earth should be a common treasury,' and added that 'if the Earth were set free from kingly bondage, so that everyone might be sure of a free livelihood, ... then many of the secrets of God and his works in nature would be made public, which men nowadays keep secret to get a living by ...' 'when the commonwealth's freedom is established, then knowledge will cover the earth as the waters cover the seas and not till then' (Needham 1948: 81).

Needham claims that many of Winstanley's remarks were reminiscent of the Soviet constitution. Basing himself on St Paul's saying 'He that shall not work neither shall he eat', Winstanley held: 'If any refuse to learn a trade, or refuse to work at seed time or harvest, or to be a waiter at the storehouse, and yet will feed and clothe himself with other men's labours, the overseers will first admonish him privately; if he continue idly, he shall be reproved openly before all men.' In April 1649, Winstanley and other levellers 'appeared armed with spades to dig up uncultivated land at Georges Hall with the intention of growing corn and other produce'. They felt the common land was the common property of the people. Needham suggests that Winstanley established many such 'collective farms' in England but as experiments they were short-lived (Needham 1948: 81).

Winstanley saw knowledge not only as part of the commons, but was clearly aware of the role of 'the new or experimental philosophy for social welfare'. Needham felt, like Winstanley, that 'to know the secrets of nature is to know the secrets of God, and to know the works of God within creation is to know God himself for God dwells in every visible work or body'. Winstanley's work was the Puritan equivalent of *The Social Function of Science* and Needham celebrates in Winstanley the *swadesi* Marxism of the British. In a poetic passage he asks, is it not of value to 'English socialists tired of hearing communism identified with foreign sounding names and doctrines, to know that communists of the seventeenth century had names like

English villages—John Lilburne, William Walwyn, Gerrard Winstanley, Robert Lockeyer, Giles Calvert, Anthony Sedley?' (Needham 1948: 91).

Needham also realized that a new relationship between science and religion had to be worked out and elaborated in the domains of politics, religion and ethics. Commenting on the changes in religion he claims that 'we reach the paradox that Marx and Engels would have been more acceptable to the martyrs and fathers than to the comfortable 19th century theologians contemporary with them' (Needham 1948: 53).

Needham's defence of communism is a theological one. Christianity, he claimed was the grandfather of communism, and it was communism that could provide the moral theology of the time. In a statement reminiscent of the *Communist Manifesto* (on the hand mill and the steam engine) he claimed that 'scientific ethics should be to communist society, what catholic ethics was to feudal society and protestant ethics to capitalist society' (Needham 1948: 55).

The challenge before communism was a religious one. The great difference between medieval theology and modern science, he felt, was that while the former incorporated a system of ethics in the form of moral theology, the latter has not yet produced one. Communism, he felt, could provide 'the moral theology of our time'. The problem was that the official church had become associated with the instruments of exploitation. Needham attributed this insensitivity to class oppression 'religious opium' and claimed that there was an equally dangerous 'scientific opium' (Needham 1948: 65). This was a worldview that was indifferent to the numinous, to the holy in one's life. For Needham, a sense of the numinous was as much a part of being man as music or poetry. In this context he states that a class analysis of the relation between man and god as being based on the relation of the primitive exploiting lord with the slave would be futile.

For Needham, Christianity was necessary to help redeem the ruthlessness of science. The increasing domination of the scientific mind creates a sense of ruthlessness. Power over nature gives one a feeling that evil could be reduced to a problem of technique, of social engineering. When one abstracts in science, one loses a sense of the individual and the unique. Needham emphasized that the principle of ethical neutrality leads to an anarchy among traditional systems of morality. Scientific domination without a sense of the numinous is what created the *Brave New World.*

Needham's distinction between the numinous and the fetish is relevant here. The numinous is an elusive quality of mystery, of poetry. For religion, 'a God comprehended is no God; one does not know why the universe is as it is; but only there is a holiness in it'. Needham senses the numinous feeling in the works of Auden and Mayakowski (Needham 1948: 59). He claimed that Christianity located the numinous in love. The fetish reduces an idea to an object in a way that eliminates mystery or emergence. The relation between the numinous and the fetishistic is in a way the opposition between Comtean positivism and Needhams's dream of scientific socialism.

Comtean positivism became obsessed with liturgy. It fetishized science, becoming eventually a catholicism without clergy. Nehruvian socialism had a Comtean touch to it. In making dams, the temples of modern India, we fetishized the dam, creating a meaningless liturgy around it. For scientific socialism, religion provided this ecological check, a sense of the sacred, of what one must not touch or touch only gently. Unlike many radicals who wanted the erasure of religion, Needham made a creative plea for it. In this he was closer to Lunacharsky than Lenin. Lenin felt 'one can't make a revolution without breaking heads' (Needham 1948: 69). But Lunacharsky pleaded for the preservation of certain buildings, art treasures in the heat of the revolution arguing that it is always necessary for Christian man to plead for the retention of Christian principles in treating people (Read 1987). Christianity thus becomes a check, an ethical thermostat both for the abstractions of science and the ruthlessness of the revolutionary. .

Needham observed that there is a ruthlessness in the statistical character of the scientific method. Noting that the ruthlessness with which the biologist throws out an anomalous embryo useless for his immediate purpose, the ruthlessness with which an astronomer rejects an aberrant observation may be too easily applied to human misfits and deviations in a socialist regime.

The Confucian/Taoist perception of science is articulated in his perceptions of biology. The organicist perception which he articulated was fundamental. If operated both at the level of biology and his perception of the scientific worker, Needham argued, there will be no such thing as a scientist without a philosophy. His position was similar to Einstein's, who stated that the scientist takes in his concepts like mother's milk without asking about its origin or genealogy. Needham held 'the purer the scientist thinks he is, the more surely he is in the grip of some unconscious system, perhaps a garbled form of idealism which he received as a virus in the milk of religion on which he was nourished as a child'. The purity of science, in this sense, is an illusion. He suggested that if one must have a philosophy then let us have a good one and among his preferred philosophies were the dialectical materialism of Marx and Engels and the organicism of Alfred North Whitehead. He contrasts both to a Cartesian worldview. He follows Whitehead in emphasizing the connection between economic individualism, seventeenth-century atomism and the individualistic ego of Descartes. All three he claimed 'lead from private worlds of experience to the private world of morals' (Needham 1948: 203). This results in the sterilization of the scientific worker, and to a disembedded notion of pure science.

For Needham, the individualism of the scientific self was compounded by the reductionism of science. Reductionism was an aspect of fetishism, a loss of the numinous. The idea of emergent levels was at the core of a Needhamian science. The notion of the level of organization, each irreducible to the other, yet each illiterate without the other was fundamental to his numinous notion of biology. The idea of the system provided for God at every level of organization. What the biological theory of emergence also provided was a sense of surprise (Needham 1948: 183). The Pascalian Demon was the demon of predictability. The Needhamian Demon was one of mystery, surprise, of a world where each level was logically unpredictable in terms of the level lower than itself.

While the Whiteheadean philosophy provided an ethic, Needham felt it was inadequate as a politics (Needham 1948: 203). Whitehead had little to say about politics, which Needham argued was the only system of mass ethics available. Needham finds this wider code in China. The Chinese intellectual philosophical systems were much more congruent with modern science than the world outlook of Christendom. He emphasizes three particular aspects in this context. First, the mechanical view of the world did not develop in Chinese thought. The organicist view that everything was connected to every other according to a hierarchical order was universal among Chinese thinkers. In addition to the notion of emergence and organicism, the Taoist ethic added what in modern terms would be called an ecological ethic. In the Chinese view, man was not the lord of the universe.

There was, says Needham, a conception of a *scala naturae* in which man was thought as the highest form of life. However, this gave him (man) no authorization, no licence to deal with the rest of creation as he liked. In the Tao, the universe did not exist to satisfy man. There was a triadic relation between man, heaven and earth, and the basic emphasis was on harmony. Finally, Chinese culture never suffered from the post-Renaissance obsession with scientism. Needham claims there was never any need for a William Blake in China, 'a Blake to oppose Newton's single vision, nor did China need a Kropotkin to oppose a Huxley'. Chinese culture had never really been tempted to regard science as the sole vehicle of human understanding. Needham suggests that scientism, the idea 'that scientific truth alone gives understanding to the world is nothing but a Euro-American disease' (Mansel Davies 1990: 62).

It is in this sense that I wish to suggest that the Marxism of Joseph Needham was a psychoanalytic project, a psychoanalysis of a civilization, facing the problems of narcisissm and reductionism. To solve this what one had to do was to exorcize its dominant myth.

Civilizationally it was Greece that was the origin and source of myths. Prometheus, Narcissus, Oedipus, Athena dominate the imagination, becoming the equivalents of Maxwellian demons in the way they determine our narrative. Greece was origin, Greece was a cornucopia of myths. Yet Greece was a kind of ethnocentricity. Ethnocentricity is a style of pre-empting narratives. When you talk of the origin of intellect, you begin with Greece. When you write a genealogy of science, ego is Greece. The Arabs, the orient, the tribal are also-rans. What Needham realizes inventively is that what one needs is not an unravelling of one myth, but an exorcism of Greece itself. The category Greece had to be relativized through the idea of siblingship.

The nature of the scientific problem was as follows. Needham realized that hegemony and dominance and its great sibling, the reductionist mode, were impoverished discourses, diminishing creativity in science and in politics. Western science in relation to the other has been hegemonic, particularly in relation to traditional knowledges. The other was never scientific. Whether it was systems of medicine or architecture or agriculture, the other always fell short. The very being of the other was only possible as a primitive, secondary, derivative, rudimentary, uneconomic form of Western science. Such a notion of the other impoverished history by restricting the origin of science to Greece.

The rest then belonged to a museum, to ethno-science or to some incarnation of orientalism. With the suppression of the history of the other, there is an equivalent suppression of the other within, of the witch, the patient, the madman, the peasant, the tribal. Each was subject to a disciplinary tactic, whether as museum, clinic, asylum, inquisition in the name of science. Science as a dominant mode of consciousness contained an epidemic of repressions. But the aetiology of the problem does not end here. There is the further problem of reductionism, of the denial of the numinous to other domains of reality.

Let us return to the methodology of the diagram we began with. The four crosscutting approaches, internalist–externalist, exoteric–esoteric carry behind their antiseptic heuristics, a demonology, a victimology (see Figure 2). Often the victim is seen as a demon.

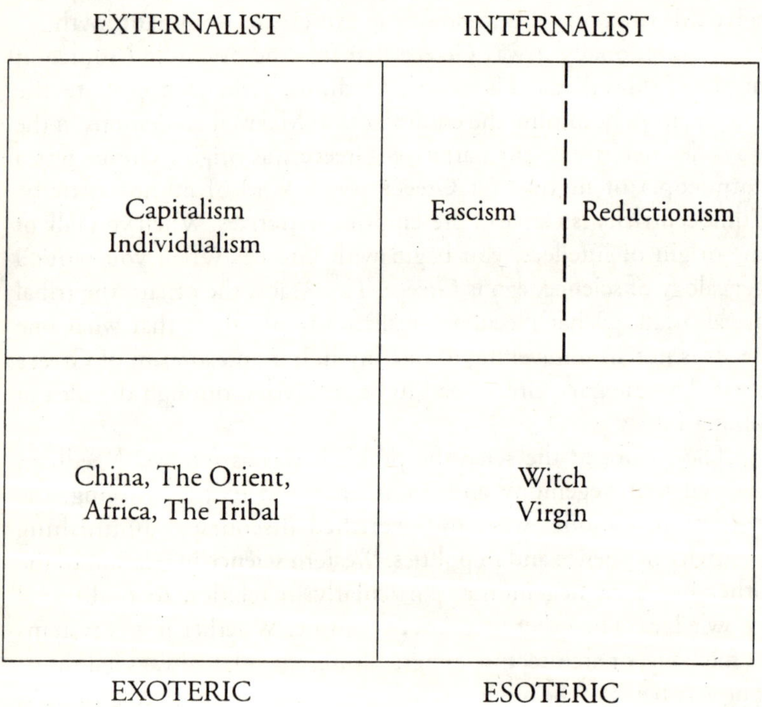

Figure 2

Needham's work as historian, scientist, sinologue, embryologist, poet and theologian employs the following repertoire of tactics.

For Needham, the scientific worker was the ascetic monk of today. But there was a danger that this monk would repress himself with an array of pollution rituals which ascribed a purity to science. Needham emphasized that the 'as if' neutrality of method demanded an engagement in politics. The specialization of science was a kind of idiocy. It is in this context that one must read the great debates between Needham and Bernal with Hill and Polanyi. Needham argued that science was work, and as a form of labour that must be liberating. This part of his therapeutics was the most publicized.

Equally public was his celebration of China. The rhetoric of strategies he uses here is fascinating. In fact, *Science and Civilization in China* reminds one of *The Golden Bough* in terms of contrast. Frazer, in his classic, emphasizes the continual error of the primitive. The books become an archive of how the primitive got it wrong. In contrast, Needham wishes to persuade us that China was a source of fecund ideas, of inventions, and he presents a monument of archives to sustain this. In his work, China is the lost sibling, twin, other, mutant. China is also that which will redeem the West. A classless society by itself is not liberating enough. Equality also needs fraternity. China provides the ethic, the style of consciousness that the modern West needs. If Marxism is the diagnosis of the wound within, Confucius is the balm for the wound without. Needham's is a therapeutic of civilizations, a breakthrough from orientalism which created the other as exotic, pornographic, ascetic, esoteric to a relationship of intimacy.

Beyond the idea of titration, there are two metaphors that mark his essays. One is that of the river. The other is of the bridge. He sees science as an ocean into which all the rivers of knowledge flow. 'All peoples have been testators, all are now inheritors each in his own way.' This is a picture of oneness. The metaphor of the bridge seeks hyphens across difference. And to me this remains more fascinating, as a vision of conversation, of dialogicity, distance, intimacy and translation between China and the West. The victim as other now returns to redeem the West as a sibling.

I would suggest that J.D. Bernal's *Science in History* and *The Social Function of Science* be seen in tandem with *Science and Civilization in China*. The two scientists were intellectual partners and their alchemical wedding produces a Martin Bernal. Breaking the narcicissm of Greece by inventing China was not enough to psychoanalyse the West. One needed to redeem Africa. Africa was the heart of darkness, the Conradian black hole. The earlier attempts to restore Africa were aesthetic ones or anthropological. One talked of the impact of primitive art on Picasso, one demonstrated the rationality of magic as in Evans-Pritchard, or one whispered about the Negritude movement. Present in Negritude was only an aesthetic and in fact Aime Cesaire in *Return to my Native Land* celebrates 'those who have never invented anything'. It is not incidental that Sartre, in talking of Leopold Senghor, dubs him the Black Orpheus. The aesthetic still needs the scientific as cognitive. Martin Bernal fills this gap with *The Black Athena.* There is equivalence but the mythology is still quaintly, overtly Greek!

Step three in this historical therapeutics is internalist historiography. This concentrates on two forms of reductionism. Needham holds that reductionism is the worst form of repression and is as repulsive as ethnocentricity. Here Needham discusses two forms—racism which has its roots in biologism and the mechanistic pathologies of science. Racism is a form of political reductionism which Needham exposes time and again. I would like to add one suggestion here. In a cultural sense the figure to contrast with Needham here is Wyndham Lewis. I don't think they met. I believe Needham repudiated Lewis's book *Time and Western Man* (Jamieson 1979). But the contrast was deeper. Here the poet-fascist meets the scientist-Marxist. One was assertively patriarchal and racist. The other was sensitive to the feminine and anti-racist. Both were close to being polymaths. Needham was philosopher, embryologist, Marxist, sinologue, even poet. Lewis too was a creative author, philosopher, painter. Their opposition reveals the real confrontation of an era. In counterposing them in this heuristic sense, one misses in Needham, meditations on language and sexuality. There are reflections on love. After all Needham sees the communist project as an elaboration of the Christian notion of love. There are interesting asides on language. For instance in showing that democracy is not essentially Greek, Needham refers to Chinese terms of address. He refers to the complete absence of special 'built in' forms of address between superiors and inferiors and notes its stark contrast with Japanese society. We also have his discussion of Vernant's studies of Greece. Needham builds on Vernant to argue that, 'while the demonstration of geometric propositions in the agora in the form of logical mathematics was particularly congruent with an assembly of equal disputing participants, arithmetic and algebra were more specialized, secretarial techniques, less susceptible to public demonstration' and therefore more preferred by bureaucratic systems of society (Needham 1969: 244). There are such interesting footnotes and asides right through but one wonders whether Needham really explores the question of language. He claims China had no problem of incorporating scientific terms into its own language. The question one must ask is: Did it work the other way? Could Chinese words be translated without carrying along the philosophical matrix they are embedded in? Was change so easy or did language, through continuous rectification, hide the traumas that Chinese modernism went through? We now move to the third level of therapeutics.

At the level of the esoteric as magic and religion one confronts, for want of a better location, the feminine. The woman as witch and virgin has been one of the great problems and victims of Western patriarchal science. It is in his handling of these that Needham remains banal. One sees this most clearly in his essay 'William Harvey and the Witches' (Davies 1990: 305). For Needham, 'nothing is more interesting in the history of science than the survival of belief in witchcraft' and yet he has little to offer on it. He is more intrigued by the problem of the even distribution of scientific men for and against witchcraft. He observes that some of the foremost scientists of the time, including Henry More, Thomas Browne, and Joseph Glanville, among the earliest fellows of the Royal Society, were defenders of witchcraft. Needham notes that witchcraft disappeared with the extension of the scientific method but says little of why the witch craze appears with the rise of science.

He has also a fascinating paragraph on the virgin, in which he cites from Henry Adams' classic work *Mont St Michel and Chartres*. It is in his excerpt on the virgin that one glimpses what the virgin and the witch may have to offer. He states

> Mary concentrated on herself the whole rebellion of man against fate, the whole protest against divine law. The virgin was above the law, she delighted in trampling on every social distinction in the world and in the next. She knew the universe was as unintelligible to her, as any theory of morals as it was to her worshippers, and she felt like them, no sure conviction that it was any more unintelligible to the creator of it. The virgin was above the law. She could save the supplicant from a ruling, a decretal, or a codex but not from the laws of gravitation or thermodynamics. (Needham 1948: 49).

What Needham observes is the decline of the power of the virgin with the rise of the power of science.

Needham borrows this powerful excerpt from Henry Adams. Henry Adams raised this problem of science in a more fundamental way. Adams began as an orthodox Cartesian, who felt that the laws underlying values were similar to the laws of physics. He saw in the virgin and the dynamo, symbols of the human condition (White 1968). The virgin who was once the symbol of life, of fecundity, of fertility was now powerless. The real defeat of the West seems to be in the inability of the two symbols to coexist. The virgin was fecund and powerless. The dynamo powerful and sexless. There is a feeling eventually, in Needham, that the West lacked this balance, this harmony of categories. He attributes it to the people of the book, who were inclined to patriarchal solutions to nature. He sees a rescue coming only from outside, in the Chinese balance between male and female, Ying and Yang, Confucian and Taoism which could create 'the spirit of humility and adaptability', replacing the masculine ordering, the sociological determinism that dominated the West. It is in that sense that the psychoanalysis of the West will never be complete without the intimacy of the other, of China.

* * *

It is relevant that this volume, produced towards the closing years of this century, seeks to explore the biography of a man and his work which spanned the century. Needham's *Science and Civilization* stands as one of the great monuments of the twentieth century. While it is monumental, it cannot be treated as a monument. It is not a statue on which an intellectual sparrow can inflict itself and move on. It is more like a Henry Moore sculpture, to be explored, opened up, shuffled again, gossiped about. It is a celebration of life and the plurality of cultures. It is an open work full of ambiguities, contradictions, and yet, because it was always crossing boundaries, it became a protean work, an archive of a civilization that becomes an invitation to the future.

Bibliography

Bray, Francesca and Joseph Needham, 1984, *Science and Civilization in China,* vol. 6, Cambridge.

Collingwood, Robin George, 1946, *Idea of History*, Oxfrod.

Davies, Mansel, ed., 1990, *A Selection from the Writings of Joseph Needham*, Sussex.

Eco, Umberto, 1989, *The Open Work*, London.

Holorenshaw, Henry, 1939, *The Levellers and the English Revolution*, London.

Jamieson, Frederic, 1979, *Fables of Aggression: Wyndham Lewis, The Modernist as Fascist*, Berkeley.

Kenyon, Timothy, 1989, *Utopian Communism and Political Thought in Early Modern England,* London.

Malraux, Andre, 1974, *The Voices of Silence,* Frogmore.

Nakayama, Shigeru and Nathan Sivin, 1973, ed., *Chinese Science*, Cambridge, Mass.

Nandy, Ashis, 1995, *Alternative Sciences*, Delhi.

Nasr, Seyyed Hossein, 1976, *Man and Nature,* London.

Needham, Joseph, 1948, *Time, the Refreshing River,* London.

——— 1969, *The Grand Titration*, London.

——— 1976, *Science and Civilization in China*, vol. v, no. 3, Cambridge.

——— 1976b, 'History and Human Values: A Chinese Perspective for World Science and Technology', in Hilary Rose and Steven Rose, eds, *The Radicalization of Science*, London.

Otto, Rudolf, 1958, *The Idea of the Holy*, New York,

Read, Chris 1987, 'The Godbuilders in Pre-revolutionary Russia: Social Democracy as a Religion', in Jim Obelkevich, Lyndal Roper and Raphael Samuel, eds, *Disciplines of Faith* (History Workshop Series), London, pp. 435–50.

Rose, Hilary, 1994, *Love, Power and Knowledge*, Cambridge.

Rose, Hilary and Steven Rose, 1969, *Science and Society*, Harmondsworth.

——— eds, 1976, *The Radicalization of Science*, London.

Solla Price, Derek de, 1973, 'Joseph Needham and the Science of China', in S. Nakayama, and Nathan Sivin, eds, *Chinese Science*, Cambridge, Mass.

Tanizaki, Junichiro, 1984, *In Praise of Shadows,* Tokyo.

White Jr, Lynn, 1968, *The Dynamo and Virgin Reconsidered*, Cambridge, Mass.

Williams, Bernard, 1981, *Moral Luck*, Cambridge.

Zimmerman, Francis, 1986, 'Why Haldane went to India', in Steven Marglin and Frederique Apfel-Marglin, eds, *Decolonizing Knowledge,* Oxford.

8

Karine Chemla

The Rivers and the Sea: Analysing Needham's Metaphor for the World History of Science*

The purpose of this essay is to analyse in a critical way the metaphor of rivers flowing into the sea, employed by Joseph Needham to depict the emergence of modern science. As he put it:

> What metaphor then can we use to describe the way in which the medieval sciences of both West and East were subsumed into modern science? The sort of image which occurs most naturally to those who work in this field is that of the rivers and the sea. There is an old Chinese expression about 'the Rivers going to pay court to the Sea', and indeed one can well consider the older streams of science in the different civilisations like rivers flowing into the ocean of modern science. Modern science is indeed composed of contributions from all the peoples of the Old World, and each contribution has flowed continuously in it, whether from Greek or Roman antiquity, or from the Arabic world and from the cultures of China and India. (Needham 1967, p. 397).

He pushed the metaphor forward to suggest the theme of his paper: 'First let us consider the fusion points, those estuaries in time when the rivers flowed into the sea, and when full mixture took place' (ibid.).

* I thank S. Irfan Habib and Dhruv Raina for having invited me to the conference where an earlier version of this paper was presented. The discussions held in Delhi on science and its history, and the possibility of reviewing Needham's contributions from the perspective of the participants was of tremendous interest to me. I also wish to thank the anonymous referee of the paper.

In fact, in this paper, Needham restricted himself to considering the merging of the traditions from the West and the ones developed in China, a merging which, in his opinion, took place after the introduction, into China, from the seventeenth century onwards, of knowledge coming from Europe, through the efforts of the Jesuits. Within this framework, he sought to show that the merging occurred at different times, depending on the discipline one considers. I have discussed elsewhere several aspects of the thesis presented by Needham.[1] I would like here to concentrate on the metaphor itself and on the problems it raises.[2] If the metaphor proves inadequate in some respects, it still captures interesting properties of modern science and fulfils crucial functions. From this perspective, it appears not to be an easy task to replace it. Hence, rather than suggest a simple rejection of the metaphor, I shall outline a research programme whose completion, I believe, should put us in a better position to find a substitute that would not lose the benefits of the Needhamian metaphor.

The Metaphor and its Problems

A basic assumption underlies the picture: history of science must be practised with an international perspective in mind, and not at any local scale. My own research convinced me of the importance of this premise. A world-wide perspective appears to be the most relevant one to fully apprehend a given phenomenon in the history of science, both conceptually and historically.[3] Moreover, the metaphor suggested by Needham is interesting in that it emphasizes that modern science was constituted with elements coming from everywhere on the planet. This designates the merging of different traditions as a fundamental

[1] I (Chemla 1992, 1993 and 1996) have examined the modalities of the fusions which merged distinct mathematical traditions, and whether this actually constituted an international body of knowledge.

[2] In fact, later on, Needham introduced another metaphor to refer to the same process: 'By a thousand capillary channels, like venules joining together to form a *vena cava magna*, influences came from all parts of the world' (Needham 1973b, p. 3). In his introduction to the same volume, Sivin, p. XXXI, attributes to Needham the following similar picture: he 'has always portrayed modern science as a river in which the waters of many streams are inextricably mingled'. The aspects of the metaphor that I want to discuss are common to all these images.

[3] Chemla (1992).

event for the constitution of scientific knowledge. However, this picture raises four kinds of problems that I would like to examine here on the basis of two concrete examples:

Problem 1: What phenomena occur when rivers meet? What are the dynamics that actually constitute knowledge at such points? If the metaphor presents the advantage of being striking, it lacks accuracy in this respect. Are we speaking of elements of knowledge remaining invariant when circulating, and eventually being piled on top of each other, or of another sort of mixture? This problem itself raises two distinct sets of issues. First, how do bits of knowledge circulate, in space or time? What transformations do they undergo on the way? Second, how is knowledge reconstituted by the process of merging? Can we provide basic patterns for how concepts, bodies of knowledge are transformed when they encounter each other?

Two such patterns are discussed on the basis of two examples (Chemla 1992). The first one shows how, while two concepts elaborated by two traditions were available for what we consider today as the same mathematical object—a fraction—one eliminated the other. Hence, concepts may disappear from modern science, as a result of the merging of different traditions. The other example starts from similar initial conditions: various concepts for what we consider today as the same mathematical object—an algebraic equation—were available in different traditions, whose merging led to the synthesis of distinct concepts. Through such a process, all concepts undergo transformations and new products appear. A synthesis of the kind radically changes previous ways of dealing with a topic. We encounter other such cases in this essay. The existence of at least two patterns indicates that the idea of a simple merging cannot be an accurate description of how such processes constituted modern science.

Problem 2: Is the image of rivers flowing into the sea relevant? The topography of such a landscape could be understood to imply that the flow of knowledge from various traditions presents the pattern of an inverted tree, the branches of which, in the end, join the trunk.[4] However, this is not the only possible kind of map for circulation constituting knowledge, and we shall present an example of another topological pattern which can be encountered, that of a circle.

Problem 3: The overall idea of rivers flowing does not clearly capture

[4] The other metaphors indicated above confirm such an intention.

a fundamental phenomenon in the history of science, as well as, at least by way of consequence, in the history of the history of science, namely the possible forgetting of knowledge.[5] And it is the possibility of such forgettings that can change the appearance of the whole picture. *Problem 4:* The fact that bits of knowledge coming from various traditions merged together must be distinguished from the fact that they become part of common international knowledge, that they circulate and join the sea. In between, several kinds of factors can intervene—inner factors as well as political, economic ones—and prevent this availability of knowledge from being widely achieved.

Let us hence exemplify these problems within the framework of two concrete examples: the rule of false double position and the solution of systems of simultaneous linear equations, before we come back to the metaphor itself.

Circulations of the Rule of False Double Position

This first example will illustrate the transformations that knowledge can undergo through transmission, including forgetting, and indicate that the roads of circulation do not always form the picture of an inverted tree. I do not want to get into technical details here, and shall rather stress the general features of the situation. The reader can find some basic information about the rule of false double position in the appendix.[6]

The 'Constitution of Modern Science' Narrative

The story of the circulation of the rule of false double position presents all the ingredients needed: it can be recounted so as to bring to light how Chinese knowledge became part of modern science.[7] Indeed,

[5] This has to be distinguished from an 'elimination', in the sense alluded to above, even though the former can be induced by the latter.

[6] I analyse the rule within the theoretical framework of the Chinese tradition (Chemla 1991), and give more details concerning its international circulation (Chemla 1997). I shall not reproduce all the relevant bibliography here, since it can be found in these two articles.

[7] Strictly speaking, the rule has disappeared from today's mathematical knowledge. If we were concentrating on the historical elaboration of the various components of contemporary mathematics, we would not take its history into account. A history that proceeds retrospectively and focuses on the constitution of knowledge 'available' at a given time can miss many an aspect of the actual scientific practice of the past.

the first known source which attests to it is the book which was to become the canon of the Chinese mathematical tradition: *The Nine Chapters on Mathematical Procedures* (hereafter *The Nine Chapters*). Compiled in the first century BC or AD, on the basis of knowledge previously available in China, this classic devoted its seventh chapter, entitled 'Excess and Deficit', to strange algorithms which were later to be called 'rules of false double position' and which it applied to solve all kinds of problems.

A few centuries later, these rules could frequently be found in Arabic manuscripts, the earliest of those which came down to us, being a treatise by Qusta ibn Luqa, dating from the 860s.[8] As Suter noticed (1908–9, p. 113, footnote 1), Qusta expresses himself in such a way that he clearly appears not to be the author of the procedure, which he calls 'of the two errors' and which he introduces to allow the solution of all computational problems without root extraction. Moreover, the terminology Qusta uses is the same as that found in *The Nine Chapters*, and the way of describing the algorithm is similar.[9] In addition to this, there is no evidence so far that such rules can be found in the extant Indian mathematical corpus, despite important scientific connections between China and India. The rules are peculiar enough so that the possibility of an independent re-elaboration of the same procedure appears to be improbable.[10] All these elements give credence to the hypothesis of a transmission, in this respect, directly from China to the Arabic-speaking world, after which the rule seems to have been widely circulated in Arabic westwards.

The treatise composed by Qusta on the rule of the false double position shows that the transmission is accompanied with a mutation. The procedure is no longer presented in relation to problems, but stated outside of any context of application and now proved in the Euclidean way. Let us stress that the new point here is not that the rule is provided with a proof. The third-century Chinese commentator, Liu Hui, had already accounted, in an algorithmic way, for the rules

[8] (Suter 1908–9) translates and analyses this treatise as well as two others on the same topic.

[9] According to Suter's translation, Qusta speaks in terms of 'suppositions', 'excess' and 'deficit'.

[10] In most of the cases, such a complex procedure is not necessary to solve the problems to which it is applied.

of false double position as given by *The Nine Chapters*. In contrast to it, the proof produced by Qusta connects the algorithm with Euclid's *Elements of Geometry*. The Greek book from the third century BC did not contain such demonstrations for arithmetical procedures. Qusta can apply Euclidean geometry to such contexts, thanks to an innovation. He associates a drawing with the problem solved by the rule (Suter 1908–9, pp. 114–15). Let us briefly describe the drawing (see Figure). On a horizontal line, one end point of which is *a*, he considers segments of arbitrary length *ax* as representing the value of a number which can be tried, in the problem, instead of the actual value of the unknown (a 'supposition'), whereas, vertically, he draws a segment corresponding to the end result associated with the trial.[11] The key point is that the end points of these vertical lines are said to generate a straight line.[12] Qusta then completes the drawing, considers it as a geometrical figure such as the ones dealt with in Euclid's *Elements of Geometry* and applies known theorems to it to establish the rule of false double position. In other words, each problem is seen from the

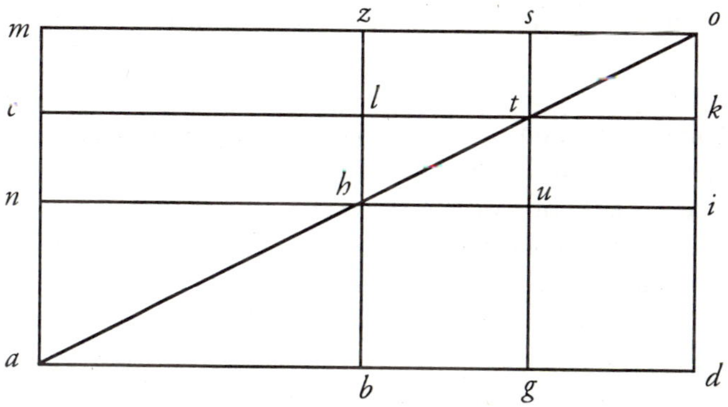

Figure 1

[11] Indeed, to each trial, the terms of the problem allow us to associate an end result. In the first problem of the appendix, that would be the amount of husked grain in the bucket at the end of the process, for a given 'supposed' amount at the beginning.

[12] This is another way of stating that the relation established by the problem between a trial and the end result is linear. Suter notes that Qusta proves the rule in this way only for some of the cases to which it can be applied. What is important for us is the new nature of the proof so devised.

perspective of the quantitative relationship that it establishes between a trial and an end result. This algorithmic dependency is represented graphically, after which the picture is considered independently from what gave rise to it and worked out as an ordinary geometrical figure.

It is difficult for me to decide whether this is the first example of such new uses of Euclidean geometry or not. Similar relationships were established, more or less at the same time, for instance when Thabit ibn Qurra (827–892) referred the proof of algorithms to solve quadratic equations to Euclid's *Elements of Geometry.*[13] However, this endeavour does not involve the elaboration of graphical representations of quantitative dependencies, as Qusta's does. In conclusion, it seems that the introduction of rules of false double position in a mathematical context, where proofs of the Euclidean kind were becoming a standard, led to two transformations. The algorithm was to be associated with the illustration of a relation of dependency, thanks to which it could be proved in a new way. Such figures, crucial for the representation of functions, may hence have emerged because of the demand of a Euclidean kind of proof. But, conversely, Euclidean geometry extended its classical domain of efficiency, as it could now be used to prove arithmetic algorithms. One should stress that, thereby, through being applied to figures of a new nature, generated in new ways, it was undergoing a mutation. The transmission seems to have produced an enrichment both for the rule and for the context within which it was introduced, a kind of synthesis of an algorithmic and an Euclidean practice, thanks to the production of a bridging type of diagram.[14]

Qusta's treatise was followed by several Arabic texts which came down to us, part of which also adopted an Euclidean presentation of the subject. This progression westwards of the procedure was relayed, from the thirteenth century onwards, by a transmission northwards. One finds the rule of false double position in Fibonacci's writings, but also in many mercantile arithmetical texts written in France in the fourteenth and fifteenth centuries under the influence of Italian mercantile textbooks, except that we do not seem to find here the

[13] See Luckey (1941). I am grateful to Jens Hoyrup for giving me a copy of this paper.

[14] Let us stress that in contrast to the synthesis alluded to above, which bore on two ways of conceiving what we now consider as the same object, this one merges two mathematical practices.

Euclidean proofs to establish the rules.[15] Later on, several German and Dutch books from the sixteenth century (Smeur 1978) also included similar rules. The stability in the way of expressing and using the rules seems to indicate that this was due to the continuation of a process of diffusion, besides which, again, some new improvements can be noted. For instance, the signs '+' and '–' appear in these texts dealing with the rule of false double position as algorithmic marks to distinguish 'positive' and 'negative' numbers, and allow the unification of what was until then presented as different computations. And this seems to have more generally been one of the contexts of elaboration of such signs. In an appendix, Smeur (1978, pp. 95-100) analyses the process of diffusion of these marks from the context of the rules of false double position (end of the fifteenth century), to algebra (beginning of the sixteenth century) and only later on to arithmetic (seventeenth century).

So would run the narrative if we were to concentrate on the constitution of modern science. It would tell the story of the priority of a Chinese invention, and of a transmission, as a result of which knowledge advanced by the bringing together of different ingredients—the rivers joining the sea, a typical story in which Needham was interested, typical of the processes which occurred in the medieval Arabic-speaking world. And such was, indeed, the kind of account Needham offered, in the part dealing with mathematics that he, together with Wang Ling, elaborated for Volume III of *Science and Civilization in China* (*SCC*, Volume III, 1959). Following a section devoted to the exposition of the rule and to its circulation (pp. 117–19), the conclusion reads as follows (p. 147): 'The Rule of False Position, found in the Han Chiu Chang (i.e. *The Nine Chapters*), appears in +13th century Italy under such names as "Regula Elchataym", indicating that the Arabs had transmitted it.'[16]

However, I shall argue that the actual picture is slightly more complex. On the one hand, the pattern of diffusion can be described

[15] For these statements, and the ones below on these texts, I could rely on the essay written by Stéphane Lamassé (1994) for his Master's degree. I am very grateful to him for having communicated to me the passages of his sources dealing with the rules of false double position.

[16] They add: 'They probably derived it from India but China may well have been its origin', but I do not understand which Indian sources they referred to. The footnote makes clear that 'double false' is meant.

more accurately. On the other hand, other events transpired in the story beyond the mere constitution of new knowledge.

Another Narrative

The beginning of the story does not change. The first source of the rules of false double position remains *The Nine Chapters*. But, there, the algorithm is in fact subtler than we said. As a matter of fact, the Chinese classic mentions it to solve two kinds of essentially different problems, examples of which are offered in Chapter Seven. Let me skip all the technical details here and call them, for the sake of simplicity, problems of the first type and problems of the second type.[17]

In fact, only problems of the first type can be found in all the Western sources, as far as I know, be they Arabic or European, all the way from the ninth up to the sixteenth century. This provides a supplementary argument underpinning the hypothesis concerning a continuous process of diffusion. Hence, the transmission did not only enrich the procedure, as we emphasized above. It impoverished it at the same time, as it reduced the efficiency which it was known to have in the earliest source that has come down to us. And, here, it is not only that a given algorithm deteriorates, but the whole subtlety of the algorithmic practice which produced it was not to be passed on. In any case, the transmission did not imply that circulated contents remained the same. A complex phenomenon modified the objects transmitted. They seem to have become richer, through their encounter with other practices, rather than by mere internal development, and to have also become poorer, due to a partial forgetting.

However, the story is not—and should not be—limited to what happened in the West. Transmission also occurred in China, and it took an entirely different course. After the compilation of *The Nine Chapters*, books were written in China, among which we mainly know those which were selected to be included in an imperial encyclopaedia dating back to the seventh century, the *Ten Classics in Mathematics*. These writings, for the most part, were treading in the footsteps of the classics, as far as the topics dealt with were concerned. Of these books, only two, the *Sunzi's Mathematical Classic* (*c.* 400) and the *Zhang Qiujian Suanjing* (*c.* second half of the fifth century) mention

[17] The reader can find a sketch of the two kinds of problems in the appendix. See also Chemla (1991 and 1997).

the procedures of false double position. The former contains a problem of each type, but the general procedure that enables their classification under the same heading is no longer stated as such. The latter, on the other hand, contains only problems of the first type. The possibility of using a similar procedure to solve problems of the second type has disappeared. And with this reduction in the variety of the kinds of problems dealt with, the specificity of a certain algorithmic practice has also disappeared.

The situation changes in the thirteenth century, namely the next period from which several books came down to us, after a long period hardly documented. Both kinds of problems re-emerge to be possibly solved by the rules of the false double position, although not in all the extant books. Hence, we are confronted with a pattern for transmission different from the westwards transmission. After what appears in our sources as a deterioration of part of the knowledge contained in *The Nine Chapters*, it seems to have been recovered. But, again, later transmission will be partial, though in another way, since the possibility of solving problems of the first type by such procedures seems to have been forgotten. Here, it is worth mentioning that, as some sources seem to indicate, this forgetting may be due to the fact that, for such problems as those of the first type, better procedures had been developed in China. It would hence be a process similar to the one responsible for the disappearance of the rules from today's mathematics, when better solutions are available. In any case, such is the state in which we find the topic in 1592, in the *General Source of Mathematical Methods* (*Suanfa tongzong*), a book by Cheng Dawei which reasonably embodies the body of knowledge contained in the Chinese mathematical tradition and still available for the literati at the end of the sixteenth century. Only problems of the second kind are included in connection with the rules of the false double position.

In conclusion, whether occurring in the same language, or involving translations between different languages, transmission appears as a complex process that can both enrich and deteriorate the knowledge passed on. Such a fact which we could grasp here, by paying attention not only to the presence or absence of the rules of false position, but also to the kinds of problems for which they were used, gives material to deal with problem 1. In this case, the two processes of transmission, one westwards and the other one in China itself, produced two completely different end results, whether one looks at sixteenth-century Europe or China. The first process was the subject of the narrative relating to the constitution of modern science. Should the second process, comparable to the first in all respects, be overlooked by history? This could well be the result of the emphasis on the European future of all ancient knowledge.

However, in addition to providing material for the analyses of transmission as a non-deterministic process, the previous analysis is fruitful in that it sets the stage from where we acquire a better understanding of other phenomena related to modern science. And the following development, which illustrates this remark, will provide the example of another topography for the constitution of knowledge—a circle, rather than rivers flowing into the sea.

Modern Science and Traditional Knowledge? Making Universal Science?

At the end of the sixteenth century, as is well known, European missionaries arrived in China, and among them, the Jesuits designed a specific strategy of evangelization wherein so-called Western science and technology were used in order to seduce Chinese literati and high officials.

Matteo Ricci, who seems to have been the one who implemented this policy in China, introduced his Chinese counterparts to two kinds of European mathematical books. The first one, embodied by Clavius's edition of Euclid's *Elements of Geometry*, introduced a new branch of mathematics to China. The second one, such as Clavius's *Epitome Arithmeticae Practicae*, was dealing with topics that Chinese books, such as the *General Source of Mathematical Methods*, mentioned earlier, were also addressing. Among these topics, we find, and Li Zhizao, a Christian convert who was to work with Ricci on Clavius's *Epitome*, could find the rules of false double position.

No wonder that Xu Guangqi published with Ricci a partial translation of the former (1607), whereas Li Zhizao, after collaborating with the Jesuit, composed a Chinese adaptation of the latter, the *Arithmetical Indicator Gathering the Cultures*, where elements from both Chinese and Western sources were brought together. Hence, there occurred a kind of synthesis, as Needham (1967) emphasizes. However, this synthesis was carried out by Chinese scholars and not by missionaries; and its products remained in China without circulating westwards (Chemla 1996)—but we come back to this point later.

Among the three volumes constituting Li Zhizao's summa, the *General Part*, published in 1613, devoted a chapter to the rules of false double position. How did he merge what appeared to him as two sources of mathematical knowledge? Indeed, in this respect, Li Zhizao was confronted with a perplexing situation. On the one hand, like all Western sources, as far as I know, following the process of transmission westwards that we described, Clavius's *Epitome Arithmeticae Practicae* was using such rules to solve various problems, even complex ones, but all of the first type. Li Zhizao was most probably unaware of the historical process which produced such a situation. If he knew of Chinese mathematical books, it seems that he had access to such contemporary productions as the *General Source of Mathematical Methods*, but not that he could read *The Nine Chapters*. Hence, on the other hand, after the second process of transmission that we described, the one which took place in China, Li Zhizao could find, in Chinese sources, procedures cognate to those coming from the West, but used only for problems of the second type. The two distinct end results of the processes we analysed arrived on his table. Here the circle is completed, hence Problem 2.

What in ancient times had been a unique procedure applied to solve two kinds of problems now appeared to Li Zhizao, who had no access to *The Nine Chapters*, as two cognate procedures having different geographical origins. He recognized the similarity between the 'two procedures', and he included both in the same chapter. Bits of knowledge that had been kept apart for centuries were again reassembled. By translating Clavius's terms with the appropriate Chinese words, Li Zhizao revealed the obvious similarities between the 'Western methods' and the 'Ancient ones'. And in Chapter 7 of the *Arithmetical Indicator Gathering the Cultures: General Part* (pp. 209 *sq.*), he not only provided the translation of Clavius's treatment of the rules of false double position, but he also included problems and procedures he had found in Chinese sources.

Li Zhizao probably thought that he was constituting new knowledge. We can see that he was reconstituting it. Again, this provides staple material to deal with problems 1 and 3. Such a process is the same as the one Joseph Needham wished to capture with the metaphor of rivers flowing into the sea. However, given the global picture, this acquires a completely different meaning. In addition to this, Li Zhizao does not seem to have understood very well the connection between the two types of problems. And, as a result of the comparison he made, he came to the conclusion that the 'Western' procedure was superior. The forgetting that occurred during the transmission, which, let us recall, might have been due to the emergence of better procedures in thirteenth-century China, resulted in a forgetting in the history of knowledge. Since this was to be later followed by a general decay of the Chinese mathematical tradition, it created a situation where 'Western mathematics' could appear superior to 'Chinese'. This feeling was set against a political and religious background which, in turn, may have had a religious and political impact.[18]

In conclusion, if knowledge was thereby constituted, would we not call this the 'constitution of modern science', *even though, again, the processes involved are the same*? A comparable international dimension is present in this case and in our first narrative, and they both involve a synthesis, although of distinct types. Moreover, the global pattern of circulations of the rules presents a circle, the completion of which involves many ingredients of diverse nature—political factors, the reality of the transmission and of the forgetting it caused. The first kind of narrative ignores the other side of the story and its relation to the political uses of knowledge.

This highlights Needham's main emphasis on the question of the constitution of modern science, where he resorts to the metaphor. But, if we only look at the making of contemporary knowledge, we cut a tree from the complex map of all circulations, which deletes from the narratives a range of events. Should not historians of science deal with such events?

Needless to say, Needham was himself aware of, and interested in, such events, and we should be cautious about simplifying his actual stand in our presentation of his endeavour. Indeed, Needham, Wang Ling and Price (1960) produced a narrative somewhat similar to our second one, with respect to the history of clockworks. They underpinned the hypothesis that a key step in the history of the escapement had been taken in China. They then showed how, after a time when the Chinese clockwork tradition peaked in the twelfth century, it declined in such a way that,

[18] Whether Li Zhizao, a Christian convert, was genuinely convinced of the superiority of Western mathematics in this respect or whether he himself took part in playing up the idea that Western science be deemed superior, such was the way in which it was presented to the Chinese literati. On his recurring attempts to provide a synthesis between Western and Chinese sources, see Leung Yuen-Sang (1990-1).

when the Jesuits arrived 250 years later there was extremely little to show them that mechanical clocks had ever been known in China .... This situation was naturally exploited by the Jesuits in several ways. To the Chinese they emphasised as much as they possibly could the superiority of the natural sciences of early seventeenth-century Europe because by the aid of it they wished to convince them of the superiority of European religion .... The general upshot was that the indigenous scientific and technical tradition of China was very much underestimated in the West. (p. 141)

Hence, in this case as in ours, it was on the basis of a forgetting, in China, of knowledge previously available that the Jesuits could impress the Chinese scholars and officials of the superiority of their own tradition.[19] Later, Needham, Wang and Price stress again:

The state of science and technology in Ming China (i.e.: 1368–1644) was in fact decadent, and what was left of it the Jesuits were not at all disposed to emphasise. They wanted to be thought of as converting a highly cultured people, but the greater the superiority of the 'new, or experimental, science' of the West over the traditional empirical science of the Chinese, the greater, by implication, was the superiority of the Western religion. Not until the time of Antoine Gaubil, and then (it must be admitted) only in a rather patronising way, did the Jesuits make a serious attempt to study the history of the sciences of the Chinese culture-area. (p. 167)

Irrespective of whether Needham was right in asserting a Chinese priority in this matter or not, he was aware of how the routes of the circulation of knowledge can imitate circles. However, these incursions in later periods are, as far as I am aware, rare in his writings, his main emphasis remaining on the Chinese 'contribution' to modern science and technology as they emerged in the West. Moreover, it seems to me that the two types of events are not conceived of as being part of the same narrative. If the prior period of the history of clockworks in China belongs to the history of science and technology, the second one would rather have to do with the history of their history, explaining the formation of Western misconceptions about the history of science and technology in China. In the excerpt quoted above, the authors, when dealing with the Jesuits' activity, lay more emphasis on the opinion the latter formed of Chinese history than how a state of

[19] Needham (1958, pp. 205–6) restated that, if not for this forgetting, the clocks brought by the Jesuits should have appeared less new and impressive than they actually have.

knowledge was constituted in China, after the introduction of knowledge from the West. History of science seems to be mainly, in their account, the history of the events affecting the research front.

Our first example already showed, as the case of clockworks alluded to did, how forgetting may create differences that can be used to various ends and disturb the neat pattern of tree-like transmission. I would like now to come back to another aspect it highlighted. Whether Li Zhizao constituted or reconstituted knowledge, his new presentation of the rules of false double position do not seem to have been circulated outside Asia. As outlined in Problem 4, bringing knowledge together does not always, or automatically, make it international, common knowledge.[20] Various factors can contribute to its remaining local, among which, as we already saw, the political and religious context within which bits of 'Western' science were introduced to China. Here, another example will provide an even more striking evidence of such aspects.

Not Forgotten Knowledge: Systems of Simultaneous Linear Equations[21]

Li Zhizao could find in the *General Source of Mathematical Methods* not only rules of false double position, but also various kinds of procedures that had been elaborated in China previously. Among them, one deserves special attention because it was not until the beginning of the nineteenth century that one finds it presented in Western sources. Hence, at the time when the Jesuits arrived in China, some bits of mathematical knowledge from the Chinese past that were still extant could have interested European mathematicians of the seventeenth century.

Li Zhizao was aware that the procedure to solve systems of simultaneous linear equations, as we now call them, was not to be found in the Western books brought by the Jesuits. He devoted a

[20] This case gives an example where, although a synthesis, a merging of different bodies of knowledge may occur, it can remain known only locally, and not circulate at the international scale. Contemporary medicine constitutes an opposite case. Although various traditions circulated, such as Chinese medicine being practised in the West, they did not in fact interact with each other, and have not so far been synthesized, a fact emphasized by Needham (1967).

[21] Again, as previously, I do not want to present technical details here. The reader can find them in Chemla (1996).

chapter to it in the adaptation of Clavius's *Epitome Arithmeticae Practicae*, on which he collaborated with Ricci, as mentioned earlier. In contrast to the rules of false double position, such a procedure was thus complementing Western sources. Li Zhizao understood this, but Ricci did not transmit the method to Europe. Here is a striking case where the rivers did not join the sea, although they could have. The constitution of a body of knowledge, created by Chinese scholars, remained local.

Why, one would like to ask. Naturally, it is difficult to answer such questions. In the case of clockworks, Needham, Wang and Price stressed how detrimental to the Jesuits' interests it could have been to emphasize a Chinese contribution to the history of clockworks, let alone advertise the existence in Chinese sources of mathematical knowledge new to European scientists of the time. Again, the question appears to be both about the history of the history of science *and* about the history of science. Indeed, as Needham, Wang and Price made clear for clockworks, we have long lived under the influence of a history of science which served the interests of the Jesuits at the beginning of the seventeenth century. After them, we have repeated that there was no mathematics worth mentioning in China when they arrived. This highlights the fact that for rivers to join the sea requires the interplay of several factors, not all scientific. In this last case, rivers joined an inner lake.

How does modern mathematics know the procedure today? It is called the method of Gauss's pivot. Gauss rediscovered it, and the political context of European empires rendered it international knowledge. This adds one more to the several political backdrops which, at different times and in various ways, played a part in the construction of the idea that science was a Western phenomenon.

Conclusion

The metaphor thus has its limits. However, Joseph Needham's focus, in *Science and Civilization in China*, on the constitution of modern science, thanks to the convergence of streams coming from everywhere in the world, presents decisive advantages to deal with what is called the 'history of non-Occidental sciences', as if such a divide could be relevant. And these advantages should not be underestimated when attempting to provide another picture.

First, the simple fact that it proposes a general framework within which to integrate results about the scientific and technological contributions from all over the world is crucial. In the absence of such a framework, the so-called 'non-Occidental sciences' remained on the margins of what, in most historians' awareness, constituted the main path for the constitution of science—the one that went from Greece to Europe, that is to say in their understanding: from Europe to Europe. The absence of such a representation of world history that would be commonly adopted by the profession could possibly explain why most of the research done in these fields has not yet become common knowledge, let alone why these results did not produce the changes in the old picture that they should have entailed.

Second, with such a picture of the constitution of modern science in mind, Needham could not but lay emphasis on those aspects of the scientific achievements made in China that were of universal dimensions, beyond singularities. Needham has often been wrongly criticized for having refused to take into account in his work the specificities which a given philosophical, social, political context imprinted on science in China at a given time. One must insist on the fact that he did not deny a certain form of relationship between a scientific result and the cultural domain with which this activity interfaced, as is clear in the following quotation: 'the ancient and medieval sciences (though bearing an obvious ethnic stamp) were concerned with the same natural world and could therefore be subsumed into the same oecumenical natural philosophy'.[22] But, if he admitted that scientific practice had everywhere its singularities, he wished to bring to light that which, in each achievement, was likely to become shared. The existence of specificities to him did not contradict the potential universality of the results produced, as is clear when he asks: 'When in history did a particular science in its Western form fuse with its Chinese form so that all ethnic characteristics melted into the

[22] (Needham 1967, p. 418). The religious connotation of the term 'oecumenical' is to be taken seriously. The development about religious oecumenism to be found at the end of Needham (1973a) is parallel to what is here stated about science. Various religions elaborated the same values, and Needham urges Christians to learn to recognize them under their different forms in the history of distinct civilizations, rather than trying to impose their views. I thank G. Métailié for having provided me a copy of this paper.

universality of modern science?'[23] And when this universality could all the more be highlighted, then one could show the ability of the results to circulate among different cultures and mix with results obtained elsewhere[24]: '... this science is an ultimate value and with its applications forms today a unity into which the comparable contributions of different civilizations (not isolated from each other as incompatible and mutually incomprehensible organisms) all have flowed and flow' (Needham 1973a, p. 16). In his emphasis here, one might infer his concern that, alleging such specificities, it was possible to conclude that there was an irreducible incommunicability between scientific cultures. And having lived through the discussions about national or racial characteristics of scientific practice, the Nazi ban on 'Jewish science', one imagines that he thought of how to prevent such deviations from happening, how to prevent the history of science from providing legitimacy to these causes. There, the property of the metaphor of rivers flowing into the sea appears clearly. Concentrating on the contributions the Chinese made to modern science specifically highlighted that which was done in China and was proved *de facto*, by history itself, to be of universal value.

Today, on the contrary, the tendency is to pay greater attention to local specificities. We should not lose sight of the limits of the relativistic stances to which this tendency could lead.[25] We should also keep in mind that they might offer little help in combating certain political forces, if only they do not foster them. In this respect,

[23] (Needham 1967, p. 397) One can question whether such a problem can be addressed with respect to only a part of the world (Chemla 1993). Moreover, as appears here, Needham often opposed the universality of modern science, to the fact that ancient and medieval sciences cannot be deemed universal. His main argument for this, as far as I know, is that modern science is expressed in the universal language of mathematics and, hence, is itself universal. Both claims seem to need further analysis. Personally, I cannot find convincing arguments in favour of such a divide. I would rather argue that the question of universality should be raised in the same terms for ancient and modern times, for mathematics and the other scientific disciplines.

[24] In the interview J. Needham gave me in September 1989 (Chemla 1989, p. 103), he restated this relationship between the fact for an item to be culture-free and its ability to circulate. Moreover, he explicitly expressed his belief that a study of the impact of cultural dimensions on science, a study of 'Chinese science' only from inside would lead to Spenglerian pessimism.

[25] If they might have some validity to account for some events in the history of science, to keep only to them would produce a truncated account. (Chemla, forthcoming).

it is helpful to remember the benefits we might draw, following Needham, from an international perspective.

These functions assigned to the metaphor are important, and we should retain them. It played another part in Joseph Needham's discourses, which does not seem to me to be relevant. The international nature of modern science echoed, in his presentations, the actual practice in science and technology in the contemporary world.[26] Aant Elzinga (1996) and Gregory Blue, in this volume, both stress the influence which his activities in international science policy had on his adopting such a stand on the world-wide nature of science and its history. Indeed, Needham did not restrict himself to noting the international practice of the science of his time. He wanted to emphasize the fact that all cultures had contributed to modern science, that the rivers had joined the sea, to plead for the equal rights of all peoples to have access to scientific knowledge and to develop its consequences (Needham, *Within the Four Seas*, 1955, pp. 15–16). And this, in his opinion, had to be taken into account in the making of international institutions of scientific research.[27] As long as the awareness of the multiple origins of scientific knowledge enables one to challenge those who claim that science is essentially European, there is no doubt that the metaphor is crucial. But, should we follow Needham in believing that history would provide an argument with which to plead that contemporary science belongs to all people that contributed to its making? If claims to the benefits of science could be grounded on past contributions, besides the fact that this would reintroduce essentialism in a dubious way, we would be promoting in a new way a very inegalitarian order. It seems to me rather dangerous to admit that history can grant rights, in this respect as well as in others. And, I would rather give up this contemporary perspective.

Instead of suggesting another metaphor, which is a delicate task, let me conclude by propounding lines of inquiry that could be pursued

[26] 'Already we entrust our lives to each other freely, flying with an Arab pilot, a Madagascan copilot, a New Zealand navigator, and an Ethiopian in charge of ground computer control. Let us have at least the generosity to salute all that their past cultures did toward the science and technology of the ecumenical world' (Needham 1973b, p. 5).

[27] See Gregory Blue's essay in this volume.

to prepare an alternative picture. However useful and fruitful the metaphor of the rivers and the sea may be, not least for the international framework it provides, it might be that it led Needham to emphasize aspects in the dynamics of science that cannot satisfactorily constitute the whole picture. The faith in a cumulative knowledge that it conveys is contradicted by ample evidence. The focus on priorities which it values leaves whole parts of the history of science untouched. The conception of transmission which it delivers, the notion of merging that it promotes, lack precision. The concentration on the constitution of modern science which it presupposes diverts attention from the observation of international dynamics of other kinds. By way of consequence, studying transmission not only as a fact, but as a process, and trying to identify the modalities through which it transforms both what it transmitted and that into which it is transmitted, understanding the various patterns that the merging of traditions can follow, looking at how science as an activity, always and everywhere present on the planet, is practised, develops, decays, circulates or not, all this could enable us to produce another metaphor which, without simplifying historical realities, would still capture *how* 'modern science is indeed composed of contributions from all the peoples of the Old World'.

Appendix

The rules of false double position solve a problem by making two suppositions of the value of the unknown that is to be determined. As a consequence of the fact that the supposition differs from the actual value of the unknown, the value of a magnitude differs from the value given to it in terms of the problem. Let us take the example of Problem 9 of Chapter VII of *The Nine Chapters:*

> We have an unknown quantity of husked grain in a 10 *dou* bucket. The bucket is filled to the brim by adding grain, which is then husked. We then have 7 *dou* of husked grain. How much husked grain was there in the bucket to begin with?

The procedure to solve the problem starts as follows, according to the third-century commentator Liu Hui:

> The bucket contains one *hu* [i.e. 10 *dou*]. If we suppose that we started with

2 *dou (a)* of grain, we had to add 8 *dou* to fill it. With 8 *dou*, we get 4 *dou* 8 *sheng* of husked grain. Compared to 7 *dou*, that produces a deficit of 2 *sheng (b)*. If we suppose that we started with 3 *dou (a')* of grain, we had to add 7 *dou* to fill it. With 7 *dou*, we get 4 *dou* 2 *sheng* of husked grain. Compared to 7 *dou*, that leaves us with 2 *sheng (b')* ....

Then, the actual rule of false double position, on the basis of the two trials (*a* and *a' dou*) and of the deviations from the value 7 *dou* they entail (a deficit of *b sheng* and an excess of *b' sheng*) produces the value of the unknown.

This constitutes in fact the first type of problem, the one which transmitted itself westwards and for the solution of which the general procedure received the name of 'rule of false double position'.

However, *The Nine Chapters* presents a second type of problem that can be solved by the same procedure, an example of which reads as follows (first problem of the chapter):

A purchase is entered into by several persons. If everybody pays 8 (*a*), there is a surplus of 3 *(b)*. If everybody pays 7 *(a')*, there is a deficit of 4 *(b')*. What are, respectively, the number of people involved and the price of the thing purchased ?

Note that what is taken to be the suppositions here (*a* and *a'*), instead of referring to trials made on the unknowns, link them together. This is what characterizes problems of the second type. And the general procedure, as above, will use *a, a'*, *b* and *b'* to produce the value of the unknowns.

The general procedure which *The Nine Chapters* gives to solve both types of problems, but which is described within the framework of problems such as the last one, is as follows[28]:

[28] In order to avoid discussing subsidiary problems, we do not translate the general algorithm here, but give a slightly simplified version of it. For a commented translation, see the sources already mentioned or our forthcoming translation of *The Nine Chapters* and of its commentaries. The symbols I, II, III indicate the starting points of the parts of the algorithm that we describe. We also give the configurations according to which the numbers were probably arranged on the counting board on which the computations were made. These configurations are not part of the text, they are reconstructed on the basis of hints made here and there in the text or in later texts. As they are not at the centre of our argument, we shall not comment any further on them here.

I

Put a and a' on the computing surface; the corresponding excess and deficit are placed under them respectively

$$\begin{matrix} a & a' \\ b & b' \end{matrix}$$

with them, cross-multiply a and a'

$$\begin{matrix} b'a & ba' \\ b & b' \end{matrix}$$

and add to get the dividend

$$\begin{matrix} b'a+ba' \\ bb' \end{matrix}$$

add excess and deficit to get the divisor

$$\begin{matrix} b'a+ba' \\ b+b' \end{matrix}$$

II

divide

$$x = (b'a+ba')/(b+b')$$

III

'If the excess and the deficit are linked with one another in a case where the numbers of persons who buy the thing are equal',[29] $(a-a')$ dividing the dividend gives y, the price of the thing

$$y = (ba'+ab')/(a-a')$$

$(a - a')$ dividing the divisor gives x, the number of persons involved

$$x = (b+b')/(a-a')$$

In fact, problems of the first type are solved by the part I+II of the general procedure, whereas problems of the second type are solved by

[29] This sentence, the meaning of which is difficult to catch but which is located at a key point of the general procedure, seems to introduce the fact that the subsequent computations are to be performed in a second range of problems, which fits the mathematical analysis of the procedure. This second range of problems seems to designate the problems the algorithm refers to, namely our second kind of problem. It may be intended to describe the particular status of excess and deficit here, in contrast to the first kind of problems. To make things simpler, I have left aside the computations to be performed if fractions occur.

I+III. Both problems are gathered under the heading of the same general procedure, on the basis of the fact that the procedures for their solution share the common part, I. This needed a rather sophisticated algorithic work, and it is this aspect which was destroyed by both processes of transmission.[30]

References

Chemla, Karine, 1989, 'Joseph Needham. Les apports chinois aux sciences: le grand titrage', *Préfaces,* 15, pp. 101–5.

——— 1991, 'Theoretical Aspects of the Chinese Algorithmic Tradition (First to Third Centuries)', *Historia Scientiarum,* 42, pp. 75–98 and errata in the following issue.

——— 1992, 'De la synthèse comme moment dans l'histoire des mathématiques', *Diogène,* 160, 1992, pp. 97–114. English translation: *Diogenes,* 160, pp. 95–111; Spanish: *Diogenes,* 160, pp. 92–107. I could check none of these translations.

——— 1993, 'D'Ouest en Est et retour. Les tribulations d'un biologiste européen en Chine', *Alliage,* 16–17, pp. 161–9.

——— 1996, 'Que signifie l'expression 'mathématiques européennes' vue de Chine?', in C. Goldstein, J. Gray and J. Ritter (eds), *L'europe mathématique. Histoires, Mythes, Identités,* Editions de la Maison des Sciences de l'Homme, Paris, pp. 220–45.

——— 1997, 'Reflections on the World-wide History of the Rule of False Double Position, or: How a Loop was Closed', *Centaurus,* 39, pp. 97–120.

——— 'Qu'attendre de l'histoire des sciences dans les aires non-occidentales ?', seminar on science studies, EHESS, 1996, 8, (forthcoming).

Elzinga, Aant, 1996, 'UNESCO and the Politics of International Cooperation in the Realm of Science', in Patrick Petitjean (ed.), *Les sciences coloniales.*

[30] Besides the nature of the computations themselves, a second fact supports an analysis in terms of a bifurcation in the general algorithm here: in *The Nine Chapters,* the description of procedures is generally such that, once the division of step II is done, $ba'+ab'$ should have disappeared from the counting board, leaving only the result of the computation. That would mean that the computations to be performed in part III of the algorithm could no longer be performed, were they to belong to the same flow as part II. The only time when they can be actually performed is at the end of part I of the procedure, when the values of x and y are still available.

Figures et institutions, series edited by Roland Waast, 'Les sciences hors d'Occident au XXe siècle', Orstom éditions, pp. 163–202.

Lamassé, Stéphane, 1994, *Les règles de proportions dans les manuels d'arithmétique marchande à la fin du Moyen Age (XIVe-XVe),* Mémoire de maîtrise (Paul Benoit adviser), Université de Paris I, p. 162.

Leung Yuen-Sang, 1990–1, 'Towards a Hyphenated Identity: Li Zhizao's Search for a Confucian–Christian Synthesis', *Monumenta Serica,* 39, pp. 115–130.

Luckey, Paul, 1941, 'Tabit b. Qurra über den geometrischen Richtigkeitsnachweis der Auflösung der quadratischen Gleichungen', *Sächsische Akademie der Wissenschaften zu Leipzig. Mathematische-Physische Klasse, Berichte,* 93, pp. 93–114.

Needham, J., 1955, 'The Dialogue of East and West', quoted according to its publication in *Within the Four Seas,* London, pp. 11–30.

——— 1958, 'The Missing Link in Horological History: A Chinese Contribution', quoted according to its reimpression in J. Needham, *Clerks and Craftsmen in China and the West,* Cambridge University Press, Cambridge, 1970, pp. 203–38.

——— 1967, 'The Roles of Europe and China in the Evolution of Oecumenical Science', quoted according to its reimpression in J. Needham, *Clerks and Craftsmen in China and the West,* Cambridge University Press, Cambridge, 1970, pp. 396–418.

——— 1973a (Henry Horolenshaw, pseudonym), 'The Making of an Honorary Taoist', in Mikulas Teich and Robert M. Young (eds), *Changing Perspectives in the History of Science,* Heinemann, London, pp. 1–20.

——— 1973b, 'The Historian of Science as Ecumenical Man: A Meditation in the Shingon Temple of Kongosammai-in on Koyasan', in Shigeru Nakayama and Nathan Sivin (eds), *Chinese Science: Explorations of an Ancient Tradition,* MIT, Cambridge Mass., pp. 1–8.

Needham, Joseph, in collaboration with Wang Ling, 1959, *Science and Civilisation in China,* volume III, section 19: Mathematics, Cambridge University Press, Cambridge (UK), pp. 1–168.

Needham, J., Wang Ling and Derek de Solla Price, 1960, *Heavenly Clockwork. The Great Astronomical Clocks of Medieval China,* Cambridge University Press, Cambridge (UK), 1986: second edition with a supplement by John H. Combridge, p. 266. I quote the book on the basis of the second edition.

Smeur, A.J.E.M., 1978, 'The *Rule of false* Applied to the Quadratic Equation, in Three Sixteenth Century Arithmetics', *Archives Internationales d'Histoire des Sciences,* 102, pp. 66–101.

Suter, Heinrich, 1908–9, 'Die Abhandlung Qosta ben Luqas und zwei andere Anonyme über die Rechnung mit zwei Fehlern und mit der angenommenen Zahl', *Bibliotheca Mathematica* 3, Folge, 9, pp. 111–22.

9

Pascal Crozet

Modernization of Science and its History Outside Europe: Egyptian Projects in the Nineteenth Century

Introduction

The field addressed in this essay is one which, until recently, has been relatively ignored in the history of science. My intention here is not to deal with the contribution of non-Western societies and civilizations to what Needham called 'the ecumenical science', but rather with the spread of 'modern science' throughout the world, as well as the role attributed to it by a given society in its projects of modernization.[1] Such an approach entails examination of the way some non-Western scientists effectively contributed to modern science, but more generally by the way they opened a space for it in their society, i.e. how they legitimized it.

This is not the place to give an account of all the recent studies which have dealt with this matter. Several seminars on this topic have elaborated different approaches, paving the way for comparative studies. But we could contemplate the relevance of such a perspective,

[1] It is not my intention here either to define modern science, or to discuss the difference between modern science and classical science. Using the words 'modern science' and 'classical science' (or 'traditional science'), I merely want, for the purpose of simplification, to distinguish the science which resulted from the massive transfer of European knowledge in some countries during the nineteenth century, and the science which existed before this transfer.

given the fact that the reported experiences often seem to be far too dissimilar. What would be the terms of an efficient comparative approach? And lastly, with what type of historiography and with what kind of representation do we embark on studying such a field?

In order to find answers we will concentrate here on one case study from nineteenth-century Egypt. We will attempt to understand the basis for modern science's legitimacy on the banks of the Nile at that time, which may well constitute a regional–cultural specificity.

Scientific Development and its Representation in Nineteenth-century Egypt

Pre-colonial Egypt and the European Challenge

The Egyptian case is not universally applicable when dealing with the transfer of European science outside European boundaries. Unlike countries such as India, this transfer was not initiated by Western powers, but instead occurred, in some respects, contrary to European expansion.

When he came to power in 1805, Muhammad Ali had to confront the challenge of military and, above all, economic dominance by Europe. The influx of European products had disrupted the Egyptian economy since the second half of the eighteenth century, and subsequent French occupation had exposed the fragility of the country's defence system.

To meet this challenge and construct an independent empire around the Nile, Muhammad Ali implemented a programme of radical modernization in various sectors. Moreover, the Egyptian state set itself new tasks, in fields in which there had been little prior intervention, such as public health, industry, irrigation, public works and education.

To appropriate 'the tools of the modernity', a local elite was trained, notably in the scientific and technological spheres, and educational missions were sent to Europe. As early as the 1820s, institutes of higher education were established in Egypt —School of Engineering, School of Medicine, School of Artillery, School of Arts and Crafts, Naval College, etc. In these schools, Arabic was the language of instruction. In order to facilitate the translation of numerous European textbooks, a school of languages was also created.

In reality, Muhammad Ali's economic and military projects were doomed to fail in the face of European pressures in the 1840s. A local elite of foreign origin formed, which tended to compete with the Egyptian state in its role as a modernizing agent. To this penetration was added increasing national debt, the Anglo-French supervision of state finances and, ultimately, British occupation in 1882.

However, during the nineteenth century, Egyptian experts continued to be trained in the same schools and in the same language of instruction, Arabic, and, until the British take-over in the 1890s, they continued to control sectors such as irrigation, public works, scientific and technological education, and some scientific institutions such as the Cairo Observatory.

Historiography and Representations

What representation can be formed of the first period of the development of modern science and technology in Egypt?

Until very recently there was no specific historiography to address this issue. There were, nevertheless, historians' representations to be found within general works on Egypt or the Arab World or among studies of more specific fields, such as education, publishing, translation or eminent figures of Egyptian political or cultural life. The bulk of these writings do not deal with scientific development per se, and their representation of it is more implicit than explicit. Despite differences in the methodology and perspective of these works, be they by Western orientalists or Egyptian historians, the intellectual background is sufficiently congruent to permit a brief account of the dominant representation of the phenomenon.

For instance, these studies have often stressed the existence, in the nineteenth century, of a binary opposition in Egyptian society between the traditional way of thinking, spread by the traditional institution, the al Azhar, on the one hand, and the scientific and technical way of thinking, generated by institutions linked to the modernization movement on the other hand. According to Tariq al-Bishri:

> Muhammad Ali did not succeed in putting together these two modes of thinking—that of technology and industry and that of al-Azhar. Such a harmonization of thoughts and material evolution could have occurred if Muhammad Ali's experiment had followed its normal course. But the 1840 defeat and the western intervention to end it by imposition on Egypt under

Muhammad Ali's reign, then Said's and Ismail's, the policy of the 'open door', contributed to widen the gap and reinforce dualism, nipping in the bud all attempts at conciliation between traditional thinking and material evolution.[2]

Moreover, the majority of authors stress the lack of scientific content in traditional curricula, while others equate the interest shown in writers such as Euclid and Ptolemy with an interest in Western civilization,[3] as if the works of Euclid and Ptolemy had not been among the most important bases of scientific thinking during the golden age of Islamic civilization.

Thus, since the beginning of this century, and in an almost uniform way, the transfer of European knowledge and methods is represented as the introduction of an exogenous learning in a kind of scientific vacuum, and the asymmetry between this exogenous science and the society of that time compromised the effects of the transfer.

This representation, obviously, was quite useful for the justification of British occupation. Thus, it can be found in use among some British officials of that time, who wrote, for instance, that 'the Arabic language, in its poverty of technical phraseology and its rigidity and complexity of construction, is peculiarly unadapted for scientific studies',[4] or that 'European skill is useless without European authority'.[5] But, this same representation is to be found in the writings of most historians and commentators of this century, right up to the present. To make this point more precisely, I shall take two examples, which I will come back to later.

My first example is that of the career of Mahmud al-Falaki, an Egyptian astronomer and engineer who spent nine years in Europe in the 1850s, and who published several research papers on geomagnetism in various European scientific reviews. From the several biographies

[2] Abd al-Basit Abd al-Muti, Tariq al-Bishri, Ali al-Din Hilal, Husam Isa et Sayyid Yasin, 'L' Etat: transformations et devenir', *Egypte—Recompositions, Peuples Méditerranéens*, nos. 41, 42, October 1987–March 1988, p. 44.

[3] See, for example, Albert Hourani, *Arabic Thought in the Liberal Age, 1798–1939*, Cambridge University Press, Cambridge, 1983, p. 41.

[4] Douglas Dunlop, in *Report by Her Majesty's Agent and Consul-General on the Finances, Administration and Condition of Egypt and the Sudan in 1906*, London, 1907, p. 109.

[5] Alfred Milner, *England in Egypt*, Edward Arnold, London, 1892, p. 286.

of and monograph on Mahmud al-Falaki,[6] it is tempting to imagine him as a modern scientist, trained in the European way, but unable to pursue the research undertaken earlier in Europe, due to the indifference and fickleness of rulers in Egypt. Moreover, he is shown as an 'isolated phenomenon', cut off from what preceded him and with no disciple or follower,[7] and his work seems to display a kind of discontinuity since studies devoted to astronomy and geomagnetism were succeeded by those on meteorology, geography and history in Egypt. Thus, it is represented as the sad story of a scientist who fell victim to unsuitable social conditions.

My second example concerns the language elaborated by Egyptian translators for scientific and technological textbooks. It is worth noting that, until now, historical investigations by specialized lexicographers have systematically ignored these books. Further, from the studies devoted to the translation movement under Muhammad Ali, one suspects that no effort was made to express in Arabic most of the concepts elaborated in Europe since the seventeenth century. Most of the specialized terms would simply have been borrowed from European languages and transliterated,[8] thus showing the fundamentally exogenous nature of imported knowledge.

Here, I do not want to replace one representation by another. For instance, in some ways, Tariq al-Bishri's opinion seems to be fairly realistic. However, it seems to me that, to really understand what was going on in nineteenth-century Egypt, it is necessary not to overemphasize dichotomies such as modern–traditional, or indigenous–exogenous, but to focus more on the relationship between the two terms. In particular, I shall try to show that Mahmud al-Falaki's career was not so unfortunate and that it shows more unity if we

[6] Two of the most extensive studies on this astronomer are Ibrahim Hilmi Abd al-Rahman, 'Mahmud al-Falaki', in *Ismail bi-munasiba murur khamsin aman ala wafatihi*, Matbaa Dar al-Kutub al-Misriyya, Cairo, 1945, pp. 331–47; Ahmad Said al-Dimardash, *Mahmud Hamdi al-Falaki*, al-Dar al-Misriyya li-l-Talif wa-l-Tarjama, Cairo, 1965.

[7] Abd al-Rahman, op. cit., p. 346; al-Dimardash, op. cit., p. 45.

[8] See, for example, Jamal al-Din al-Sayyal, *Tarikh al-tarjama wa al-haraka al-thaqafiyya fi Misr fi asr Muhammad Ali*, Dar al-Fikr al-Arabi, Cairo, 1952, pp. 213 14. This idea is much more explicit in some works resting on al-Sayyal's study; see, for example, Anouar Abdel-Malek, *Idéologie et renaissance nationale; l'Egypte moderne*, Anthropos, Paris, 1969, p. 136.

interpret it in terms of a dialectical relationship between imported and native knowledge.

Moreover, such an analysis will be needed in order to grasp the relationship between scientific development and its historiography.

'Traditional' and 'Modern' Scientific Activities

Survival of Traditional Scientific Activities

Obviously, at the time of the transfer of European knowledge, Egypt was far from a scientific vacuum. In saying this, I do not want to refer to only the scientific inheritance of the golden age of Islamic civilization, but also to the traditional activities which continued to be practised throughout the nineteenth century.

The survival of such activities is not surprising. In fact, the main applications of classical scientific studies, as regards some characteristic features of Islamic civilization—such as computations for the calendar, for the prayer schedules or for the *qibla* (Mecca's direction), computations of inheritances, construction of sundials and astrology—did not disappear with the transfer of the new learning for the simple reason that the transfer was not aimed at such applications.

Thus, traditional training in arithmetic, algebra and astronomy, which probably constituted a very limited part of the studies in the traditional university of al-Azhar, but which contributed to keeping alive some of the scientific writings in use in previous centuries, continued in Egypt till the end of the nineteenth century. Moreover, some of these treatises, such as the famous *al-Risala al-fathiyya fi al-amal al-jaybiyya* of Sibt al-Mardini (1423–1506), which contains the majority of trigonometric terminology, were republished several times in the second half of the century.

The same topics, corresponding to the same applications, prompted some *shaykh* of that period to write original treatises. Thus, apart from religious writings, an author like Khalil al-Azzazi composed a short textbook on sexagesimal computation for his students. He also wrote a short treatise, on the *al-muqantar* quadrant, a traditional instrument of astronomy. Finally, he is the author of many more voluminous treatises on sundials. The case of al-Azzazi is not an isolated one.

This traditional science could have interested only a narrow circle of scholars. However, the preservation of scientific teaching, even in narrow circles, constitutes an important fact. Could 'modern scientists' have ignored these scientific activities? I do not think so, and this is due to at least two reasons.

The first reason arises from rather practical considerations. Since it was decided to construct a modern scientific language in Arabic, resorting to classical texts allowed some lexical difficulties to be resolved more simply. Obviously, from this point of view, the textbooks still in use in al-Azhar were a prime target.

The second reason is closely akin to the first one in its most symbolic aspects. It was in the interest of at least some 'modern' Egyptian scientists to turn towards traditional activities because of the possibilities offered in legitimizing both the new knowledge, as well as their own place in an Egyptian society largely dominated, from a cultural point of view, by the al-Azhar institution.

To develop these two points, I shall very briefly take up the two examples mentioned above.

The Trajectory of Mahmud al-Falaki[9]

A close examination of his scientific trajectory shows that, along with an involvement in the modern sector of the state's activities (he taught astronomy in the Egyptian School of Engineering, managed the Cairo Observatory, and was in charge of preparing the first astronomical map of Egypt, etc.), Mahmud al-Falaki was always interested in fields connected with the environment into which modern science was introduced.

Thus, as early as in the 1840s, before his journey to Europe, he undertook the making of the Egyptian calendar. In particular, he composed almanacs, with lunar and solar ephemeris, for mosques and even astrologers, which show their form as well as their content. Moreover, it was his own identity, in the most prosaic sense of the word, which was at stake here, since it was after the publication of these almanacs that, in the words of a contemporary, 'he received *from the public* the title of al-Falaki, that is the astronomer'.

When he was in Europe, he continued dealing with calendar problems, in parallel with his research on geomagnetism. He then

[9] For further information and references on this point, see my article 'La trajectoire d'un scientifique égyptien au XIX siecle: Mahmud al-Falaki (1815–1885)', in *Entre réforme sociale et mouvement national. Identité et modernisation en Egypte (1882–1962)*, A. Roussillon, ed., CEDEJ, Cairo, 1995, pp. 285–309.

published studies devoted to Islamic and Jewish calendars in orientalist reviews. Later, he went on to publish, in the same reviews, works on metrology in order to highlight the traditional system of measures of Egypt.

But he was also interested in the other great field of application of traditional astronomy, the creation of sundials. Thus, as early as the 1840s, he created three sundials which were placed in the yard of the Egyptian School of Engineering. Later, he conceived of a sundial inscribed on a paraboloid of revolution, and installed it on the western facade of his house in al-Azhar square.

Though nothing remains of these sundials, it is useful to mention a book by his friend, Mustafa Shawqi, devoted to the *al-muqantar* quadrant. This book was published in 1872, when its author was an engineer in the public works department and is written in a modern style, with modern notations and references, but without any historical or ethnological intention. Mustafa Shawqi only wanted to give an account of a contemporary instrument and to explain its use and purpose.

While there is no doubt that the bulk of the scientific activities of nineteenth-century Egypt were aimed at quite different targets, notably the occupations of engineers or physicians, we must consider that some scientists like Mahmud al-Falaki did not wish to ignore the traditional system, and tried to articulate modern scientific learning with categories, objects and customs which existed before the transfer of European knowledge.

Before pondering on the meaning of such developments, let us now move on to my second example.

The Reconstruction of a Scientific Language[10]

Contrary to the widespread representation I mentioned above, the analysis, historical as well as linguistic, of the textbooks in use at the School of Engineering, clearly shows a resolute willingness to inscribe these texts as best as possible in the existing scientific and linguistic fabric.

[10] I am referring here to my two papers, 'Langue scientifique et fait national: le cas de l'Egypte à partir du XIXè siècle', in *Sciences hors d'Occident au XXe siècle, Sciences coloniales: figures et institutions* (vol. 2), P. Petitjean and R. Waast, eds, ORSTOM, Paris, 1996; and 'Les mutations de la langue crite au XIX siecle: le cas des manuels scientifiques et techniques', *Egypte/Monde Arabe*, nos. 27-8, Cairo, 1996.

If we consider the actors of the translation's movement of the 1840s and 1850s—i.e. after the 1840 defeat—we find that, besides the scientific teachers of the School of Engineering and the graduates of the School of Languages, who were in charge of the standardization of vocabulary and syntax, some ulama of al-Azhar were also involved. These ulama were mainly in charge of correction work, but they also had to choose the titles of books and write introductions in the traditional rhyming prose (*saj*), as if it was important to inscribe scientific writing in a tradition to which they appeared to be the last legitimate heirs.

If we now consider the lexis, we notice a willingness to root scientific terminology in that of classical science.[11] It is clear—and this point was corroborated by the actors themselves—that an intensive research programme was undertaken on classical treatises in order to solve lexical difficulties. Which then were the treatises consulted? Where no precise reference was given by the translators, it seems quite obvious that the texts were those still in use in the traditional system, or in use in a recent period, and not texts abandoned long before in traditional studies. Thus the translations contain almost all the classical lexis in arithmetic, geometry, trigonometry and astronomy, whereas some differences can be noticed in the theory of numbers or in statics, which were absent from traditional curricula for a long time. Everything suggests that the traditional system, contemporary and still alive, supplied the references to be employed.

Finally, if we consider modern scientific concepts, unknown to classical science, we notice a clear willingness to find solutions using the mould of the language and avoiding as far as possible borrowings from European languages. These borrowings are in fact very rare, especially in theoretical disciplines such as mathematics, rational mechanics and astronomy. For instance, the translation of Belanger's *Hydraulics* (1848) contains no terms for this discipline which have not been translated with Arabic roots.

End-of-century Shifts

How, then, can the meaning of these scientific activities in nineteenth-century Egypt be grasped? Comparing this scientific development

[11] It is important here to remember that, in the case of Egypt and unlike many other countries, this classical science was very similar to the science cultivated in Europe up to the middle of the seventeenth century.

with the scientific developments in the following century and in analysing the nature of the shift that occurred between the two centuries may yield the solution.

The contrast between the two periods is rather striking. In contrast to the twentieth century, in the nineteenth century, fundamental research on an international level was not really undertaken. In the 1890s, the language of instruction for science and technology suddenly and definitively moved from Arabic to English. In order to demonstrate this contrast, I would like to quote Ali Mustafa Musarrafa, a prominent physicist of the first half of the twentieth century. It is hardly conceivable that what he wrote in 1942 could ever have been written by predecessors of the nineteenth century. Musarrafa writes:

> In Egypt today, we are transmitting the knowledge of others, then we let it drift without link to our past, without contact with our soil; it is a foreign commodity tinged with strangeness, with strangeness in its words and strangeness in its concepts. If one touches on theories, one associates them with foreign words, which one can hardly recognize; if one expresses the concepts, it is by fear-inspiring words which chase thinking away and trouble the imagination.[12]

Comparing the 'modern scientists' of the nineteenth century with their followers of the twentieth century, or juxtaposing, for instance, the trajectories of Mahmud al-Falaki and Ali Mustafa Musarrafa, it appears that Egyptian scientists of the last century were more inclined to link the new learning and traditional activities, as we saw above, or to organize its diffusion in an ill-prepared environment, than try to build a system for fundamental research in Egypt, as Musarrafa did.[13] Without doubt, the difference rests on many factors, notably the evolution of the colonial situation and the emergence of the national movement.

But the link with classical science also seems to have been modified between the two centuries. In the two cases, classical science offers the possibility of legitimization, but in each period, it is for different reasons and through different ways and means. The major difference

[12] Atiya Mustafa Musarrafa, *Ali Mustafa Musarrafa*, Markaz Kutub al-Sarq al-Awsat, Cairo, 1966, pp. 153–4.

[13] On Musarrafa, see Roshdi Rashed, 'Recherche scientifique et modernisation en Egypte: l'exemple de Ali Mustafa Musharafa (1898–1950). Etude d'un type idéal', in *Entre réforme sociale et mouvement national*, pp. 275–84.

is, perhaps, that it was still alive in the nineteenth century, whereas it was only an object of history in the following century. This point is particularly clear when one deals with the publication of classical scientific writings. In the explicit task of finding historical roots for scientific development in Arabic, Musarrafa tried, for example, to promote the publication of the greatest writings of classical scientific literature, such as the writings of al-Biruni, al-Buzjani and al-Battani.[14] He undertook a critical edition of al-Khwarizmi's *Algebra*. At the same time, one of his contemporaries, Mustafa Nazif, undertook the publication of important works on Ibn al-Haytham's *Optics*, whereas, in the previous century, the publication of classical writings was related to al-Azhar's curricula and exclusively undertaken by traditional scholars. They did not present any historical perspective, and only concerned late authors such as Sibt al-Mardini, Baha al-Din al-Amili or Ali ibn Ghanim. It must be kept in mind that a book such as Mustafa Shawqi's did not have any historical intention.

In describing the project visualized for modern science by those who were in charge of it in the nineteenth century, I am tempted to introduce the word 'naturalization'. Rifaa al-Tahtawi, the intellectual guide of the first generation of translators, seems to outline this project in one of his books. The rational sciences, he explains, which now *seem* to be foreign, were cultivated in Egypt until recently, and it is necessary, if not to resurrect their vestiges, at least to inscribe the new knowledge in a kind of historical continuity,[15] that is to reconstruct a scientific language fitting as well as possible with the existing linguistic and scientific fabric, to see to it that modern science should take charge of some of the problems formerly dealt with by classical science, and to be careful not to leave aside traditional society from which legitimacy is being sought, not only for knowledge itself, but also for the very existence of 'modern scientists'. Thus, for al-Tahtawi, as well as for the Egyptian teachers or translators, if modern science appeared foreign, it was more contingent than necessary.

Towards the end of the century, with British occupation and the emergence of reformist movements, the very foundations of this project seem to have been contested. While modern science was still considered

[14] Atiya Mustafa Musarrafa, op. cit., p. 154.

[15] Rifaa al-Tahtawi, *Manahij al-albab al-misriyya fi mabahij al-adab al-asriyya*, 2nd edn, Sharika al-raghaib, Cairo, 1912, p. 373.

to be of Western origin, for an increasing part of Egypt's political and intellectual scene the nature of this origin was seen in quite a different light. Rather than being contingent, this origin could now be seen as essential. More than ever before, the very nature of modern science would have appeared foreign to the Egyptian society, as if the East and West were opposed not only as different geographical places, but also as different 'historical positivities'.[16]

Such a representation, as we saw before, offered arguments for the justification of British occupation. But it also appeared in non-British discourses. The point was to explain the difficulties of Egyptian society in gaining access to modernity. To reach this modernity, the invocations to Western knowledge became recurrent in the discourses of the first Egyptian reformists, who were mainly non-scientists. However, these invocations signalled the difficulties in introducing modern science into Egyptian society, which ought to have been entirely re-envisaged, almost as if the nature of science was fundamentally Western.

Egyptian reformist discourses and aims, which basically constituted a reaction to the colonial situation, began with the feeling that, up to that time, the Egyptian reforms had been initiated in accordance with foreign interests and in foreign ways.[17] The reformist purpose was then to formulate a specifically endogenous 'social-historical project', focusing on the Egyptian identity in order to prepare to receive modernity: language reform, review of the belief system, selection of pertinent traditions.[18] Thus the whole emphasis was laid on the problems posed by the *Western origin* of modernity and its tools.

Perhaps then, it is possible to attempt the hypothesis of a *re-Westernization* of modern science to qualify the shift between the two centuries, a *re-Westernization* of which the change from Arabic to English as the teaching language would only have been a sign. Other

[16] I am quoting here Roshdi Rashed, 'La notion de science occidentale', in *Entre arithmetique et algèbre. Recherches sur l'histoire des mathématiques arabes*, Les Belles Lettres, Paris, 1984, pp. 301–18, especially p. 305.

[17] See Alain Roussillon, who presents the starting point of the reformist discourses as 'underlining the fact that the Other's reformist practices impose upon the Self a prior model and establish the modalities for accessing modernity' (A.Rousillon, 'La modernité disputée: réforme sociale et politique en Egypte', *Entre réforme sociale et mouvement national*, pp. 9–35).

[18] Ibid. See also Roussillon, 'Le partage des savoirs: effet d' antériorité de l'image coloniale en Egypte', *Annales islamologiques*, vol. XXVI, 1992.

investigations will of course be necessary to confirm this hypothesis. It will be necessary to measure precisely the influence of nineteenth-century orientalists on the way Egyptian intellectuals of that period conceptualized their own society. This point is all the more important since these orientalists were, as Roshdi Rashed showed, primarily responsible for the idea of the Western origins of science.[19]

It is worth noting that the emergence of the reformist movement, at the end of the nineteenth century, and the shift of the intellectual landscape which occurred at that time, coincided with the sudden emergence of a new way of visualizing the history of science, that did not exist before, either in nineteenth-century Egypt, or in eighteenth-century Europe.

The dominance of the idea of a Western science that survived the arguments which sustained it originally, will, perhaps, provide insights into not only the prevailing representations of the historians of this century, but also the devaluation of Rifāa al-Tahtawi's naturalization project, both in principle and in its achievements, such as the reconstruction of a scientific language in Arabic.

However, let me return to the end of the century to cite another reason for the failure of this first project of scientific modernization. The situation of the Egyptian group which had taken charge of modern science is of prime importance here. This group, that emerged under Muhammad Ali, saw its position deteriorating through the second half of the century, with the increasing domination of European powers and companies over the economy and management of the country. In the 1890s, it progressively lost its teaching mission. Moreover, the social prospects of a career in engineering were so unattractive that the study of science was discouraged. Finally, when after six years the British were obliged to re-establish free education in the School of Engineering, it was the less wealthy people, having few links with the most prominent intellectuals and politicians, who were recruited.

The intellectual landscape was recast, possibly at the beginning of the twentieth century, as if science had no role to play. Along with a shift towards a colonial management of scientific development, this produced a different perception of the relationship between science and history on the one hand, and Egyptian society on the other.

[19] Roshdi Rashed, 'La notion de science occidentale', pp. 301–18.

Conclusion

The case of the scientific modernization of Egypt in the nineteenth century seems basically to illustrate two things. First, the importance of a closely related study of civilizational factors and scientific content, and second, the exertion of a determining influence, specifically through the lack of consideration of this scientific content, by a vision of history inherited from the nineteenth-century history of science. This vision of history could have led, as for more ancient periods, to the distortion of some historical perspectives, as also to the exertion of influence upon scientific development itself. Furthermore, in the case of Egypt, the whole question of 'modernity' deserves to be reformulated, and perhaps the history of science could be of some help there. Perhaps then we might have an example of what Roshdi Rashed called an *applied* history of science,[20] i.e. a discipline which could step out of its narrow reference framework in order to contribute to an explanation of some of the main questions posed today.

Egyptian projects and achievements of the nineteenth century seem to be characterized by two essential features. First, scientific development which took place in a national and not in a colonial framework and hence involving local cultural specificities which were often erased by colonial discourses in other situations; and second, the existence of a living scientific tradition which played quite a determining role. Thus, at least for one part, the legitimacy of modern science in Egypt seems to have been built on *naturally* fitting the new learning in an *existing* framework, whether related to language or the applications of science.[21]

Colonial or, more generally, Western discourses and actions do not seem to have exerted a determining influence in this case. For example, it is difficult to find any discourse on the universality of science in Egypt before the 1890s. Similarly, strategies of expansion, aimed at establishing a cultural hegemony, did not appear to have

[20] Roshdi Rashed, 'Science classique et science moderne a l'époque de l'expansion de la science européenne', in *Science and Empires*, P. Petitjean, C. Jami and A.M. Moulin, eds, Kluwer Academic Publishers, Dordrecht, 1992, p. 19.

[21] It seems to me that such a phenomenon has little to do with the movement for Islamicization of science observed nowadays, where one is searching to link science and Islamic civilization, but often in an artificial way and without any real historical continuity.

much influence, unlike in other countries or in Egypt post-1890.

If the two characteristic features just mentioned are truly essential for giving an account of Egyptian projects and achievements in the nineteenth century, then one must understand why it would be pointless to build a universal model for the expansion of European science, or to compare situations which cannot be compared. That is permissible only within a historiography that minimizes differences. A comparative perspective will be useful to describe precisely the mechanism of modernization of science, but then, the Egyptian case should only be compared with situations that are analogous on these two counts.

10

Catherine Jami

Joseph Needham and the Historiography of Chinese Mathematics*

The Needham Problem and China's Modern History

As he was building up his monumental work on the history of Chinese science, Joseph Needham raised a crucial question: Why did modern science, the mathematization of hypotheses about Nature, with all its implications for advanced technology, take its meteoric rise *only* in the West at the time of Galileo?[1]

This question, now famous as 'the Needham problem' ('the Needham puzzle'), is not about China. Needham's contribution was not only to state it, but to turn it into a motivation for writing a comparative history of science. Instead of repeating the usual clichés about 'the Greek miracle' and the 'uniqueness' of Western civilization, he challenged them by arguing that China's scientific tradition was by no means inferior to that of pre-Renaissance Europe. China, he went on, might therefore have been a plausible milieu for the

* This was a paper prepared at the Institute for the History of Natural Science, Chinese Academy of Sciences, Beijing. I would like to express my gratitude to my Chinese colleagues for their help. Wang Qianjin provided a wealth of material, and He Shaogeng valuable information on the Chinese translation of Needham's work. Discussions with Liu Dun and Liu Bing helped me interpret the material I was reading. The responsibility for the views in this essay is mine.

[1] Needham (1969), p. 16.

development of modern science. The Needham problem can then be restated as: 'Why did not the scientific revolution take place in China?' He went a step further. Once the Eurocentric bias is discarded, the fact that 'between the first century BC and the fifteenth century AD, Chinese civilisation was much *more* efficient than occidental in applying human natural knowledge to practical human needs', has to be accounted for.[2] Thus formulated, these questions are of obvious relevance to the comparative history of 'sciences and civilizations'.

Needham did not propose a full solution to the why-not problem. He postponed a full treatment of it to Volume VII of *Science and Civilization in China*,[3] which was not completed during his lifetime. However, the much-discussed chapter, 19(k), the last one in the section on mathematics, concludes with a tentative answer to the more specific question of the necessary conditions for the merging of mathematics and knowledge of nature to take place—in China 'there came no vivifying demand from the side of natural science'.[4] He goes on to say, 'apparently a mercantile culture alone [that of Renaissance Europe] was able to do what agrarian bureaucratic civilisation could not—bring to fusion point the formerly separated disciplines of mathematics and nature-knowledge'.[5]

Many Western historians of traditional Chinese science have criticized the way in which Needham stated his central question.[6] They have chosen different approaches, more concerned with a deeper understanding of the patterns of thought than with adding to the list of China's contributions to today's scientific knowledge. What is currently regarded as the most innovative research in the field focuses on what is found in the Chinese scientific tradition, rather than on what it missed. However, assessments by the latter criterion are still somewhat of a reflex. In trying to get rid of it, recent research has been striving for positive characterizations, for alternative questions to those put forward by 'occidentalist' colleagues, usually phrased as: 'Did Chinese science have ...?', or 'How did the Chinese deal with ...?',

[2] Ibid., p. 16.

[3] Needham (1954), hereafter *SCC*.

[4] Among discussions, see for example, Cohen (1973), Koslow (1975), Du and Mei (1982).

[5] *SCC* III, p. 168.

[6] The most influential essay on this issue is Sivin (1982).

and the like, where the suspension points can be replaced by anything that is held to be an essential and universal object, method, or question constructed in writing the history of 'Western' science. Looking for such alternative questions does not imply cultural relativism: the search for universally valid models cannot dispense with a systematic questioning of our research tools.

There is no denying, however, that the list of Chinese contributions started by Needham has served to open the field not only of Chinese science, but of 'non-Western' science, or rather, of the global history of science. This partly explains the enthusiastic reception of his work not only in China, but also in many Third World countries. Moreover, the Needham problem is in tune with a concern that has been shared by intellectuals of these countries, that is understanding the historical process of scientific as well as economic development.

Let us confine ourselves to China in dealing with this wide issue. Since the opium wars, a number of Chinese intellectuals were brought to interpret their country's failure to meet the military and technological challenge posed by the West as a failure of its scientific, or even of its whole intellectual tradition, to provide the necessary framework for modernization. The conviction that science is a driving force of economic development, rather than one of its symptoms or outcomes, is still conspicuous in today's China, where slogans promoting a practice of agriculture based on science and technology, or claiming that 'science is the first production force' can be seen in both rural and urban surroundings.

China's 'failure' since the middle of the nineteenth century has often been contrasted with Japan's 'success', all the more because of the brutal shift in the balance of power between the two countries that ensued. By 1900, Japan had become both a model and a threat for China. Given the abundance of literature retracing Japan's success story and trying to explain how and why the Japanese were clever enough to prefer Western modernity to Chinese tradition, the comparison between the two countries has contributed to inscribing the failure of the whole Chinese intellectual tradition into its recent history. This trend has found an unexpected rationale in the 'Needham problem' and in the broad social and intellectual factors put forward to solve it.

Science and Civilization in China, Section 19[7]

The widely accepted characterization of modern science as 'the mathematization of hypotheses about nature' points to mathematics and its relation to scientific disciplines as crucial in the transition to modern science—a transition which China missed, if we follow Needham. Accordingly, mathematics is the first scientific discipline he analyses:

> Since mathematics and the mathematisation of hypotheses has been the backbone of modern science, it seems proper that this subject should precede all others in our attempt to evaluate Chinese contributions in the many specific sciences and technologies.[8]

In contrast with this statement, the section on mathematics is strikingly short. Those inclined to believe that mathematics is the queen of sciences might feel disappointed by its brevity—168 pages seem very little, amidst the thousands of pages already published in the *SCC* series.[9] One question then is, what elements (if any) do we find in the internal analysis of Chinese mathematics, which forms the core of this section, that can throw light on the 'Needham problem'? In other words, what is the link between his general approach and the way in which he tackles mathematics in *SCC*?

In keeping with *SCC's* general structure, section 19 is mainly divided according to modern topics. It opens with a discussion of the secondary literature, which acknowledges, among others, the contributions of Li Yan[10] and Qian Baozong,[11] who are nowadays regarded as the two founders of the field in China. The second chapter discusses numeral notation, place-value, and zero. It offers an extensive comparative discussion, relying on a wealth of secondary sources, as well as on original Chinese material. Here as in the following chapters, notions, notations

[7] Section 19 was written in collaboration with Wang Ling (1917–94). Wang was one of the scholars who contributed substantially to *SCC*. His Cambridge Ph. D (Wang Ling 1956) was the basis of the section on mathematics.

[8] *SCC* III, p. 1.

[9] The section's first sentence: 'With this section we enter the second half of the present work', suggests that when it was written, Needham still had no idea that the *SCC* series was to expand so impressively.

[10] 1892–1963; transcribed by Needham as Li Nien.

[11] 1892–1974; transcribed by Needham as Chhien Pao-Tsung.

and methods are assessed according to their importance in a retrospective view of the global history of mathematics rather than according to their significance within Chinese mathematics. For example, whereas place-value notation is indeed one of its crucial features, it is difficult to assess to what extent various ways of representing an empty position might have affected this place-value system.

Next comes a survey of the main mathematical works up to 1600. This date is in tune with Needham's much-questioned periodization, according to which 'with the coming of the Jesuits to Beijing at the beginning of the seventeenth century, the period of what may be called "indigenous mathematics" for the purpose of the present book comes to an end'.[12] The works are divided into three periods: from Antiquity to the Three Kingdoms (third century AD), from the Three Kingdoms to the early Song (tenth century AD), and finally, the Song, Yuan, and Ming dynasties. This periodization is slightly different from the received one. Most authors (including those discussed below) propose periodizations which highlight the institutionalization of the study of mathematics by the Tang dynasty (seventh to tenth centuries AD), and single out the Song and Yuan dynasties (tenth to fourteenth centuries) as the golden age of mathematics.

Each of the following six chapters is devoted to one particular mathematical topic: arithmetica and combinatorial analysis, logistic of natural numbers, mechanical aids to calculation, artificial numbers, geometry, and algebra. This tells us what can be found in Chinese mathematics (and what cannot), but it is difficult for the reader to put together a global picture from all this information. For example the central role of counting rods, listed as one of the 'mechanical aids to calculation', in the elaboration of the 'logistic of natural numbers', is not especially brought to light. Also, whereas counting rods have been the instrument on which most of Chinese mathematics (including the famous *tianyuan* algebra) relied until the fourteenth century, much more space is devoted to the abacus, usually considered a more recent device, important mostly for the development of popular arithmetic. There is an extensive discussion on the origin of the abacus and its transmission to other cultures. This is in keeping with the point made above concerning place-value notation and zero. The importance given to the abacus is mainly due to its role in world history.

[12] *SCC* III, p. 52.

A short chapter on 'Influences and Transmissions' lists fourteen 'mathematical ideas [that] seem to have radiated from China southwards and westwards', and five that seem to have travelled in the other direction.[13] Recent research has confirmed Needham's central point. In the study of ancient and medieval science the Eurasian continent should be regarded as a whole rather than as a set of impermeable cultures. On the other hand he provided no evidence of transmission (which is systematically hypothesized whenever he finds similarities), and seems to have underplayed the possibility of independent inventions or discoveries. Many documents brought to light since Section 19 was written rule out the chronology on which he hypothesized transmission.

The section's final chapter, 19(k), is a general discussion of the relationship between mathematics and other sciences, proposing a partial solution to the Needham problem. The dissection of mathematical knowledge undertaken in the previous chapters has prepared the ground. It enables Needham to analyse the ingredients of the Galilean method, and to point to those missing from traditional Chinese mathematics. He does not limit himself to an analysis of social and economic factors. In particular, he points to some specificities of Chinese mathematics, such as the calculating devices on which it relied and the algebraic notations derived from those during the Song dynasty (tenth–thirteenth centuries AD), as possible inhibiting factors.[14] In retrospect, the analysis of the various branches of 'modern' mathematics appears as the best way to display evidence for the argument constructed in 19(k).

Histories of Chinese Mathematics since 1959

In order to assess Needham's influence on later research in the field, I would like to give a brief survey of the main general histories of Chinese mathematics published since 1959. This will not do justice to the wealth of literature on the subject, but will allow us to compare the general approach of various authors.[15]

[13] *SCC* III, pp. 146–8.

[14] In the whole discussion, as in many other passages of *SCC*, a chemical metaphor provides a model to account for the presence or absence of some historical development. The title of one of Needham's books, *The Grand Titration*, clearly indicates that he deliberately used this metaphor.

[15] See for example, the bibliography in Martzloff (1996), and the *Newsletter for the History of Chinese Science* (edited in Taiwan by Huang Yilong since 1990).

A word should be said about Chinese research institutions. The Institute for the History of Natural Science (Ziran kexueshi yanjiusuo) in Beijing specializes in the history of Chinese science. It is a constituent of the Chinese Academy of Sciences (Zhongguo kexue yuan), and was founded in 1957 (under the name of Research Office for the History of China's Science and Technology—Zhongguo keji shi yanjiu shi) and first directed by Li Yan. Apart from that, scholars are scattered in various research institutes and mathematics departments. Recently, two other centres have been producing considerable work. These are the team set up by the late Professor Li Jimin at China North-West University's Mathematics Department, Xi'an, and the Institute of History of Science, founded and led by Professor Li Di, at the Inner Mongolia Normal University, Huhhot.

In 1961, Li Yan and Du Shiran published *Chinese Mathematics—A Concise History*, the only Chinese work on the subject to have been translated into English.[16] Structured mainly according to chronology, it is essentially an abridgement of an earlier work by Li Yan, and does not seem to take into account the questions put forward in Needham's section on mathematics. It covers the whole period of imperial China.

The most widely used reference work on the subject is the volume edited by Qian Baozong first published in 1964, *History of Chinese Mathematics*.[17] It is structured according to chronology (from the origins to the end of the Qing dynasty), and divided into four parts: (1) from Antiquity until the Qin unification (221 BC); (2) from unification until early Tang (756); (3) from early Tang until late Ming (1600); (4) from late Ming until 1911. 221 BC is a date of importance in China's history; 756 is the year when the examination system was set up; 1600 is one year before the Jesuit missionary Matteo Ricci (1552–1610), who translated Euclid's *Elements* into Chinese, settled down in Beijing. The fourth part thus covers the period during which, according to Needham, Chinese mathematics had already been merged into modern, universal mathematics. The book's first part describes the basis of traditional mathematics (numeration, counting rods, etc.), some elements of knowledge that can be traced back to the Warring States period (475–221 BC), and the mathematics of some philosophical

[16] Li and Du (1961, 1988).

[17] Qian Baozong (ed.), *Zhongguo shuxue shi* ('History of Chinese Mathematics'), Beijing, Kexue chubanshe, 1964. I have used the second edition (1981) on which the Japanese translation by Kawahara Hideki (1990) is based.

schools of that period. For the second period, the division is mainly according to authors and works, whereas for the third and fourth, some specific mathematical topics are dealt with in separate chapters. There are enough sources to get a sense of the scholarly community interested in mathematics during both these periods. Each of the four parts opens with a short discussion of the historical and social context. This may have owed more to the political context in which the book was written than to its author's conviction: the Marxist discourse is thus put in a conspicuous position, but at the same time kept well apart from the main body of the text, which is mostly traditional internalist history.

Whereas Needham had drawn heavily on Qian Baozong and Li Yan's earlier research, it is difficult to see any influence he might have had on either of the two works just discussed. The absence of references makes it difficult to tell what, if any, secondary material has been used in them. They seem to have been meant as textbooks rather than reference tools for research.[18] One might also wonder whether Li Yan and Qian Baozong had actually had a chance to read Section 19 of *SCC* in the early 1960s.

Hardly any research was carried out in China during the Cultural Revolution. Hence there is a gap of about ten years in the literature. In the meantime, a small book by the great historian of Chinese astronomy Yabuuti Kiyosi appeared in Japanese. This popular book is structured according to chronology, closing with an epilogue which briefly evokes the 'second entry' of Western mathematics into China after the opium wars. Unlike the two works just mentioned, it does combine internal analysis with discussion of social factors. Yabuuti seems to have been influenced by Marxist historiography, which was quite an important trend among Japanese historians after the Second World War. However, contrary to Needham, he does not describe Chinese society as a 'feudal bureaucracy'. Discussing the social background of Song mathematics, he designates the period 'early modernity'. He quotes Li Yan and Qian Baozong, as well as several Japanese authors, a number of times. Needham's name, on the other hand, appears only once (in fact the reference is to Wang Ling who co-authored Section 19), when Yabuuti discusses the origin of the use of a circle to represent zero in the place-value notation of numbers.

[18] The educational purpose is stated explicitly in Qian Baozong (1964).

There is no doubt, however, that he had read *SCC.* He often addresses the same issues, but from a different viewpoint. In many passages, he discusses the development that some of the mathematical knowledge transmitted from China to Japan underwent. Moreover, in his introduction, he underlines the necessity of studying Chinese mathematics as one important source of Japanese mathematics. This emphasis on what the Japanese scientific tradition owes to China has political implications. Disregard for China had been the dominant attitude during Japan's militarist period, but a number of scholars have striven to make its culture and history better known since the end of the Second World War. A comparison with Needham's approach to Chinese science, and with Li Yan's and Qian Baozong's works brings to light the close link between the standpoints of various scholars, their motivations and their approach in dealing with the same topic.

In 1981, there appeared a revised edition of Qian Baozong's work. Three years later, Li Di published *A Short History of Chinese Mathematics.*[19] Its structure is mainly chronological and has two peculiarities—its mention of 'national minorities' and its discussion of the Republican period. The 'national minorities' contributions to mathematics consist first and foremost in Minggantu's work on power series expansions of trigonometric functions. Minggantu was an eighteenth-century imperial astronomer, a Mongol of the Banners. His scientific work, fully integrated in Chinese institutions, bears no specific link to his descent. Some other Manchu and Mongolian scholars are mentioned. Tibetan and Dai (Yunnan province) calendars and calculating devices are also discussed. The paragraph on Tibet opens with the following statement: 'Tibetans are one of our country's minority nationalities who have a long history and a relatively high cultural level.' This, as well as the characterization of Minggantu as a Mongolian astronomer, points to the ambiguous implications of research on 'national minorities' undertaken in the People's Republic of China. It is a way of acknowledging that minorities have there own culture, while reasserting that they are an integral part of the Chinese empire in history, and therefore of the People's Republic of China in our times. In the past few years, Li Di (whose career was spent in Inner Mongolia), has edited a series entitled *Research on the History of Science and Technology of the Minority Nationalities.*[20] Studies published

[19]Li Di (1984).

[20]*Shaoshu Minzu Keji Shi Yanjiu.* The seventh volume was published in 1992.

in it are a sample of what is an important trend of the history of science in China today.

The book's last chapter depicts the beginnings of modern mathematics in China, with the setting up of institutions and the role played by those who returned after completing higher education in foreign universities. It concludes with an outline of research on the history of Chinese mathematics, from Ruan Yuan (1764–1849) to Li Yan and Qian Baozong (at the very end comes one page on the world history of mathematics). This last chapter gives the reader a sense of the continuity between the last years of the Manchu dynasty and the Republic. Needham's name does not appear in the book.

A volume compiled by a team of eleven historians, *A Concise History of Chinese Mathematics*,[21] which appeared in 1986, is organized along the same lines, and goes a step further towards the contemporary period. The last chapter deals with the mathematical research done in the People's Republic before the Cultural Revolution. Needham is quoted once, in reference to his statement that 'between the third and the thirteenth centuries AD, China maintained a level of scientific knowledge with which Europe was too far behind to compete'.[22]

In contrast to these works, Shen Kangshen's *Introduction to Chinese Mathematics*, also published in 1986, is structured according to modern mathematical topics, except for a short presentation of various mathematical traditions (European, Chinese, Indian, Islamic, and Japanese) at the beginning. This book is obviously oriented towards mathematical education, as each chapter concludes with a series of related problems posed to the reader.

In 1987, just before the English translation of Li and Du's book appeared, the only comprehensive work on Chinese mathematics written in a Western language since Needham's was published by Jean-Claude Martzloff.[23] It is divided into two parts, 'Context' and 'Content'. The author discusses historiography, and is rather critical of Needham's approach, and especially of his dating the end of 'indigenous mathematics'. However, the work also concentrates mostly on pre-1600 mathematics.

[21] *Zhongwai shuxue Jianshi Bianxie Zu*, 1986.

[22] Op.cit., p. 70.

[23] Martzloff (1987, 1996). The only previous work in a Western language on Chinese mathematics is Mikami (1913).

Two works by Chinese historians of the younger generation (born around 1949) have appeared in the 1990s. Li Zhaohua's *History of Chinese Mathematics* is structured strictly according to chronology and to various authors and works.[24] The words used in the titles of the various parts (sprouts, development, flourish, decline, revival) strongly convey an organic metaphor, suggesting that the historical development of mathematics is the result of an internal dynamics.

Liu Dun's *How Great the Discourses on Mathematics Are* has some remarkable features.[25] It is divided into four parts, the last three of which discuss mathematical topics that can be rendered by 'arithmetic', 'algebra', and 'geometry' respectively, while the categories thus put together take into account those of the *Nine Chapters on the Mathematical Art* (*Jiuzhang suanshu*, first century AD), the founding classic of China's mathematical tradition. In other words, the author has attempted precisely what Needham has been criticized for not doing—adopting the categories of Chinese scientific knowledge rather than modern ones. The underlying issue is the possibility of matching traditional and modern categories.

The first part opens with a list of authors and works, then describes the numeration and calculating devices, and finally discusses 'mathematics and society'. In this chapter, the historical evolution of mathematics is analysed in terms of the society's structure and of some important intellectual trends. The author goes further and refines some of Needham's ideas, distinguishing between several historical periods, and proposes a positive characterization of what in *SCC* appears as an absence. However, this is not done with explicit reference to Needham. Rather than taking up his ideas, it seems that Liu Dun takes into account recent research on the social history of science.

Several works on the history of mathematics written since *SCC* appear to have taken China into account. An example is Yushkevitch's *Geschichte der Mathematik im Mittelalter*.[26] Here we have limited our discussion to authors who worked directly on primary sources.

One conclusion of this quick survey is that historians of Chinese mathematics do not consider Needham's work on the subject as a

[24] Li Zhaohua (1994).

[25] Liu Dun (1993), 2nd edn, 1996. The title is a quotation from Confucius' *Analects*.

[26] Yushkevitch (1964).

central reference, and that none of them finds it relevant to discuss the 'Needham problem' in the framework of a historical analysis of Chinese mathematics. However, although it is not apparent from histories of mathematics, Needham's work did have some influence on Chinese historians of science.

The Reception of Needham's Work in China

SCC was first translated into Chinese by Taiwanese scholars. The first volume appeared in 1972.[27] This seems to have stimulated research in the field, which has undergone a remarkable development since the 1980s.

In mainland China, the first volume of *SCC's* Chinese translation appeared in 1975, under the title *History of Chinese Science and Technology*.[28] Section 19, translated by Mei Rongzhao, was published as a separate volume in 1978. It has the same short preface as the first volume, presenting the author and the work, and stating that 'concerning periodisation and historical assessment, some [of Needham's] viewpoints deserve discussion'. Readers are invited to carry out this discussion from a Marxist viewpoint.[29] The translation project was terminated at the end of the Cultural Revolution, apparently because of its coordinator's political stand. A revised edition of this translation is being done under the supervision of Lu Jiaxi.[30] Besides, a number of books such as translations of essays and biographies have contributed to introducing Needham's work into China.[31] All these have given rise to many discussions and comments; the literature on the subject is much too vast to be listed here. I will limit myself to a few articles, mostly relevant to mathematics and to the 'Needham problem'.

Before discussing them, I should like to emphasize that, on the whole, the reception has been extremely favourable. Needham's general

[27] *Zhongguo zhi kexue yu wenming,* Taipei, Shangwu yinshuguan, 1972.

[28] *Zhongguo kexue jishu shi.* This is the Chinese title found in each volume of *SCC.*

[29] One sentence which refers to the specific content of vol. 1 has been left out in vol. 3.

[30] The first volume appeared in 1990.

[31] See, for instance, Wang Guochong (1992).

views, both his high appraisal of the Chinese scientific tradition and his Marxist analysis of scientific development, could not but meet with a positive echo in the People's Republic of China, where Needham himself was extremely popular. Witness to this are the many visits he made there and the honours conferred on him by various Chinese institutions.[32] Thus, in 1994, he was one of the first foreigners elected academician of the Chinese Academy of Sciences, and one of the very few among them who were not of Chinese descent. If one considers these honours together with the seemingly little influence on the historiography of science evidenced in the case of mathematics, one is tempted to conclude that the main value of Needham's enterprise from the viewpoint dominant in China is that, as a distinguished member of the Western scientific establishment, he granted recognition and legitimacy to China's scientific tradition.

A review of the translation of Section 19 by Du Shiran and Mei Rongzhao appeared in 1982.[33] Their overall assessment is quite positive. After rectifying a few dates and details, they devote most of the essay to the 'Needham problem', which they subdivide into three different questions, all of which are stated in *SCC:* (1) Why did Chinese mathematics experience no further progress after the Song-Yuan period (tenth to fourteenth centuries)? (2) Why did traditional Chinese mathematics not develop into modern mathematics? (3) Why did modern science develop only in Europe and not in China?[34]

The first of these three questions, as restated by them, is a *locus classicus* of research on the history of Chinese mathematics: why did Chinese mathematics decline after the Song-Yuan period? Interestingly, Du and Mei use this decline to argue for the relative autonomy of mathematics as a discipline vis-à-vis social factors. According to them, the fact that mathematics underwent a retrogression at a time when 'sprouts of capitalism' were coming out illustrates the importance of internal factors. This argument is quite interesting, because the 'decline' of mathematics and more generally of science at that time is usually blamed on either the Mongolian invaders' lower cultural level[35] (an argument which overlooks the formidable and rather well-documented

[32] See, for example, Wang Guochong (1992), especially pp. 555–61.

[33] Du Shiran and Mei Rongzhao (1982).

[34] Ibid., p. 3.

[35] Dong Guangbi 1990, p. 45.

circulation of knowledge that took place under the *Pax Mongolica*), or on the conservative attitude of the Ming dynasty (a point made by Needham).

Concerning the internal weaknesses of Chinese mathematics, pointed out by Needham to explain the decline, the reviewers find his judgment excessive. Taking up his line of analysis, they identify all the necessary prerequisites for the development of calculus, and, following a deterministic pattern, see the failure of the 'sprouts of capitalism' to fully blossom as the ultimate cause of the failure of calculus (and therefore of modern mathematics) to appear in China. They also point to astronomy as one important case in which mathematics and nature-knowledge did successfully merge in the Chinese scientific tradition, thus questioning the general absence diagnosed by Needham.

Other discussions of the 'Needham problem' in more recent Chinese literature which I have consulted do not focus on mathematics. Thus, among the many discussions of China's 'scientific backwardness' since the sixteenth century, is Lin Wenzhao's[36] assertion of the links between science and society, laying the blame for this backwardness on the obstacle posed by feudal economy to the development of capitalism.[37] This view, expressed in 1985, then allowed for some optimism as to the development of science in a socialist society.[38]

The relevance of Needham's approach is not agreed upon by all. Thus some scholars have pointed to contradictions in the statement of the 'Needham problem', and to its problematic definition of modern science.[39]

Since the beginning of the 1990s, some interesting papers on the topic have been published in the *Journal of Dialectics of Nature*. As examples, let us discuss three of them, in chronological order.

As the title indicates, in 'Transplantation, Merging, or Revolution? A Review of the Modernisation of Traditional Chinese Science',[40]

[36] It was originally read at a conference entitled 'Causes of Backwardness of Chinese Science in the Modern Period', in Chengdu (1982).

[37] The term 'feudal' is used uncritically by most Chinese historians to characterize China for two millennia.

[38] Lin Wenzhao (1985).

[39] See, for example, Zhang Binglun and Xu Fei (1993).

[40] Dong Guangbi (1990).

Dong Guangbi chooses to tackle what did happen to Chinese science in the early modern period, rather than what it did not produce, and tries to characterize it. He chooses to call it a 'scientific revolution', and redefines the notion more broadly than the Kuhnian sense used by Sivin.[41] To this end, he gives a general outline of the relationship between traditional Chinese science and its philosophical and political context, and points to a dual origin of science's modernization: the philosophical trend of 'concrete studies' (*shixue*, which developed in the late Ming) on the one hand, and 'Western learning' (*xixue* as introduced by Jesuits from 1600 on) on the other hand. Unlike most authors dealing with this period, he brings together not only mathematics and astronomy, but also medicine, which, he argues, underwent some institutionalization in the late Ming, and technology. The two sources of modernization, he argues, were brought together in a process of integration (*huitong*). Even the mid-Qing trend of 'Revival of Antiquity' (*fugu*), gave rise to some innovations. It is only after the opium wars that this process was replaced by a massive transplantation. The argument is sketchy, but it emphasizes the 'endogenous factors' of the transformation which Chinese science underwent in the late imperial period.

Another 'positive' study of Chinese science in the same period is that of Hu Yongquan, who grounds his argument in a quantitative study of scientific literature between 1550 and 1750.[42] He uses the *Siku quanshu* corpus, the library commissioned by the Qianlong Emperor in the second half of the eighteenth century. He discusses mathematics and astronomy, medicine and agronomy separately, points to the importance of regional differences (with the well-known predominance of Jiangnan), and suggests that the social upheaval that accompanied the change of dynasty affected scholarly science more than it did popular knowledge. While noting the correlation between scientific and economic growth, he does not establish a strong causality between the two. He points to the distinction that should be made between the intellectual impact of 'Western learning', which was significant, and their influence on the scientific community, which, he argues, was minor. Finally, he emphasizes the importance of 'concrete studies' as the general framework in which this scientific growth makes sense.

[41] Sivin (1982).

[42] Hu Yongquan (1991).

A more recent paper by Xiao Wei exemplifies attempts to apply sociological theses like those of Merton and Ben-David to the analysis of Chinese science in the early modern period.[43] Emphasizing the significance of professionalization, and of the scientific community's role and status within a society, he falls back on the usual comparison with early modern Europe, which leads him to the stereotyped description of the Chinese case in terms of absence and failure.

The common feature of these three studies is that they propose alternatives to the repetitive Marxist discourse predominant until the 1980s in Chinese history of science. While the first two papers avoid value-loaded judgments, the last emphasizes the importance of social and institutional analysis. In different ways, all three point to the value of the 'Needham problem' as a source of inspiration, if not as a question that can yield a straightforward answer.

Concluding Remarks

The relationship between Needham's work and the Chinese historiography of mathematics in the past forty years appears to be more a convergence of interests and methods than a direct influence. Let me draw attention to some of the points of convergence.

The 'Needham problem' matches the endeavour of many Chinese intellectuals to understand, and thereby influence, the economic as well as scientific evolution of their country. Even as the notion of a general failure of the Chinese tradition is subsiding, with the present economic growth, Needham's high appraisal of the Chinese scientific tradition is still equated with Western acknowledgement of the 'scientificity' of Chinese civilization. This partly explains why his work has been furthered (and sometimes caricatured) by a mass of literature listing China's 'firsts'. Nationalist feelings often lay behind such endeavours.[44]

As regards methodology, Needham's systematic quest for priority meets a tradition of Chinese scholarship mainly set up during the last dynasty, namely textual criticism (*kaozhengxue*). This mainly philological approach to ancient texts first aimed at challenging the authority of

[43] Xiao Wei (1994).

[44] This approach is also found in Western literature. It is epitomized in Temple's popularization work (Temple 1986).

some classics, and of Song dynasty commentaries. The continuation of this method to the present day has been made possible by its 'factual', 'objective' character, which has made it relatively safe and compatible with the dominant discourse.

Another feature of Needham's work, his 'vertical approach' (that is his separate treatment for each scientific discipline)[45] corresponds to the dominant approach of PRC scholarship to this day, where most research specializes in the history of one particular scientific field rather than being cross-disciplinary. This is changing slowly, following the general evolution of history of science.

On the question of periodization, Needham and the historians of mathematics whose work has been discussed in this essay agree in viewing the Song and Yuan dynasties as the golden age of Chinese mathematics. Whereas Needham does not deal with post-1600 developments, Chinese authors often emphasize their limitations. The continual reference to modern mathematics as a yardstick, a reference still shared by most historians of mathematics around the world, imposes limitations on their analysis of mathematical texts as historical material. Giving up the systematic use of progress (a simplistic and linear notion insufficiently questioned by historians of mathematics) as the sole criterion might yield more fruitful assessments. This is already apparent from some recent research. It will probably take some time to redraw the global picture. However, it is a necessary step if one aims at understanding the viewpoints and concerns of those who wrote the texts, and of those who read and used them in different periods.

The above discussion is by no means exhaustive. The material under scrutiny is only a small fraction of what is relevant to the history of science. Needham's work has been discussed, in China as elsewhere, by scholars of other disciplines—history, philosophy, history of religions, comparative culture studies.[46] In this respect, Chinese scholarship does not differ from Western scholarship. It has acknowledged the importance and significance of his work beyond the discipline of history of science.

[45] Ho Peng Yoke (1977), p. 9.

[46] His assessment of the respective roles of Taoism and Confucianism, for example, has been much debated.

References

Cohen, Robert S., 1973, 'The Problem of 19(k)', *Journal of Chinese Philosophy* 1: 103–17.

Dong Guangbi, 1990, 'Yizhi, ronghe, haishi geming? Lun Zhongguo chuantong kexue de jindaihua' (Transplantation, Merging, or Revolution? A Review of the Modernisation of Traditional Chinese Science), *Ziran bianzhengfa tongxun* 12, 1: 43–9.

Du Shiran and Mei Rongzhao, 1982, 'Ping Li Yuesi zhu, Zhongguo kexue jishu shi yi shi de shuxue bufen', (Review of Needham's *SCC* section on mathematics), *Kejishi wenji* 8: 1–9.

Ho Peng Yoke, 1977, *Modern Scholarship on the History of Chinese Astronomy*, Canberra.

Hu Yongquan, 1991, 'Ming Qing zhiji Zhongguo kexue de zengzhang: dingliang miaoshu he dongyin fenxi' (The Growth of Science in China During the Ming and Qing Dynasties: Quantitative Description and Analysis of its Causes), *Ziran bianzhengfa tongxun* 13, 6: 25–33.

Huang Yilong, ed., 1990, *Zhongguo kejishi tongxun* (Biannual Newsletter for the History of Chinese Science and Technology).

Koslow, A., 1975, 'More on 19(k)', *Journal of Chinese Philosophy* 2: 181–96.

Li Di, 1984, *Zhongguo shuxue shi jianbian* (A Short History of Chinese Mathematics), Shenyang, Liaoning renmin chubanshe.

——— ed., *Shaoshu minzu kejishi yanjiu* (Studies on the History of Science and Technology of Minority Nationalities), 7 vols published, Nei Menggu renmin chubanshe.

Li Yan and Du Shiran, 1961, *Zhongguo shuxue jianshi*, Beijing, Zhonghua shuju, *Chinese Mathematics: A Concise History*, Oxford, OUP, 1988.

Li Zhaohua, 1994, *Zhongguo shuxue shi* (History of Chinese Mathematics), Taipei, Wenjin chuban.

Lin Wenzhao, 1985, 'Lun jindai kexue mei you zai Zhongguo chansheng de yuanyin' (On the Causes that Prevented the Development of Modern Science in China), *Kejishi lunwen* 14: 14–29.

Liu Dun, 1993, *Da zai yan shu* (How Great the Discourses on Mathematics are), Shenyang, Liaoning jiaoyu chubanshe, 2nd edn, 1995.

Martzloff, Jean-Claude, 1987, *Histoire des mathématiques chinoises*, Paris, Masson

——— 1996, *A History of Chinese Mathematics* (English translation of a revised version of Martzloff [1987]), Berlin, Springer Verlag.

Mikami Yoshio, 1913, *The Development of Mathematics in China and Japan*, Leipzig, G.B. Teubner.

Needham, Joseph, 1954-, *Science and Civilization in China*, Cambridge, Cambridge University Press.

———1969, *The Grand Titration: Science and Society in China and the West*, Toronto, University of Toronto Press.

Qian Baozong (ed.), 1964, *Zhongguo shuxue shi* (History of Chinese Mathematics), Beijing, Kexue chubanshe, 2nd edn, 1981. Japanese translation by Kawahara Hideki based on this 2nd edn, Tokyo, Misuzu Shobo, 1990.

Shen Kangshen, 1986, *Zhongsuan linglun* (Introduction to Chinese Mathematics), Shanghai, Shanghai jiaoyu chubanshe.

Sivin, Nathan, 1985, 'Why the Scientific Revolution Did not Take Place in China—Or Didn't It?', *Chinese Science* 5: 45–66.

Temple, Robert K., 1986, *China Land of Discovery and Invention*, Wellingborough, Patrick Stephens.

Wang Guochong, 1992, *Li Yuesi yu Zhongguo* (Joseph Needham and China), Shanghai, Shanghai kexue puji chubanshe.

Wang Ling, 1956, 'The "Chiu Chang Suan Shu" and the History of Chinese Mathematics in the Han Dynasty', Ph. D thesis, Cambridge University.

Xiao Wei, 1994, 'Tizhihua' yu qi fanli' (Institutionalization and its Opposite Case), *Ziran bianzhengfa tongxun* 16, 5: 54–62.

Yabuuti Kiyosi, 1974, *Chûgoku no sûgaku* (Chinese Mathematics), Tokyo, Iwanami shoten.

Yushkevitch, Adolf P., 1964, *Geschichte der Mathematik im Mittelalter*, Leipzig, G.B. Teubner.

Zhang Binglun and Xu Fei, 1993, 'Li Yuesi nanti de luoji maodun ji kexue jiazhi' (Logical Contradiction in and Scientific Assessment of the Needham Problem), in *Jinian Cheng Dawei, Mei Wending, Dai Zhen, Wang Lai Guoji xueshu hui taolun wenji*, Hefei: 60–71.

Zhongwai shuxue jianshi bianxie zu, 1986, *Zhongguo shuxue jianshi*, Jinan, Shandong jiaoyu chubanshe.

11

Dhruv Raina and S. Irfan Habib

The Missing Picture: The Non-emergence of a Needhamian History of Sciences of India

Almost four decades after Needham's *Science and Civilization in China* first appeared, it is fitting to ask a question that has a Needhamian twist to it: Why has an Indian equivalent not been produced? In seeking an answer to the question, this essay proposes, rather provocatively, that an Indian variant of the Needhamian oeuvre cannot be instituted. In hoping to answer one question, we raise a counterfactual. The proposition that a Needhamian Science and Civilization in Ancient India cannot be written may be falsified. However, the proposition could still be saved, for if *Science and Civilization* is a cultural monument of a specific socio-political milieu; then the meanings associated with science and civilization no longer obtain in the same way.

This essay is divided into four parts. The first very fleetingly outlines the social history of the history of sciences in India. The second raises some questions and problems in Needham's historiography that are problematic for the history of science in India. In the third, the conditions contributing to the non-emergence of the Indian equivalent of *Science and Civilization in China* (hereafter *SCC*) are discussed. A 'New Big Picture' (hereafter NBP) of the history of science has emerged from sociological approaches to the history of science. What this picture holds for the history of science in India is then considered.

Stages in the Institutionalization of the History of Science

Long before Whewell's *History of the Inductive Sciences* (Whewell 1857) appeared, British and French orientalists had begun documenting the traditional knowledge forms of India within the idiom of modernity (Lettres 1810, Colebroke 1873, Dharampal 1971).[1] This documentation did not merely cover the mixed sciences, of interest to the Whewellian historian of science, but also consisted of detailed studies on indigenous systems of jurisprudence, philosophical thought, including redactions of the schools of atomism (such as the Vaisesika), the evolution of logic (Nyaya), as well as comparative linguistics and philology. Naturally, orientalist scholarship prior to the *History of the Inductive Sciences* could not have been structured around the key themata that were ingrained in history of sciences writing after Whewell's paradigmatic opus.

History of science writing of the last three decades of the nineteenth century inasmuch as it addressed the non-West, celebrates the emergence of modern science in Europe, almost bemoaning its non-emergence in the non-West. For example, the influence of the orientalist, Ernest Renan on the French chemist Marcelin Berthelot was decisive (Rosu 1986), and the latter's *Histoire de l'Alchimie* (Berthelot 1885), was presentist in that it hoped to recover from alchemy the traces of a positivist science, homologous to the chemistry of the turn of the century.[2] This was Whigish in the sense that alchemy was for Berthelot nothing more than a low rate chemistry (Guillemain 1992, Bensaude-Vincent 1992). In any case, Berthelot himself was associated with two projects on the history of sciences related to India.

In the first instance, he and Ernest Renan were partially responsible for the inauguration of the French tradition of Indology, in particular those bearing on the traditional medical sciences of India (Rosu 1986). There is an entire lineage of scholars extending from Cordier to Filliozat (Filliozat 1964, 1974) (notable differences between their

[1] For a collection of British orientalist writings on the sciences of India, see Dharampal (1971).

[2] In an interesting paper Brush points out that while scientists are more likely to judge the history of science in terms of present-day science, a presentist predisposition to history is not identical with Whig history (Brush 1995, 21–220).

epistemological commitments notwithstanding). On the other hand the founder of the Indian school of chemistry, Prafulla Chandra Ray, was formally ordained in the discipline of the history of science through his association with Berthelot (Ray 1932). Acknowledging his indebtedness to the latter, Ray in his subsequent writing was to break away from the central axioms of the orientalist historiography of science, in the same breath laying the foundations of the social history of sciences in India (Raina 1997).

The break with orientalism was facilitated by a trend in economist thinking, gaining ground in India during the last decades of the nineteenth century, that sought to explain India's underdevelopment in economic terms and shelved the essentialist interpretive modalities of the orientalists (Pandey 1992). A reflection of this strain is found in Ray's externalist history of chemistry, as well as a large number of other writings (Ray 1918). At about the same time, drawing upon the exemplar of the inductive sciences and the positivist conception of science, B.N. Seal was to provide the complementary internalist account in his work *The Positive Sciences of the Ancient Hindus* (Seal 1915). Both these books must be seen as a response to the exaggerated depiction of India as a spiritual civilization, devoid of a modern scientific or industrial tradition. It is not coincidental that these complementary works were published around the time that the nationalist movement gathered momentum, acquired a cohesive ideology, and important cultural figures like Seal and Ray advocated the critical assimilation of modern science (Raina and Habib 1995).

This stage in the history of science studies in India coincided with the period when Indians commenced lobbying for the institutionalization of postgraduate teaching and research within the university system. In addition, there were independent efforts by the Indian Association for the Cultivation of Sciences devoted single-mindedly to the inauguration of a research system, followed almost three decades later by the founding of separate institutions under national control and management by the National Council of Education (ibid.). At about the same time, Jamsetji Tata strove to found the Indian Institute of Science, Bangalore (Subbarayappa 1992) as a postgraduate research institution, through a tripartite agreement with Princely Mysore, and the Government of India.

The major concerns of the history of sciences of India during this period were: (1) establishing the existence, in the Indian subcontinent, of proto-sciences, or elements of science based on empirical procedures, a methodology of validation and logical reasoning (Ray 1918), (2) explaining the inability of these systems to develop further, in terms of either the separation of theoretical knowledge from the crafts and practices, or the suppression of the sciences themselves, (3) deriving a principle of quasi-continuity from the traditional sciences to modern science, that facilitates the legitimation and adoption of modern science. Further, in these histories of science, science is represented as 'transcendental', and the history of science is visualized as the progress of scientific knowledge culminating in its modern incarnation.[3]

Between this stage and the next, was an intermediary one, marked not so much by the institutionalization of the discipline, as by the notable studies of A.N. Singh and B. Datta, undertaken between 1927 and 1940, on the history of mathematics in India from the ancient until the medieval period (Singh 1936, 1936a; Datta 1929, Datta 1933). These studies belong entirely to the internalist genre of the history of mathematics. The next stage in the social history of the history of sciences, in the decade beginning 1950, was marked by both a cognitive and an intellectual watershed. This was the first decade of modern science in independent India. This high-tide of the Nehruvian era witnessed the proliferation of industrial research laboratories under the aegis of the CSIR, the emergence of the Tata Institute of Fundamental Research, Bombay, as the centre for advanced scientific research in the Third World, and the inauguration of India's atomic energy programme. In this decade, when the frame of Big Science was being introduced in India, two scientific academies contended for initiating and institutionalizing the study of the history of science in India—the National Institute of Sciences of India, now the Indian National Science Academy, and the Council of Scientific and Industrial Research (Bose 1956, Bose 1963).

The former drew its sustenance from the initiators of the discipline in India. The programme suggested by the latter was inspired by Needham's *Science and Civilization.* Both factions saw the Needhamian project as worthy of emulation in India. In the newly independent

[3] The parallel with the prehistory of STS is most striking. Fuller points out that during this phase, in Europe, the theorists of science and its history 'sought ways of expressing scientific claims that would move appropriately educated audiences to support emergent scientific institutions over their competitors for cognitive authority ...' (Fuller 1993, p. 4).

republic, a new set of questions acquired currency for historians of science, and, more importantly, older questions were rephrased. The historians in both cases belonged to the scientific community, though D.M. Bose of the National Institute of Sciences of India had suggested involving the Asiatic Society and Bhandarkar Oriental Research Institute, Pune, for studies on ancient India. The dominant vision for the theorists of science and society during this period was Bernalist-Nehruvian. An outline of the elements of this vision will be followed by a brief discussion of Needham's historiography that echoed in the projects of historians of science in India.

The 'Grand Question' and the First Decentring

By the end of, what Hobsbawm (1993) calls, 'the Age of Catastrophe', a number of scientists and scientists-turned-historians of science attempted to project science as a cultural activity that enjoined humanity—this moral vision was to fill the vacuum left by the two wars. From Sarton (Pyenson 1993) to Bernal and Needham an attempt was underway to redefine humanism, wherein science would provide a cultural affirmative for the West ruined by two world wars. This 'moral' vision of science was to strike a sympathetic chord within the nebulous scientific communities of newly independent nations like India where the bonds between state and science were mutually reinforcing (Salomon and Lebeau 1993).[4]

Within this frame, science came to be characterized in three different ways (Cunningham and Williams 1993). The first was 'philosophical', where science was characterized by a method that generated knowledge structured by 'causal laws, preferably mathematical, as in the physical sciences' (ibid.). This perspective is evident in the programme of the Vienna Circle. The second characteriszation was a 'moral' one, where science embodied the ideals of freedom and rationality, of truth and goodness, and its internalization within the heart of contemporary culture would ensure social and material progress, and possibly end all suffering and disease. The third, and possibly the most important, from the point of view of the study of

[4] In the writings of Sarton, Needham and others a non-hegemonic idea of science, for the first time begins to appear in the historical literature. Mendelsohn suggests that this idea was '... overlooked by the enthusiasts of the European image of the scientific community' (Mendelsohn 1995, p. 63).

science and civilizations, was the ascription to science of the quality of a cultural universal (ibid., p. 411). This meant that across all civilizations, scientific activity was embodied in the 'grand universal desire to understand the world'. This activity embraced all of humanity, and its origins were traceable to the age of the prehistoric megaliths (ibid., p. 408). This vision of science, in different combinations of these characterizations, and best epitomized in Vannevar Bush's attempt to situate science at the centre of modern culture (Holton 1993) echoed in a multitude of nations entranced by the rhetorical promise of science.

The Needhamian project proved alluring for Third World scholars since it provided a counterpoint to the Eurocentric discourse on the history of sciences. Cunningham and Williams speak of the decentring of the Old Big Picture in the light of contemporary researches in the sociology of science. However, it may be argued that the first decentring of this discourse, within the Western tradition, was to occur in the work of Needham.[5] From the postulate that science is a cultural universal, he proceeds to examine various 'other' ways of pursuing the study of the history of science in Chinese civilization. China for him is in some absolute sense the civilizational other of Europe, since India and the Arab world had been trading in goods and ideas with Europe over centuries. The China project offered him the possibility of departing from the European way of examining the history of scientific knowledge.

However, both Needham's supporters and critics have shown that he could never entirely disentangle himself from this picture (Cohen 1994). The centrality of modern science in contemporary culture, and the founding impulse of the scientific revolution of the seventeenth century that radically transformed European society and culture set the terrain of his investigations. Though the scientific revolution was central to the subsequent history of science, for Needham this was not the whole story: '... to tell this part alone is to be deeply unjust to the other civilizations. And unjust here means both untrue and unfriendly ...' (Needham 1973, p. 1). The famous river metaphor invoked by Needham suggests that while there existed several streams

[5] One of the senses of 'decentring' employed by Cunningham and Williams, drawn from Mead and Piaget is' ... the process by which children come to realize ... that other people can have different visual perceptions of the same scene, and that other people can have different knowledge ...' (ibid., p. 429).

of science that joined the stream of European science, the origins of modern science were in Europe (Needham 1969, p. 176).[6] The Needhamian decentring shifted the origins of science itself from Greece, and the universal script implied a more polycentric notion of science, of its emergence across civilizations.

The second decentring discussed in the last section of this essay is a product of the scepticism concerning science as a cultural universal script, and may thereby destabilize the foundations of science and civilizations' studies. This also reflects the collapse of nineteenth-century colonial structures, emanating from Europe and mapping the world. As a result, Mendelsohn writes that a train of revision of 'myths of roots and origins' is set in motion. These myths 'had become part of European self-explanation created during the late Renaissance and early modern era' (Mendelsohn 1995, p. 59). The idea of the scientific revolution that swept Western Europe in the seventeenth century constitutes a sort of Kuhnian hard core, around which the history of science from Whewell to Duhem to Sarton, Koyré, Butterfield and Needham was organized. These synoptic histories, most of which were written during the first half of this century, present the scientific revolution as a revolution of the mind, a revolution that reorganized the geography of the imagination, and hence could only be plotted on 'maps that were metaphysical, metaphorical' (Porter and Teich 1992, p. 3). Further, this new science was emblematic of the age of modernity, and subsequently of industrial civilization. Naturally, historians of science were concerned with the question of civilizations at the threshold of the scientific revolution. Why was it that civilizations with a significant tradition of scientific knowledge failed to generate a scientific revolution, or even an industrial revolution (Elzinga 1988)?

Needham was certainly aware of the problematic nature of his own search and the obstacles in the way of historians. This awareness is recognizable in the constant renewal of the form of the 'Grand Question' posed by him, the first formulation of which dates back to the 1930s, when his Chinese students asked him why modern science emerged in Europe. This was subsequently reformulated to why the

[6] Needham employs a range of metaphors to depict the circulation of scientific knowledge, for example, 'By a thousand capillary channels, like venules joining together to form a *vena cava magna*, influences come from all parts of the world' (Needham 1973, p. 3).

scientific revolution eluded China, to why the Chinese civilization, unlike the European civilization, did not give rise to modern science and technology, to why was there no indigenous revolution in China. In all these formulations, there is a certain unease about the projection of a European theory of science in history on another civilization—in qualifying the last formulation above he speaks of an 'indigenous revolution'. While all these formulations are situated within the framework of comparative history, the questions were reformulated as his exposure to Chinese civilizations grew. The question itself concerned what Needham considered the 'greatest problem in the history of civilization' (Needham 1969, p. 148). Modern science and, subsequently, the emergence of industrial civilization appeared as major watersheds in the history of the West. For Needham, since science was a cultural universal it was germane to the enquiry into the non-emergence of modern science in the non-West where both the state of science and techniques was comparable to, and in the case of China more advanced than, that of Europe prior to the scientific revolution.

The later formulations of the 'Grand Question' were far more focused, unconcerned about problematic counterfactuals, and, despite the inescapable schema of European history, sought to explain the phenomenon of non-emergence. The innocent question posed by his Chinese graduate students in the 1930s was subsequently rephrased as: why did modern science originate in the Semitic–occidental civilization and fail to do so in China? The more direct civilizational variant of this was: why did modern science and technology develop in Europe and not in Asia (Needham 1969, pp. 147–54)? In the more detailed studies on China a related problem is encountered: why was East Asian culture between the second century BC and the sixteenth century AD 'more efficient than the European West in applying human knowledge of nature to useful purposes' (Needham *SCC* 1, pp. 3–4).

Rather than address Needham's response to this family of related issues, we merely wish to focus upon the ambience generated by his ecumenical vision and the enormous possibilities that materialized in cross-cultural history both in the West and the non-West. In order to reverse the customary search for the shadow of Greece in Eurocentric history of science of the ancient world, he proposed a new historiographic rule. He began by postulating that cross-cultural transmission is a natural phenomenon, and that simultaneous discovery or invention is something that the historian must establish. Since he was working with Chinese documents and inventions whose antiquity pre-dated those of Europe, the onus of proof now resided with those who wished to maintain independent invention (Needham *SCC* 4, pp. xxvii, 1). Needham's sense of the constructedness of history is unmistakably evident here. Explicitly stated, if the Eurocentric account drew its potency from the idea of the Greek miracle, then a more ecumenical and inclusive history demanded the insertion of another set of historiographic rules.[7]

Non-emergence and the Interfoliation of Rhetoric and Practice

The ecumenical vision was to inspire the Indian historiography of science. Beyond the vision of a cross-cultural history of science, two crucial 'nodal events' from Needham's *Science and Civilization* are important for the study of the historiography of science in India. Before discussing these, it is essential to highlight that Needham, to begin with, projected a European schema onto the tabula rasa of Chinese society (Cohen 1994, pp. 450–1). This European schema subsequently receded into the background as he came to be possessed by China's cultural symbolism. But this received vision of European science marked by the polarities of the rational and the irrational, tradition and modernity, etc. (Thapar 1993, pp. 25–59), provided progressivist historians of science in India with a frame to explore the non-emergence of modern science in India.

As India ambled into the age of 'Big Science' in the 1950s, three important works acquired canonical status amongst a section of the Nehruvian scientific community and decision-makers—Bernal's *The Social Function of Science* (1942) and, later, *Science in History* (1954), and Needham's work on China then in progress. The exemplar for the history of sciences in the 1950s was Needhamian. It is not difficult

[7] While recognizing this possibility in Needham's oeuvre, the counterfactual history of scientific revolutions cannot entirely disentangle itself from Eurocentrism. This, it may be conjectured, has to do with the fact that counterfactual history is enfolded within, what Fuller would call, an overdeterminationist view of history. The view postulates certain nodal events 'through which all possible events would have had to have passed' (Fuller p. 210). Further, scientists reconstruct history in such a frame, paying little attention to 'contingent turning points', that reduce the number of paths available to history (Fuller 1993, p. 5).

to see why this was so in the light of the foregoing discussion. Put differently, the 'high church'[8] of science studies in India was predisposed to the Needhamian project, whilst the 'low church' drew capital from the Bernalist one.

In retrospect, Bernal's *Science in History* was not the exemplar for historians of science in India or elsewhere in the world. As Ravetz remarks, it was a four-volume history, the bulk of which discussed the history of the last three hundred years, made a 'disappointingly small contribution to the history of science', and was conceptually and politically obsolete by the time it was completed (Ravetz 1992, p. 165). Furthermore, his exaggerated faith in the rationalism embodied in science distorted his appreciation of the science of the past. *Science in History* was drafted from a triumphalist early twentieth-century perspective that could not have accommodated the non-West within its narrative. In short, it is, at best, 'externalist-Whig' history, 'differing only in detail from the internalist-Whig style of historians of science dominant in his time' (ibid., p. 169). Hence, Bernal's influence was restricted to the domain of policy, a narrow section of the scientific community (Elzinga 1988), and within social movements drawn by his positivist evangelism.[9]

However, the opportunity to respond to Needham's 'Grand Question', offered latecomers to modern science the opportunity to legitimate modern science culturally within their socio-political orders, to reinvent cognitive connections between the practice of modern science and the cultivation of traditional knowledge. It will not be an exaggeration to suggest that *Science and Civilization* provided the master narrative, to be elaborated or subverted, for subsequent research on the history of science of the non-Western nations.

[8] The high church of science, technology, society studies, according to Fuller, includes historians, philosophers and sociologists who seek to understand the hold that scientific knowledge exercises on society. The members of the 'low church', on the other hand, include policy-makers, political activists, schoolteachers, 'political scientists and economists ... civic minded natural scientists and engineers' who have to contend with the positive or negative impact of scientific knowledge on society (Fuller 1995, p. xiii).

[9] In fact, the radical agenda of Bernalism of the 1930s was appropriated 'by captains of industry and ministers of government in the post-war period' (Elzinga 1988, p. 94). Further, the science of science ideology contained prefigurations of the misconception, later implicit in practice, that 'natural scientists' were seen to be 'the repository of reason and ... the most suitable for leading the world' (ibid., p. 99).

What was the core agenda of research in the history of sciences in India as articulated by the scientific community? Why, despite the official push given to a Needhamian history of sciences in India, has such a history not emerged, even though the 'conditions' required for its production were more or less present? Floris Cohen, in his massive work on the historiography of the scientific revolution, reviews the responses to the Needham question for the Islamic and Chinese civilizations. The Indian picture, among others, is missing, more or less by default. This section outlines the vocation of the missing picture.

The second leap forward in the history of sciences in India runs parallel to the creation of new institutions of scientific and technological research in the 1950s and 1960s. The National Institute of Sciences of India, now the Indian National Science Academy, organized a symposium with UNESCO in November 1951, where the idea of writing a comprehensive history of sciences in India was mooted. Needham's plan for his six volumes was considered the exemplar for this massive project (Bose 1956, pp. 397–402). Reportedly, the most attractive historiographic principle of this scheme was 'Needham's' idea that the history of science of a region was integrated with its social, environmental, and economic history. But overall, the plan outlined by Bose was quite disparate, for it was unable to problematize, either in detail or in a general perspective, what the sciences in India either were or could have been. For Bose, as for so many other Indian scientists of the time, the history of science was to provide a humanist garnishing to science, in a milieu where the rationality of science (actually technology), considered autonomous, loomed large over daily life (ibid., pp. 396–7). Commencing with the script of science as a cultural universal, studies on the history of science, produced under the aegis of INSA, glossed over the notion of culture. The journal founded by INSA in 1966, the *Indian Journal of the History of Science (IJHS),* is divided between externalist and internalist history—science in non-Western cultures is never problematized, and the aim, although this was not the vision of the founding fathers, is to find homologues of Western scientific theories in the Indian tradition (Raina 1996). The history of sciences of India as initiated by the National Commission on the History of Sciences, and codified in the publications of INSA, has developed into a journal wherein a very strict internalist–externalist dichotomy is maintained.[10] The internalism, paradoxically, is rooted in a late–nineteenth-century positivist image of science. But even the internalist genre, an adequate epistemological account has failed to emerge.

The history of science, with a few exceptions, failed to acquire an ecumenical character. Bogged down by considerations of nationalist historiography, about a third of the publications in the *IJHS* addressed a range of priority disputes. The researches of the early generation of historians of science, who had grown up within the international idiom of science, made way for some revivalist writing in the post-1980s era. This new writing sought legitimacy in the existing caveats of the age. It was professedly a history free of ideological biases, and therefore more scientific.

The theme of the decline of science in the non-West, be it China, India or the Islamic world, dominated in the post-World War Second years. Two important themes run against each other and provide robust templates for the history of science in India. In *Science and Civilization*, Needham suggests that the Mohist texts of China are closest to the spirit of Western science. It is left to the sociology of science to answer why this school did not develop subsequently within Chinese society, or why this tradition did not survive its original suppression (Needham, *SCC* 2, pp. 171–84). For convenience, let us refer to this episode as the legend of suppression.[11] The legend of suppression is about two contending schools of thought, the one epistemically open, empirically oriented, and progressive, and the other bigoted, doctrinaire and obscurantist. The legend relates how the former is persecuted, almost eradicated by the forces of the latter. The evolution of Taoism is

[10] Almost a year before the first issue of the *Indian Journal of the History of Science* was launched, S.N. Sen, amongst the pioneers of the history of science in India wrote a piece in *Science and Culture* on 'Changing Patterns of the History of Science', where he points out the gradual shift in focus from the histories of Tannery and Comte, to the external history of science. He points out that during the early stage of the history of science there is a natural tendency to accord greater emphasis to discipline histories, while the interpretive history of science must come when the historiographic tradition has reached a sufficient level of development (Sen 1965, p. 218).

[11] The legend of suppression may be seen as the opposite of what Weingart calls the legend of abuse, according to which despite suppression, scientific truth prevails in the end and the scientific community stands vindicated (Weingart 1993), whereas in the former case, the tradition of science declines.

blocked by Confucianism, the state-ideology of the feudal bureaucracy (Needham 1969, p. 161). The resemblance of this to the incarceration of Galileo is not lost on the readership. It may be suggested that out of the diversity of the Needham corpus, the legend of suppression provides one of the most powerful themes for Needham's Marxist following in the history of science in India.

The social history of science received a fillip outside the ambit of INSA. The efforts of two historians of science, Debiprasad Chattopadhyaya and Abdur Rahman, both of whom drew their inspiration from Needham's work, will be briefly outlined. On this count, Needham's work on the Mohists, Logicians, and Taoists was of particular importance. The legend of suppression that encapsulates the history of these schools, provided Chattopadhyaya one of many frames for exploring the origins of materialist thought in ancient India (Chattopadhyaya 1959, 1976), and the evolution of Ayurvedic medicine (Chattopadhyaya 1979). Chattopadhyaya's principle thesis appears to have been that materialist schools like the Lokayata and, later, the Ayurveda of the early period provide possibly the only instances that approximate to our present conception of science. The decline of these schools is imputed in part to their suppression by the religious orthodoxy. In fact, a major corpus of Chattopadhyaya's work consists in making an inventory of the various traditions of 'science' in ancient India, traditions that were systematically and consecutively demolished by the theocracy and those with a vested interest in the politics of irrationalism.

Abdur Rahman worked from the other end. Inspired by the twin ideas of cross-cultural history and the shifting centres of world science, he proceeded to document the practice of sciences in medieval India and, in the process, revealed how the medieval socio-political context and cosmos conditioned this science (Rahman 1982, 1986). Rahman could not resist countering Needham's assertion that Taoism was the world's only mystical tradition that was favourably disposed to the sciences. Debatably, contrary to Needham and the findings of some Indian historians, Rahman went on to insist that the universalism of the Bhakti and Sufi movements in India could not have proved antithetical to the practice or the cultivation of the sciences (Rahman 1996).

Other than the broad historiographic issues involved, a number of sub-plots within the Needham text resurfaced in the research programmes of historians of science in India. These sub-plots were essential to the elaboration of a response to the Needham edifice. But this transposition to India, in formal terms, of Needham's findings for China, was based on a vital assumption—similar social processes across civilizations either promote or deter the progress of science. This assumption was central to the script of science as a cultural universal, and the cross-cultural history that it later produced.

This leads us to the question of the non-emergence of the Needhamian project in India. In a lighter vein, it may be suggested that while a Needhamian history of sciences in India has not been written, it is unlikely to be written. Nevertheless, it may be argued that the meanings associated with *Science and Civilization in China* as a cultural text between the 1950s and the 1970s no longer obtains at the end of the twentieth century. So any future rendering that may appear from India will actually be a different project, despite the Needhamian inspiration. To begin with, science and scientism no longer occupy the centrestage of human culture. We shall return to this later.

Cohen drops the discussion on the liminality of the scientific revolution in India on the count that 'there is as yet no historiographical tradition worth mentioning' (Cohen 1994, p. 382). Cohen proposes four possible reasons why this is so. The first has to do with the absence of a scholar who could tap the misunderstood sources of indigenous science. This can be taken to mean that historical scholarship within the science academies was marked by scientism of the external or internal variety. The second and third echo the customary argument of the insufficiency of the historical material for such a reconstruction, and the absence of a reliable chronology of Indian history. This is a much discussed shortcoming that need not be imputed to Cohen. Despite this, historians have done a splendid job in India. More than anything else it reflects the inability to countenance another sense of history. Fourth, Indian civilization was visualized as metaphysically rather than scientifically inclined (ibid., p. 382) in tune with the fundamental orientalist partitioning, not to mention Weber's characterization of civilizations in terms of rationality. Historians of science have challenged this characterization. Thus, out of the four historiographic points raised by Cohen, the first is the most crucial and gives us cause to probe the sociology of the discipline for a landscape of the social history of science in India, and to identify the factors that curtailed its emergence.

Needham's history of *Science and Civilization* appeared against the backdrop of two political events. The first was the crisis of European humanism precipitated by the two wars. The second was the emergence of China in the realm of international politics in the 1940s. In a period when stable structures of democratic politics in the capitalist world came apart, 'Third Worldism' offered the European left the new utopia that the world would now be emancipated by the impoverished, agrarian periphery, which had been pressed into dependency by the 'world system'.[12] Despairing of what Europe had to offer in the early 1940s, three important members of the Cambridge left turned eastward for inspiration, and possibly a new sense of purpose. Rephrasing Werskey's point, China was the 'functional equivalent' of what India was to Haldane and the erstwhile Soviet Union was to Bernal (Werskey 1978, p. 318). The search was perhaps for an alternate conceptual scheme, or theory of action, but more importantly for refashioning the science–society framework. The monumental nature of Needham's oeuvre must then be situated in the context of changing perspectives of the Third World and China in particular, in the arena of global politics.

Prior to the attempt, however unsuccessful, to give history of science in India an institutional identity in the 1950s, there already existed a fairly extensive corpus of orientalist scholarship on the sciences of ancient India. In addition, as specified earlier, a micro-ensemble of Indian scientists had already inaugurated the 'other' picture of the history of science. These histories were drafted piecemeal as histories of specialist disciplines. They differed from each other on the finer points of historiography, and exhibited depth in areas such as the history of astronomy or alchemy. Was it possible to reconfigure this archive around a different organizing principle, say the Needhamian one? It more or less appears certain, if one goes by what was accomplished rather than by what was professed, that for D.M. Bose, the commitment to the Needhamian framework was purely rhetorical. Further, the first generation of Indian scientist-historians of science were themselves situated within the cross-cultural perspective, by virtue

[12] As Hobsbawm writes, first world left theorists held that 'if the world's troubles lay not in the rise of modern industrial capitalism, but in the conquest of the Third World by European colonialists in the sixteenth century, then the reversal of this historical process in the twentieth century offered the powerless revolutionaries of the First World a way out of their impotence' (Hobsbawm 1993, p. 443).

of their own cultural diversity, but more importantly through the influence of comparative philology that had come down through the influence of the orientalists.[13]

In the post-war years, the polarization of global politics and the emergence of socialism and Third World societies resulted in the canonization of Needham's work amongst the cognoscenti of the scientific world. As scientific academies and societies promoted the Needhamian project within their respective constituencies, science continued to be conceived of as a neutral instrumentality. The community in India, as elsewhere, was still enchanted by the two worlds dichotomy (Snow 1959), that exhibited science's sense of cultural triumph. In Third World societies this was the golden age of scientism. Riding the wave of scientism, the contradictions scientism posed for a Needhamian history were possibly not evident to this generation of scientist-historians of science. Historiographically and sociologically the problem resided here.

What then of the historians? D.D. Kosambi, a mathematician-historian was, in the 1950s, laying the foundations for drafting the Marxist history of Indian civilization (Kosambi 1985, Thapar 1993, pp. 89–113). The history of science was not his primary concern. The formulaic application of Marxist historiography to the study of Indian civilization was first challenged by Kosambi, and it is more than likely that he might have reframed the science–society relationship had it been his primary concern. Bernal, methodologically defining science in his *Science in History*, commences with Kosambi's definition of science (Bernal 1954). In the newly independent republic, historians were preoccupied with redrafting the socio-economic history of India, and in the context of the history of technology, were concerned with the potentialities of capitalist development in pre-colonial India (Irfan Habib 1971).

Besides, the first generation of Indian scientist-historians of science in the early years of this century had done much to neutralize the cultural import of modern science (Raina and Habib 1996). The history of the sciences of India lacked, by the mid-twentieth century, a sense of novelty. Finally, it must be remarked that the history of science, by and large, remained, throughout the 1950s to the 1970s, a sort of amateur history pursued largely by scientists in their spare

[13] For a brief discussion of philology and the history of science, see Raina (1995).

time, or prior to retirement. Hence there was a loss of rigour among orientalists. Much of the history of science writing during this period had to do with the sciences of ancient India. Consequently, the history of science, at least for Bose, could have been construed as a branch of antiquarian studies.[14] This history was a combination of internal–external history, inflected with empiricism and often with nationalist overtones (Raina 1996). This framework of historiography was flat. In the mid-1970s, when alternate perceptions of science appeared, a historiographical renewal was required. A new image of science was being reckoned with in social movements, and policy, and a frontal attack came from academic sociology. None of these changes left a mark on the history of science supported by science academies.

A peculiar ideological prejudice prevailed in this domain of the history of science. Other than the excessive preoccupation with history of science as antiquarian studies, science more often than not came to mean the mixed or exact sciences, and to an extent the history of medicine (Raina 1996). The history of technology remained a woefully neglected discipline, which was left to economic historians. The concerns of the historians of science mirrored the perceptions and prejudices of the scientific community (ibid.). Another obstacle in the way of a Needhamian history of science has been an ideological fixation that was deeply ingrained amongst the historians of science as well as historians. The historians of science were rooted in internalist or externalist history. The historians were preoccupied with questions that were then perceived as more crucial to Indian politics and culture, but it is also likely that they were themselves partisan to the two cultures divide, and saw the externalist account as merely providing the trimmings for the internalist one.[15]

The snowballing of developmental crises in the 1970s, complicated

[14] Bose, Sẹn and Subbarayappa authored *A Concise History of Science in India.* Bose felt that since the Asiatic Society and the Bhandarkar Oriental Research Institute possessed expertise in antiquarian studies, they should be drawn into the project of the history of science (Bose 1963). Sen, on the contrary, felt that the scientist-historian was located in a far more advantageous interpretive point than the 'non-scientist historian' who 'may not only overlook the significance of a scientific work, but is more likely to misinterpret it' (Sen 1965, p. 218).

[15] See Elzinga's essay in this volume, which discusses Shapin's account of the asymmetric relation between the internal and external history of science. While the former can have an independent existence, the latter is parasitic upon the internalist account

further by a perceived dysfunction of India's scientific institutions, a lack of engagement between the scientific and technological research system and industry, a research prioritization strategy that had little to do with the demands of India's rural population, an energy crisis acquiring menacing proportions, etc., a revision of perspective of the science–society relationship was all but imperative. Amongst the many areas of revision, there commenced a search for alternatives, and in historical discourse science came to be problematized differently. The Indian analogue of the Kuhnian revolution fed on the anomalies piling up in socio-political life. The actors who subsequently enabled the reconfiguration in science–society studies belonged not merely to the scientific community. Wearing different hats they came from social movements and from the social sciences. This dimension of the story takes us away from our present problematic.

Do Civilizational Studies of Science Have a Future?

The principal concern today relates to the possibility of the project of science and civilizations in a world where post-modernity denies the likelihood of a global narrative. In the post-modern world, science itself becomes cultural discourse (Harding 1994), no longer a transcendent entity, but historically contingent. The script of science as a cultural universal is reduced to mere anachronism.

The history of science studies have announced the arrival of the New Big Picture (hereafter NBP), discussed earlier, at variance with the Old Big Picture (OBP). We shall briefly outline this picture. The NBP is a picture of contemporary society and science, but the science of the NBP is really modern science of the OBP, whose invention and not origins, since science is contingent, can be located in Hobsbawm's Age of Revolutions. These revolutions are: (i) the French revolution of 1789, a political revolution that transformed the organization of society, (ii) the industrial revolution, that commenced in Britain in the 1770s, which was a revolution in the means of production, exchange, ownership of wealth and resources, and (iii) the post-Kantian intellectual revolution centred around the German state (Hobsbawm, 1973). These revolutions enabled the emergence of a consolidated middle class that wielded political, industrial and intellectual power (Cunningham and Williams 1993, p. 425). The NBP thus shifts the question of the origins of the scientific revolution in seventeenth-century northern Italy to the question of its invention in the later half of the eighteenth century. This picture has been commissioned through the efforts of sociologists of science, who in the process have deconstructed the mythology of the scientific method, and supplanted it with a pluralist principle that states that 'there are many possible ways of knowing and studying the world, and that science is just the particular way of knowing dominant in our culture'. The character of Whewellian history of science changes radically, since, given the loss of science's epistemic privilege, the history of science is transmuted into a 'history-of-ways-of-knowing-the-world', rather than of 'one single thing at different stages of development' (ibid., p. 409).

Rather than go into the historiography of the NBP, it would be important to fathom its impact on those drawn by Needham's 'embracing vision', if not by the nitty-gritty of his method. The NBP's principle of spatial decentring that requires treating all native knowledge forms with perfect symmetry is a good anthropology of science, that might assist in destabilizing essentially political arrangements claiming epistemological sanction in the name of science or modern science. But in treating modern science as the invention of the age of revolutions are we not losing out on the sense of oikumene, or limiting the history of astronomy and mathematics to the history of those disciplines in Europe over the last three hundred years? Where do we situate the efforts of, for example, the lesser known Abraham Ibn Ezra the Spaniard (d. AD 1167), considered one of the foremost transmitters of Arabic scientific knowledge to the West, and who sympathized with Ibn al-Muthanna's desire to harmonize Greek and Indian astronomy (Goldstein 1996, pp. 9–13)? Or how do we address mankind's common experience relating to the passage of the stars, those narratives that wove communities together from Toledo and Cordoba to Marageh and Jaipur? The question to be asked is, how do particular scientific styles acquire currency and legitimacy across civilizations and nations? How does the interplay of the metaphysical, the epistemic and the political unfold in the history of ways of knowing or a history of science? The NBP could surely find a place for these questions, but its emphasis on so recent a segment of human history provides no insight into the longue durée that concerns the study of civilizations. Furthermore, it appears, that as far as the history of the non-West is concerned, the history of science, alias modern science, must commence with the age of colonialism, which makes us wonder, with Said (1994), whether colonialism as a metanarrative will ever be displaced from the former colonies.

From the point of view of history of science studies in India today, a fissure continues to run right across the discipline. There is a scientist's history of science, reflected even in an enlightened popular science magazine like *Resonance,* published through the efforts of India's leading scientists, where society is merely an add-on (Prathap 1996). On the other hand, there is a gradually growing community of professional historians of science, little acquainted with the cognitive content of science. Will a new history of science in India have to sit-out the enactment of Planck's hypothesis: the old generation cannot be converted, it just dies out? Arguing reflexively, the studies related to the production of theories and disciplines enable us to inquire into the social and conceptual obstacles in the way of not merely the non-emergence of a Needhamian history of sciences of India, but possibly why a Needhamian history of sciences of India.[16] In either case, Needham's project remains an exemplar for the non-West as much as for the West. Price carefully chronicled the evolution of many a research discipline, and there was certainly an air of prophecy in a remark, possibly based on the observation of an emerging research front, that may be about the project of science and civilizations, when he wrote: 'It may well be judged that to Joseph Needham lay the honor of contributing the last great traditional and comprehensive work of human scholarship' (Price 1973, p. 9).

References

Bayly, C.A., 1996, 'Who Discovered the Cause of Malaria', *Times Literary Supplement,* 11 October, 33.

Bensaude-Vincent, Bernadette, 1992, in Jean Dhombres and Bernard Javault

[16] In a recent review of two books on the history of sciences in India, Bayly again raises the inadequacy this essay has sought to address. Bayly writes: 'While there are many worthy tomes and articles on Indian learning, it still lacks its Joseph Needham to synthesize its mass of findings and project its importance to the forefront of international scholarship' (Bayly 1996, p. 33). When posed this way, we think the onus for the production of this Needhamian history must lie with the West. For as in the case of China, so in India, as this essay indicates, there existed a corpus on the history of sciences of India prior to the arrival of the Needhamian exemplar. It is thus for the West to produce a Needhamian history of sciences of India, for the West.

(eds), *Actes de Colloque 'Marcelin Berthelot: Une Vie, Une Epoque, Un Mythe'*, Paris, pp. 141–3.

Bernal, J.D., 1942, *The Social Function of Science*, London.

——— 1954, *Science in History*, London.

Berthelot, Marcelin, 1885, *Les Origines de l'Alchimie*, Paris.

Bose, D.M., 1956, 'History of Science in India', *Science and Culture*, 21, 8, February, pp. 395–402.

——— 1963, 'History of Science in India—How it Should be Written?, *Science and Culture*, 29, 4, April, pp. 163–6.

——— 1963, 'Asiatic Society's Contribution to Science Studies in India, *Science and Culture*, 29, 5, May, pp. 219–24.

Brush, Stephen G., 1995, 'Scientists as Historians', *Osiris*, 10, 215–31.

Chattopadhyaya, Debiprasad, 1959, *Lokayata: A Study in Ancient Indian Materialism*, New Delhi.

——— 1976, *What is Living and What is Dead in Indian Philosophy*, New Delhi.

——— 1979, *Science and Society in Ancient India*, Calcutta.

Cohen, Floris, 1994, *The Scientific Revolution: A Historiographical Inquiry*, Chicago.

Colebroke, H.T., 1873, *Miscellaneous Essays, with the Life of the Author by his Son, Sir T.E. Colebroke*, 3 vols, Trubner and Co.

Cunningham, Andrew and Perry Williams, 'De-centring the "Big Picture": The Origins of Modern Science and the Modern Origins of Science', *British Journal of History of Science*, 1993, 26, pp. 407–32.

Datta, B. 1929, 'The Scope and Development of Hindu Ganitha', *Indian Historical Quarterly*, 5, p. 479–512.

——— 1933, 'The Algebra of Narayana', *Isis*, 19, pp. 472–85.

Dharampal, 1971, *Indian Science and Technology in the Eighteenth Century: From Contemporary European Accounts*, New Delhi.

Elzinga, Aant, 1988, 'Bernalism, Comintern and the Science of Science: Critical Science Movements Then and Now', in Jan Annerstedt and Andrew Jamison (eds), *From Research Policy to Social Intelligence*, Macmillan Press, pp. 87–113.

Filliozat, Jean, 1964, *The Classical Doctrine of Indian Medicine: Its Origins and Greek Parallels*, New Delhi.

——— 1974, *Laghu-Prabandah: Choix d'Articles d'Indologie,* Lieden.

Fuller, Steve, 1993, *Philosophy, Rhetoric and the End of Knowledge: The Coming of Science and Technology Studies,* Wisconsin.

Goldstein, Bernard R., 1996, 'Astronomy and Astrology in the Works of Abraham Ibn Ezra', *Arab Sciences and Philosophy,* 6, pp. 9–21.

Guillemain, Bernard, 1992, 'Marcelin Berthelot et le Positivisme', in Jean Dhombres and Bernard Javault (eds), *Actes de Colloque 'Marcelin Berthelot: Une Vie, Une Epoque, Un Mythe',* Paris.

Habib, Irfan, 1971, 'Potentialities of Capitalist Development in the Economy of Mughal India', *Enquiry,* New Series II, 3, Winter, pp.1–56.

Harding, Sandra, 1994, 'Is Science Multicultural? Challenges, Resources, Opportunities and Uncertainties', *Configurations,* pp. 301–52.

Hobsbawm, E.J., 1973, *The Age of Revolutions: 1789–1848,* Cardinal.

——— 1993, *The Age of Extremes: The Short Twentieth Century 1914–1991,* Michael Joseph.

Holton, Gerald, 1993, 'Can Science be at the Centre of Modern Culture', *Public Understanding of Science,* 2, pp. 291–305.

Kosambi, D.D., 1985, *The Culture and Civilization of Ancient India,* New Delhi.

Lettres Edifiantes et Curieuses: Ecrits des Missions Etrangeres, Tomes 6–11, Toulouse, 1810.

Mendelsohn, Everett, 1995, 'Science and the Construction of the Idea of Europe', *VEST,* 4, 8, pp. 59–64.

Needham, Joseph, 1969, *The Grand Titration: Science and Society in East and West,* London.

——— 1973, 'The Historian of Science as Ecumenical Man: A Meditation in the Shingon Temple of Kongosammai-in on Koyasan', in Shigeru Nakayama and Nathan Sivin, eds, *Chinese Science: Exploration of an Ancient Tradition,* MIT Press, pp. 1–8.

——— 1977, *Science and Civilization in China,* vol. 1, Cambridge (referred to as *SCC* 1 in text).

——— 1977, *Science and Civilization in China* (assisted by Wang Ling) vol. 2, Cambridge. See Chaps. 10 and 11 on the Tao Chai and Taoism and the Mo Chia and the Ming Chia (referred to as *SCC* 2 in text).

——— 1977, *Science and Civilization in China* (assisted by Wang Ling and Kenneth Girdwood Robinson), vol. 4, Cambridge (referred to as *SCC* 4 in text).

Pandey, Gyanendra, 1992, *The Construction of Communalism in Colonial North India,* New Delhi, p. 166.

Porter, Roy and Mikulas Teich, 1992, 'Introduction', in Porter and Teich (eds), *The Scientific Revolution in National Context,* Cambridge, pp. 1–10.

Prathap, Gangan, 1996, 'The Origins of Science; Part 1 Thales' Leap', *Resonance,* April, pp. 67–73.

Price, Derek J. de Solla, 1973, 'Joseph Needham and Science of China', in Shigeru Nakayama and Nathan Sivin, eds, *Chinese Science: Exploration of an Ancient Tradition,* MIT Press.

Pyenson, Lewis, 1993, 'Prerogatives of European Intellect: Historians of Science and the Promotion of Western Civilization', *History of Science,* xxxi, pp. 289–315.

Rahman, Abdur, 1982, 'Science and Technology in Medieval India', in Debiprasad Chattopadhyaya, ed., *Studies in the History of Science in India,* vol. 2, pp. 805–15.

——— 1984, 'A Conceptual Framework of History of Science in India', in *Science and Technology in Indian Culture: A Historical Perspective,* New Delhi, pp. 21–43.

——— 1996, 'Science and Social Movements: Bhakti and Sufi Movements, 10th–18th Centuries', unpublished manuscript.

Raina, Dhruv, 1997, 'The Early Years of P.C. Ray: The Inauguration of the School of Chemistry and the Social History of Science (1885–1907),' in *Science, Technology and Society,* 2, 1, pp. 1–40.

——— 1995, 'Coomaraswamy and Sarton: An Interdisciplinary Exchange in the Age of Catastrophe', NISTADS-HPSS/951018B.

——— 1997a, 'Evolving Perspectives on Science and History: A Chronicle of Modern India's Scientific Enchantment and Disenchantment' (1850–1980)', *Social Epistemology,* 11, pp 13 24.

——— 1998, 'Historiographic Concerns Underlying the *Indian Journal of the History of Science:* A Bibliometric Inference', *Economic and Political Weekly,* XXXIII, 8, 21 February, pp. 407–14.

Raina, Dhruv and S. Irfan Habib, , 1995, 'Bhadralok Perceptions of Science, Technology and Cultural Nationalism', *Indian Economic and Social History Review,* xxxii, 1, pp. 95–117.

——— 1996, 'The Moral Legitimation of Modern Science: Bhadralok Reflections on Theories of Evolution', *Social Studies of Science,* 26, 1, pp. 9–42.

Ray, P.C., 1902 and 1908, *A History of Hindu Chemistry,* Calcutta, vols 1 and 2.

——— 1918, *Essays and Discourses,* Madras.

——— 1932, *Life and Experiences of a Bengali Chemist,* Calcutta and London.

Ravetz, J.R., 1992, 'The Marxist Vision of J.D. Bernal', in *The Merger of Knowledge and Power,* London and New York, pp. 153–73.

Rosu, Arion, 1986, 'Marcelin Berthelot et l'Alchimie Indienne', *Bulletin de l'Ecole Francaise d'Extreme-Orient,* LXXV, pp. 67–78.

Said, Edward, 1994, *Culture and Imperialism,* Vintage.

Salomon, Jean-Jacques, 1992, *Le Destin Technologique,* Paris.

Salomon, Jean-Jacques and Andre Labeau, 1993, *Mirages of Development: Science and Technology for the Third World,* Lynne Reinner.

Seal, B.N., 1915, *The Positive Sciences of the Ancient Hindus,* London.

Sen, S.N., 1965, 'Changing Patterns of the History of Science', *Science and Culture,* 31, 5, pp. 214–19.

Singh, A.N. 1936, 'On the use of Science in Hindu Mathematics', *Osiris,* pp. 605–28.

——— 1936a, 'A Review of Hindu Mathematics up to the Twelfth Century', *Archeion,* 18, 43–62.

Snow, C.P., 1959, *The Two Cultures and the Scientific Revolution,* Cambridge.

Subbarayappa, B.V., 1992, *In Pursuit of Excellence: A History of the Indian Institute of Science,* New Delhi.

Thapar, Romila, 1993, 'Durkheim and Weber on Theories of Societies and Race Relating to Pre-colonial India', in *Interpreting Early India,* New Delhi, pp. 25-39.

——— 1993, 'The Contribution of D.D. Kosambi to Indology', in *Interpreting Early India,* pp. 89-113.

Weingart, Peter, 1993, 'Science Abused? Challenging a Legend', *Science in Context,* VI, 2, pp. 555-67.

Werskey, Gary, 1978, *The Visible College,* London.

Whewell, William, 1857, *History of the Inductive Sciences, from the Earliest to the Present Time,* 3 vols, London.

12

Michel Paty

Universality of Science: Historical Validation of a Philosophical Idea

1

The Need for a Historical Analysis

The problem of the universality of science is polarized between the perspective that considers science as an ideal type and that of the social production of science. In the first case, science is conceived as the 'hard core' of its statements and results in a particular period. Further, its supposed universality ignores factors, conceptual and social, that relativize the contents of knowledge. Conversely, an exclusive focus on the social production of scientific knowledge discounts the objective contents of that knowledge. These include theoretical objects such as mathematical entities and descriptions of phenomena of the real, physico-biological as well as human and social worlds. These two extreme positions highlight the absence of an interdisciplinary perspective that integrates the insights of philosophy, the various sciences, the history of sciences and general history.

It is now generally admitted that science is not merely knowledge governed by the judgments of reason and experience. Science is also embedded in a cultural and axiological context, and rests on values that are often accepted by the culture in which it is embedded. In addition, science includes social practices. The consequences of the application of scientific knowledge include the transformation of society, which, in turn, could reshape science itself. Science, technology and society are closely related, and the description of contemporary science as 'technoscience' expresses this complexity.

In contemporary terms, science connotes the inseparability of science and technology. This has two dimensions, the one depicting the content of scientific knowledge and the other the practices of science. When we speak of science, we refer to a system where the very notion of science implies the inseparability of the contents of knowledge and the concrete situations in which this knowledge is embedded. Thus, even if we were to consider the epistemological dimension of science, wherein science involves a type of thinking, then its possible effects or consequences would virtually be considered properties of such a mode of thinking. This implies that human thinking, whatever be the form under which we consider it, carries within itself the signature of human practices, actions and society.

In every period of history, and in every culture area, science, or its equivalent, is embedded in a cultural totality of an organic nature and constituting a system. Given this systemic character of science, it is difficult to conceive of the possibility of the transmission of a science and a culture to another one, whether this transmission is considered in space at a given period, or in time along the flow of history, for during the process of transmission, the elements that constitute these systems undergo a shift of meaning. This difficulty in transmission, as I shall try to show, is at the heart of the question of universality.

The debate on the universality of science is closed for those situated within the Western intellectual tradition that could be labelled positivist-instrumental. Those who share this conception consider contemporary science, or what may be called 'global science', the only universal science, and they would deny any other and further consideration. The objection could however be raised that 'universality' so construed is not necessarily a feature of contemporary science, and that it has been imposed through economic domination related systematically to the power of modern technology.

Philosophical Analysis

On the contrary, radical opponents of this 'imperialist' conception of science deny the notion of universality of science, on account of the systemic relationship that joins science and domination. This view ignores that systems evolve and that elements taken from within a structure can be transferred to another structure without disappearing. In such a case, it will be interesting to investigate the changes of meaning these elements undergo, or, eventually, why certain elements remain unchanged during the process of transmission. And, indeed, if translation is betrayal, it might well, for this very reason, be endowed with some creative potential.

With this in mind, we may wonder whether there is not something in the idea of the 'universality of science' that withstands the strong criticism it has elicited. I am sure that Joseph Needham, who authored an admirable meditation on the historian of science as an 'ecumenical man',[1] would not have liked to give up the idea of universality of science, for it would mean at the same time abandoning the idea that science is valuable, and deprive us of one of our means for achieving enlightenment and the happiness of mankind. In the process, the field will be left free for obscurantism and fundamentalism. But, we certainly have to question science and the idea of its universality in order to acquire a better understanding of its nature and utility.

In the analysis and reconstruction that follows, I shall take as my point of departure well-known critiques of science. These are essentially: (1) reflections on the location of science in the industrial–capitalist society (see, for instance, the works of Max Weber, Jurgen Habermas, Jacques Ellul, among many others); and (2) reflections on the relationship of science and imperialism, which have been growing over the last two decades and of which the studies on 'Science and Empires' are a part.

I shall try, in what follows, to develop elements of an argument that in turn calls for further thematic investigations, rather than arrive at something conclusive, for the question of universality of science is primarily about the nature of science as we know it today and as it has been in the past. Regarding these aspects our understanding is far from certain. We constantly discover new dimensions and aspects of science that shed more light on the question of its universality.

Therefore, the question of the universality is twofold. On the one hand it is, undoubtedly, a philosophical question and must be approached philosophically. But we cannot, as is the case with closed philosophical systems, consider that philosophy alone will provide

[1] Needham (1993).

the answer, for science is—quite apart from any attempt at definition—a historical, intellectual and social reality, and only through historical investigation can we hope to grasp something of its evolution. The same can be said about the idea of the universality of science, the meaning of which has evolved through the centuries.

At this stage, we propose that *the universality of science is a philosophical question submitted to the proof of history.*

2

The Philosophical Argument

The path I pursue is the following: examine the concepts underlying the idea of the universality of science, and then examine whether the criticism provides us with adequate elements that meet the requirements of intelligibility. This is followed by an inquiry into history—general, cultural and anthropological history—and the history of science, in order to arrive at a concept of universality that can be reconstructed from such elements as are compatible with real and living science.

I shall restrict myself, in this investigation, to the philosophical tradition of the Mediterranean and Western worlds that is usually considered to have commenced in ancient Greece. In doing so, I ignore other traditions of our modern science, thus calling for a complementary account of the foundations of science and philosophy in other cultures that have influence on the constitution of present modern science. The underlying assumption is that Greek culture is only one of the sources of the origins of our present conception of science. Our notion of science, and modern science itself, is traced back to the founding moment of the birth of science and philosophy in ancient Greece. This was a period when ideas about the world and cosmology underwent a crucial transformation from that of *mythos* to *physis*, or from mythological descriptions regarding the creation of the world and the order of nature to a natural description based on the process of reasoning.[2] A new form of thinking thus appeared that developed from the 'positive' thought of the ancient Ionians, and from another theoretical strain emerging from the schools of Parmenides and the Eleatics and which, from Socrates to Plato, asked questions relating

[2] Cornford (1912, 1952), quoted and discussed by Vernant (1965).

to ultimate truth and the being that lies hidden behind mere appearance and a changing nature. They established the *logos* as the principle of rational thought and this, together with the requirement of intelligibility, refers to the idea of *being*.[3] Aristotle would explicate the role of the *logos* as the series, endowed with a meaning, of the words having a meaning, that led him to emphasize the idea of demonstration. It is this *logos* that makes the difference, according to him, between man and animal.[4]

In the idea of the *logos* as developed in Greek philosophy—although not so much in Aristotle, who was disposed to logic—the role of mathematical thinking is emphasized, since the mathematical method of reasoning overcomes the approximations of discourse in ordinary language and due to its ability to serve as a model. The ideal objects of mathematics make it possible to relate, through rigorous logical reasoning, the one and the many, the identical and the varied, by referring one to other elements of geometrical figures.

The *logos* permanently verifies the validity of established meanings. In other words, any utterance demands its own criticism. This function of the *logos* is, at the same time, and for this very reason, a claim of universality, whose idea was born with it.

The conception of science inherited from the Greeks, even if it has been modified—in particular through the introduction of the notion of experience and the experimental method—is closely linked with the ideas of universality, of reason, and hence philosophy.

The close link between the emergence of the philosopher and the advent of the citizen is worthy of emphasis, for these ideas emerged, as a matter of fact, together with the rational mode of arguing, that is dialogue, in a particular society.[5] In a way, these ideas are an emanation of this society. Rational dialogue is, in this sense, a formalization of that social invention of the Greek city, democracy. But, we should add, it not only formalizes but transcends the Greek restrictive notion, for its further fecundity would surpass the limits inherent in this first form of democracy, moving towards a broader 'universal' that is present in the idea of humanity (or mankind).

Due to its 'social origin', the idea of universality, together with the

[3] Vernant (1965), 1985 edn, p. 383.

[4] Aristotle, *Politics*, book I, ch. 2. Cf. Labarriere (1994).

[5] Vernant (1965), 1985 edn, p. 392.

ideas of reason, science and wisdom (*sophia*), pretending to transcend, through the notions of being and truth, the circumstances that gave birth to them—these ideas cannot, indeed, escape criticism.

But we are aware that the contingency of circumstances does not deny—on the contrary it can possibly make it manifest—the universality of other human dispositions such as art, symbolization, the ability to make objects and to develop techniques, religious sentiments and metaphysics, the demand for meaning. Rationality is such a disposition. Therefore, our task is to question this rationality and its content, from the point of view of universality.

Let us recognize that the idea of universality, as well as the ideas of reason and demonstrative (and even objective) science, with which it has a constitutive link, carries with it the requirement of its own criticism. In this, it would be adequate, when it aims at overcoming the limitations of a local culture, or enabling the communications between cultures. This idea was embedding and 'universalizing' itself with regard to the conditions of its origin. This process took several centuries of maturation, wherein philosophy, religion and theology, sciences, and social transformations as well, were closely linked.

I shall only mention the affirmation of the universality of the faculty of judgment, that appears to be well established between the European Renaissance and the eighteenth century. This idea is well expressed by Descartes in the *Discourse on Method*: '*The power of judging well* and of distinguishing the true from the wrong, that is properly what we call *good sense or reason, is by nature equal in all men.*' Worthy of emphasis is the strong relationship with the possibility of universal doubt. With Descartes, doubt is the foundation of a knowledge that is to be, at the same time, and for this very reason, universal. Here something quite new emerged. The only true knowledge is that knowledge which, for every thinking subject, overcomes the obstacles posed by doubt.

The great lesson from Descartes' *Regulae ad directionem ingenii* (*Rules for the Guidance of the Mind*), the importance of which was somehow underestimated by positive science that developed after his time, is that there is no knowledge and science if not through subjectivity, which is the proper place of intelligibility. Every mind, every subjectivity founds in itself its understanding and its judgments, and retains the ability of arriving at truth without the need to refer to any authority.

We have pointed out that later-day science ignored this lesson, and instead focused on the one concerning man's power over nature. But philosophies generally maintained something of this idea. Kant's philosophy, for instance, re-established, against the critiques of empiricism, the demand of reason, by gifting the transcendental subject with sensibility and understanding, which makes knowledge possible by allowing man to organize rationally his experience of phenomena. This scheme took science as given, in order to justify it, but without considering a possible future recasting.

Though Descartes' project of founding a science that is certain has, with the exception of geometry, largely failed, its inspiration is to be found in his philosophy, notably in his *Meditations*, where he proposed to turn to the *ego cogito* as the first evidence, eventually able to provide an absolute foundation for knowledge.[6] Edmund Husserl was to see in it 'a new type of philosophy', in which 'naive objectivism is replaced by transcendental subjectivism' (see Husserl's *Cartesian Meditations*).[7] This universal science was, for Husserl, philosophy itself, such as it is considered by phenomenology. To found knowledge, the philosopher must take everything again to his own account.

> Philosophy is a personal affair of the philosopher, ... it must be constituted as his *own*, be *his* wisdom, his knowledge that, although it tends towards universal, is acquired by himself and that he must be able to justify that since the start and at each of its stages.[8]

We turned to Husserl's phenomenology to point out the resonance of this perspective with the criticism and eventual refoundation of the universality of science. Its claim begins with a radical questioning of everything that we believe to be evident or certain, and this questioning is decidedly fundamental, as a return to thought as such, to the subject as the place and the condition of thought, action and warrant of any possible knowledge.[9] This subject of knowledge is, truly, at the heart of the question of universality.

[6] Descartes (1641).

[7] Husserl (1934), 1992 edn, p. 21.

[8] See also Husserl (1954).

[9] See Groethuysen (1995), pp. 127–34.

3

Lessons from History and Anthropology

The notion of universality that, from the philosophical point of view, accompanies that of reason and, from the historical point of view, that of humankind, is 'very recent and limited in expansion', as Claude Levi-Strauss informs us in *Race and History*.[10]

The idea of humankind, surpassing the narrow frontiers of the tribe or of the city to embrace all human beings, has developed progressively, in the Mediterranean and Western world, from Greek philosophy and monotheist religions, though not without some regression, the most recent forms being racism and a variety of contemporary ethnocentrisms. The notion of humankind, and of the universality of the humane, undoubtedly owes much to Judaism and to Christianity. I am not very familiar with non-Western philosophies, but I recall that Joseph Needham had emphasized the need to investigate what form the idea of humankind and of its unity had for instance in Confucian thought.[11] The same should also be done with the teachings of Buddhism.

However, the sense of humankind's exact relationship with nature, and its place in the universe was still missing in the cultural area of Western Europe till the end of the middle ages. It required that the idea of divinity, insofar as it represented closed thinking, be challenged. With this came the opening of the skies, the foreseen possibility of man's freedom. It might not merely have been a matter of chance that the last restrictions on the universality of the humane fell at the same time as the rigid spheres of the old cosmos came crumbling down.

Let us move from the humanism of the European Renaissance and the new vision of the world symbolized by Nicolas Copernicus, Giordano Bruno and Galileo Galilei, up to the eighteenth century, with its great travellers, whose relations speak of the encounter of man with man, and the blossoming Enlightenment—from Giambatista Vico to Charles Secondat de Montesquieu and to Giambatista Beccaria, from Jean-Jacques Rousseau to Denis Diderot and M.J. Nicolas Caritat de Condorcet—elaborating anthropological, philosophical and juridical conceptions that proclaim the equality of all. The first article of the

[10] Levi-Strauss (1952), p. 21.

[11] Needham (1993).

Universal Declaration of Human Rights of the 1789 French Revolution acknowledges this irreversible acquisition of human thinking: 'All men are born free and equal in right.'

At the same time, humankind was admitted as a conscious reality, and, maybe as an effect of it, people became aware of the great variety of human cultures, incomparably more varied than races or ethnicities.[12] Levi-Strauss observes that cognate cultures and languages tend to diversify, while those which have different origins develop common features and tend to converge. He underlines further that all human peoples and cultures, known or ignored, considered evolved or 'primitive', leave behind a history that is as long as any other. Some of these histories are cumulative, for complex, always contingent, reasons. These could be due to interactions, influences and borrowings between them, that cannot be simplified in linear terms as a function of the progress of a single one.

From prehistory to our own time, the neolithic revolution reveals all the characteristics of universalization among all the peoples, as do other important transformations, of which the most recent are the scientific and the industrial revolutions, whose propagation has a certain universality, though, as has been suggested, perhaps more through coercion than free choice. Anyhow, through these transformations, progress neither appears to be a necessary occurrence, nor does it imply continuous change. Progress seems to proceed through jumps, mutations and changes in orientation. We could possibly apply to its jerky and random but unavoidable motions the metaphor of the transforming continental drift—in the end, some movement, visible in time, what we call progress. Yet, the vantage point from which this judgment is made has to be discussed. The fact, nevertheless, remains that from a given point of view that depends on the determinants and values relevant to our culture, the progress of scientific and technical knowledge itself is asserted.

The idea of progress, stemming from the accumulation or summing up of knowledges, is inherent to our conception of science, and it is related to the critical function of reason, constitutive of science. However, the cumulative aspect of knowledge that underlies this notion of progress, notwithstanding its objective character, is relative and has to do, so to speak, with the projection on one axis only of the

[12] Levi-Strauss (1959).

various components of a culture. 'Modern culture' privileges the axis that measures scientific development. Other cultures privilege other, different, axes.

What makes for the originality of each of these cultures is

> its particular way to solve problems, to put values in perspective, problems and values that are approximately the same for all men: for all men without exception possess language, techniques, art, knowledges of scientific type, religious beliefs, social, economical and political organization.[13]

These characteristics can also manifest themselves in the genuine assimilation of elements that originate in a different culture, through the development of particular abilities that may eventually create new forms of thinking or new practices. All cultures proceed with their inheritance from previous generations and through communication with other cultures. A part of a common legacy may eventually be rediscovered through a process of differentiation–unification that is characteristic, generally, of the relations between cultures.

One of these cultures, the so-called 'Western culture' (in fact, a bunch of cultures, the origin of which is not only Western), has generated the industrial revolution and contemporary science with its particular meaning, its own cultural value. It has spread throughout the world, a situation not unprecedented if we think of prehistory and the neolithic revolution, and of the beginning of history with the invention and the diffusion of writing.

Over the centuries that followed, the question of where exactly they originated is of secondary, even anecdotal, interest, compared with the fundamental fact that all cultures responded to these changes rather quickly for their own sake and assimilated them with, perhaps, transformations of the original. The origin of the neolithic revolution is of little importance as all human cultures have been able to make it their own. Similarly, is it of importance, Levi-Strauss observed, to know in which culture the industrial and scientific revolutions began? The simultaneous emergence of the same technological movement, followed by a social upheaval in the societies that absorbed technologies, showed that these modifications did not hold to the peculiarity of the genius of a culture, but 'to conditions that are so general that they are located outside the consciousness of men'.[14] Such

[13] Levi-Strauss (1959).

[14] Levi-Strauss (1959), p. 50.

'revolutions' are not limited to the modalities of their emergence. They acquire new forms, in all cultures, whatever the conditions.

Given this background, the circumstances by which, in history, modern science has constituted itself in correlation with the idea of its universality have to be considered. In very broad terms, this history is known. The essential idea I shall retain is that the representations given by science appear to us more and more explicitly as constructions.

To admit that science and, up to a certain point, reason itself is constructed—beyond the more immediate *function of rationality*—does not affect its universality in principle, that is its potentiality to expand. In a given cultural environment, these rational elements have been progressively elaborated, woven by the endlessly intertwined threads of experience and understanding. But this universality might still be only a partial and restricted one, relative to the culture where it was born. Universality will be in a position to be safely stated only if its elements resist confrontation—at the level of knowledge, and not as a result of coercion at other levels—with other elements originating in other cultures. If they get informed by these other elements, they may achieve a larger universality.

4

A History of Transmissions

In their variety (variety of their objects and of their methods), and in their diachrony, the sciences never ceased to inform one another, producing subsequent changes in their nature and renewing themselves. It is a lesson of history that when a 'science' becomes rigid and closes in on itself, it is doomed to disappear from the field of knowledge. It gets fossilized, as is the case with astrology and alchemy, living sciences yesterday, cultural relics today.

The history of sciences also illustrates the formation of scientific traditions over a given time and in a given place, characterized by given types of problems. Filiations in works at successive periods indicate the existence of these traditions, of schools.[15] Eventually, the latter extend to a whole culture. In that sense one can speak of the Chinese scientific tradition,[16] of the scientific (mathematical,

[15] Paty (1990), ch. 4.

[16] Needham (1974, 1993).

astronomical, etc.) tradition in the Arabic language or of Islamic countries.[17] Transmissions from one tradition to another are observable, for example from the Greeks to the Arabs,[18] from the Arabs to Mediterranean Europe.

Contemporaneous science has resulted from these transfers and modifications. In particular, if science as understood today is eminently the heir of Greek thought, of its notions of *physis* and *logos*, it has also borrowed one of its fundamental constitutive components from a different source—technics and experimentation, that seem to have been foreign to Greek thinking on science, notwithstanding the remarkable inventions of Greek engineers, mostly developed in the Alexandrine period.[19] And despite the mechanical works of Archimedes that relate to statics, there is an absence, in this culture, of a relationship between theoretical thought and practical knowledge, this knowledge being considered as belonging to empiria and escaping pure rationality.

In the long run, new currents have developed from several preceding traditions, elements borrowed from which are integrated in an original construction, giving birth to fruitful new traditions.

Modern science, as it has developed in Europe since the Renaissance of the sixteenth and seventeenth centuries, was inherited from the science of Greek antiquity, from science in the Arab-speaking world and from the science of the European middle ages, all these scientific traditions being themselves in filiation one to the other, although in a non-linear manner. It was also indebted to other traditions, such as those from the East, through the complex and innumerable exchanges from antiquity and the middle ages to the European Renaissance, between the East and Mediterranean shores and from there to Western Europe. We know about caravan trade, about military and political conquests, and for sure the exchange of ideas and knowledge has accompanied them, though the history of this process is yet to be fully uncovered. In such exchanges, openness to difference and novelty from inside a tradition towards other ones is indeed a factor that

[17] Rashed (1984, 1986, 1992, 1994).

[18] Rashed (1993), ch. 1.

[19] See the argument of George Thomson against Benjamin Farrington's claim in favour of a tight link between the coming of reason and a supposed technical progress among Ionian people (Farrington 1944, Thomson 1955). See also Vernant (1965), 1985 edn, pp. 261–322.

benefits the former and enlarges the scope of universality. Modern science itself began in Italy and in regard to the science of nature, its first period culminated with Galileo. Its second period is described by J.C. Poggendorf, well-known nineteenth-century German historian of physics, as one 'where the other peoples of Europe took an always [*sic*] more active part', with British and French scientists leading the ranks (this second period began, according to him, with the foundation of the London Royal Society in 1662 and the Academie des sciences of Paris in 1666).[20] Hereafter the flame was passed on to numerous European countries, and the development of the different branches of knowledge occurred in the most varied directions, with science itself acquiring a new form.

One may evoke a more specific example that deserves detailed study—the case of 'mathematical physics', that was established in the eighteenth century, and from its start in the 1730s to its culmination in Joseph Louis Lagrange's *Analytical Mechanics,* published in 1788.[21] Mathematical physics crystallized at the confluence of two currents of mathematical and physical thought. The first resulted from the transmission and development of Gottfried Wilhelm Leibniz's differential and integral calculus, around 1700–20, thanks to Leibniz himself, the Bernoulli brothers, Johann I and Jacob in Basel, and to scholars within the mathematical milieu of the Paris Academy of Sciences, such as Pierre Varignon and Alexis Fontaine.[22] The second resulted from the transmission of Newtonian physics to the continent. The heirs of the first tradition are Alexis Clairaut, Leonhard Euler, Jean d'Alembert, followed in turn by their own successors, Lagrange and Pierre Simon Laplace.[23]

A new 'school' formed at a place and in an intellectual milieu that differed from those where the initial developments took place, and this new school opened a new branch of science, the fecundity of which would last nearly one century, after which the flame dispersed and was lit again in new places. It is in this sense that theoretical physicists from Germany and North Europe of the second part of the nineteenth century, who worked on electrodynamics and thermody-

[20] Poggendorf (1878), French translation, pp. xii–xiii.

[21] Paty (1994).

[22] Blay (1992), Greenberg (1996).

[23] Greenberg (1986).

namics, considered themselves the successors of mathematical and theoretical physics created by Lagrange and Laplace as well as by Jean Baptiste Biot, Joseph Fourier, Simeon Denis Poisson, Augustin Fresnel, and Andre Marie Ampere.

The history of sciences is constructed from such examples and it is a history of transmissions. The problem of the universality of science might be illuminated from a similar but more general consideration with respect to cultures. The transmission of a scientific current indeed involves cultural factors, even if the impact of these factors is less important inside a given culture considered on a large scale, as is the case of Western Europe.

This particular form of culture, where science has such a large part to play, and which was constituted in the context of the Christian civilization of the middle ages, arose, indeed, at the confluence of two cultures—'the science of the Greeks' and the 'wisdom of the Jews'. In *Birth of God*, Jean Bottero suggests that they represent two millenary traditions oriented in two very distinct directions. 'The science of the Greeks', he writes, 'is the fruit of one millenary of progress, of struggles, of refinements and discoveries about intellectual reflection. The wisdom of the Jews is the fruit of one millenary of progress, of struggles, of refinements and discoveries about religious feeling'.[24]

Prior to both, Babylonian writing enabled both 'Greek science' and 'Jewish wisdom' to be moulded in their respective forms. (And we should not forget 'the admirable achievements of Babylonian culture as well as the cultures of the other peoples of Mesopotamia that have preceded the Greeks')[25]. Undeniably, writing has contributed in endowing science with genuine properties that differentiate it from cultures without writing. For instance, the Mayan civilization of Central America developed well-elaborated scientific knowledge in mathematics (they, it appears, knew the zero even before the mathematicians of India), in astronomy, and possessed extremely sophisticated techniques (architecture and town planning, water supply in areas distant from rivers and with irregular precipitation).

Finally, it must be recognized that the great cultural filiations mentioned above, through which the scientific traditions that gave rise to contemporary science have emerged and developed cannot

[24] Bottero (1992), pp. 31–2.

[25] Needham (1991), p. 342.

exclude other influences, which have thus far been underestimated, but which might well have had a fundamental and decisive character. According to Joseph Needham, the openness of the skies claimed by Giordano Bruno and William Gilbert as a corollary of the Copernican vision, would have benefited from the knowledge that had recently reached Europe, of Chinese astronomical representations of celestial bodies, floating inside an infinite space.[26]

For the great historian of Chinese science, the civilizations of China and India play a part in the emergence of modern science in Europe. He saw the ancient scientific currents of the various civilizations as rivers flowing into the ocean of modern science.[27] In 'The Historian of Science as Ecumenical Man', Joseph Needham argues the need to get rid of the Eurocentric notion, still prevalent in the history of science, that the scientific revolution first occurred in Europe.[28] The origins of science are multiple, he reminds us, evoking the contributions of the various civilizations to the common patrimony of scientific knowledge—'By thousands of capillaries, as veins that converge to form a major trunk, a big cave vein, influences came from the whole world.'[29]

The collaboration of cultures, to be enriching, requires the existence of differences between them. Monolithism entails frailty, as Claude Levi-Strauss says in *Race and History*. 'It is', he writes, 'the holy duty of mankind' to beware of blind particularism and 'also to never forget that no fraction of mankind possesses formulas that could be applied to the whole, and that mankind will merge into a unique way of life is just inconceivable, because it would be an ossified mankind'.[30]

Therefore, universality requires recognition of difference from the other, and it implies exchange. Universality is not to be confused with uniformity and monolithism. We are hence left with the problem of communication and exchange, and these are in no way to be drawn from the model offered to us by the present 'world civilization', that is a commercial media that is no more than a monologue with loudspeakers. Universality calls for the other and requires dialogue.

[26] Needham (1954), vol. 3, p. 438; Needham (1993), p. 132.

[27] Needham (1991), p. 271.

[28] Needham (1991), pp. 340–8.

[29] Needham (1973), p. 3.

[30] Levi-Strauss (1952), pp. 76–83.

In dialogue, two or more entities exist (human or cultural), each with its own system of reference.

From the history of sciences, as well as from the communication of cultures it is evident that a sympathy, and even some kind of 'empathy', is needed to comprehend the intelligence of modes of thought different from one's own. This is valid for systems of thought distant in time as well as for the contemporaneous ones studied by anthropologists. In both cases, a decentring is required in order to communicate with and recognize the *other* in its own system of thought. The universal, if it corresponds to something possible and meaningful, contains among its conditions of possibility, *recognition*. As a corollary, the choice of a point of view is not only unavoidable but necessary, and demands the acceptance of difference. It demands admitting the existence of other cultures and other values, even of those which one cannot assimilate because they are too alien or too opposed to the values we consider essential. Such is the spirit of tolerance. This spirit appears to be a genuine feature of a science in its approach to its object.[31]

I would like, incidentally, to make a remark about the variety of scientific traditions and their bonds to culture. Historical reality should not be confused with extremist ideological considerations antagonistic to the universalistic claim, and, to this end, I would like to draw attention to the variety of scientific traditions and the bonds with culture. Recent history has demonstrated the unfortunate consequences produced by the idea that there could exist an irreconcilably opposed Jewish science and a so-called Aryan science (actually, a Nazi one), or a 'bourgeois science' as opposed to a 'proletarian' one. To believe in something like a 'imperialistic– capitalistic science' and, on the other side, a 'Third-World science', eventually an 'Islamic science', is nothing but obscurantism, and the same holds true with respect to the idea of reason. What is required is not exclusion but dialogue, because nothing living and evolving is closed and isolated from the rest of the world.

History, in particular history of sciences, shows that there was, in the development of civilizations, a science in China (commonly called) Chinese science, a science in India (Indian science), a science in Greece (Greek science), a science in the Arabic language or in the Islamic countries, a science of the European Christian middle ages, a science of the European Renaissance, and thereafter a science called 'modern'

[31] It is also a feature of anthropology.

or 'classical', followed by 'contemporary science'. In addition, there were other sciences, conceived of by other peoples, in particular those of the ancient civilizations of America. None of these sciences was or is universal, including the one in which we are situated, that is 'contemporary science', but, to the extent to which they expressed the dimension of human knowledge—knowledge of nature and of humans—they all aimed at something that does not differ from what we have called universality.

I shall not delve here into comparing both the claims to universality of the various sciences and the potential, or the lack of it, of the epistemological features of these sciences to make them wider in scope and more unified. What I want to retain at this stage is the idea that there is, from one science to another, some kind of a current of universality, which corresponds to a widening of universality when something has been passed on—assimilated, transformed, recreated, whatever be the case—from one culture to the other. This means that some kind of communication is established. And this is related to the question, significant in contemporary philosophy of science, of 'incommensurability', that is actually of 'incommunicability'. Let us retain, nevertheless, the recognition of an inherent tendency to universality of all sciences and all knowledges.

If such is the case, it means that a communication must be possible between these various forms of knowledge, between various sciences and various conceptions of science, notwithstanding the systemic situatedness in their respective cultures. Indeed, such communication has been effective either linearly or through partial borrowings, even if we do not always know how to fully evaluate such exchanges.

If the system of 'contemporary science'—it is improper to call it 'Western science' for it has been constituted from the most varied contributions—is to tend more towards greater universality, then it is necessary for it to know and to understand these other endeavours, parallel or related. We must, to this aim, transcend sociological or structural relativisms, and invent the means of comprehension that build bridges from a science and from a culture to another one, taking into account the difficulties inherent in any translation.

We can take as our starting point, present science and our own culture to build the universality of science. It is not necessary, nor is it desirable, to convert to another point of view—in this, indeed, we would never fully succeed. But we can measure precisely the detailed knowledge of these other conceptions that would allow us to understand to what extent present science is universal, and what is still missing in its pretence of universality.

I set aside the problem of integration, and of compatibility, of science, or of what we think in science is universal, with the various cultures that constitute humankind. Cultures are not fixed and rigid. They themselves can be transformed in part by their confrontation with science and, in turn, each of these cultures can enrich our conceptions of the science and culture problem and reconfigure the relationship by offering genuine ways of acculturating science.

The idea of universality reflects open representation or system. What is conveyed by the idea of 'universal science' is a vision of a world still being built, a world in evolution. A science aiming at this qualification permanently points to its own reform. Looking for improvements, it is critical and knows that it still remains unachieved. There is an obvious link between universality and truth, both being regulative requirements for the constitution of knowledge, and both escaping eternally any full achievement and possession. As the young mathematician Evariste Galois wrote, the night before his death in a duel: 'Science is the work of human mind, that is more fated to study than to know, to search the truth than to find it.'[32]

5

Conclusion

There is, actually, no conclusion, but only questions. We no longer have the initial simplicity, exemplified by the two mutually exclusive positions from which we had started. Having given the notions the flesh of reality the questions themselves have changed. They have evolved from purely abstract ones, addressing mere ideological symbols, to actual problems for exploration, which are not a priori impossible and, above all, are meaningful.

Methodologically, the questions raised by the idea of the universality of science and its criticism exhibit the need to be considered from the perspectives of philosophy and history. Only then can these questions acquire real content and deep meaning. From

[32] Evariste Galois, Letter to Gabriel Chevalier, June 1832, in Galois (1962).

both points of view, they become apparent as questions of fact and questions of right or principle. The questions of fact bear on the nature of science and its historical relationships to society. The question of right, informed by criticisms, leads to the question of whether *the idea of a universality of science is thinkable* and whether *a universal science is possible*? We have to probe deeper in order to detail the true nature of their contents, and to activate the science we recognize and the kind of universality we think right and valuable.

Fundamentally, indeed, these questions invite us to a renewed dialogue between philosophy and history regarding sciences (and about the history of sciences), due to the necessary imbrication of each of these two types of questioning if we wish to continue to be able to doubt and to inquire with the view of arriving at some kind of true knowledge, while simultaneously establishing the truth of this knowledge.

References

Blay, Michel, 1992, *La naissance de la mecanique analytique: La science du mouvement au tournant des XVIIe et XVIIIe siecles*, PUF, Paris.

Bottero, Jean, 1992, *Naissance de Dieu: La Bible et l'historien*, Gallimard, Paris.

Chemla, Karine and Francois Martin, 1993, eds, *Le juste nom, Extreme-Orient, Extreme-Occident: Cahiers de recherches comparatives,* Vincennes, no. 15, 1995.

Descartes, René, 1628, *Regulae ad directionem ingenii* (published in 1701), French trans., *Régles pour la direction de l'esprit*, in Descartes (1964–74).

——— 1737, *Discours de la methode* (1964–74).

——— 1964–74, *Oeuvres*, eds, J. Adam and A. Tannery, 11 vols, Vrin, Paris.

Ellul, Jacques, 1954, *La technique ou l'enjeu du siecle*, Armand Colin, Paris, 1954; reissued, Economia, Paris, 1990.

Farrington, Benjamin, 1944, *Greek science*, 2 vols, London.

Galois, Evariste, 1962, *Ecrits et mémoires mathématiques,* Gauthier-Villars, Paris.

Greenberg, John L. (1986), Mathematical Physics in Eighteenth Century France, *Isis*, 77, 1986, pp. 59–78.

Groethuysen, Bernard, 1995, *Philosophie et histoire*, ed. Bernard Dandois, Albin Michel, Paris.

Habermas, Jurgen, 1963, *Theorie und Praxis. Sozial philosophische Studien*, H. Luchterhand, Berlin, 1963, French trans. from *Theorie et pratique. Etudes de philosophie sociale*, 2 vols, Payot, Paris, 1975.

——— 1968a, *Technik und/s Ideologie*, 1968, Frnach trans. by Jean-Renéladmiral, *La technique et la science comme ideologie*, Payot, Paris, 19.

——— 1968b, *Erkenntnis und Interesse*, French trans. Connaissance et intérît, Paris, 1976.

——— 1985, *Die philosophische Diskurs der Moderne*, Frankfurt, French trans., *Le discours philosophique de la modernit*, Paris, 1988.

Husserl, Edmund, 1934, *Meditations cartsiennes*, French trans. by Gabrielle Peiffer and Emmanuel Levinas, Paris, 1934; reissued, Vrin, Paris, 1992.

Husserl, Edmund, 1954, *Die Krisis der europaischen Wissenschaften und die tranzendentale Philosophie*, La Haye, French trans. by Gerard Granel, *La crise des sciences europeennes et la phenomenologie transcendantale*, Gallimard, Paris, 1976.

Jami, Catherine, Anne-Marie Moulin and Patrick Petitjean, 1992, eds, *Science and Empires*, Dordrecht.

Kant, Immanuel, 1781, *Kritik der reinen Vernunft*, Hartknoch, Riga, French trans. *Critique de la raison pure*, in I., Kant, *Oeuvres philosophiques*, published under the direction of Fernand Alquie, t.1, Gallimard (Pleiade), Paris, 1980.

Labarriere, Jean-Louis, 1994, 'La difference entre l'homme et les animaux chez Aristote', Seminar at Equipe REHSEIS, CNRS, Paris, (unpublished).

Lafuente, Antonio, A. Elena and L.M. Y. Ortega (eds) (1993), *Mundializacion de la ciencia y cultura nacional, Actas del Congreso Internacional 'Ciencia, descubrimiento y mundo colonial'*, Doce Calles, Madrid, 1993.

Levi-Strauss, Claude, 1959, *Race et histoire*, Paris.

Needham, Joseph, 1954, *Science and Civilization in China*, 7 vols, Cambridge.

——— 1969, *The Grand Titration*, French trans., Eugene Jacob, *La science chinoise et l'Occident (Le Grand Titrage)*, Seuil, Paris, 1973.

——— 1973, 'Historian of Science as Ecumenical Man', in Shigeru Nakayama, and Nathan Sivin, eds, *Chinese Science: Explorations of an Ancient Tradition*, Cambridge, Mass., pp. 1–8.

——— 1974, *La tradition scientifique Chinoise*, Paris.

——— 1993, *Dialogues des civilizations Chine-Occident: Pour une histoire ecumenique des sciences*, Paris.

Paty, Michel, 1990, *L'analyse critique des sciences, ou Le tétraédre epistémologique*, Paris.

——— 1996, 'Sobre o estudo comparativo da historia da difusao integra cao das ciéncias', in Ana-Maria e Maia, Goldfarb, and A. Carlos (orgs), *Historia da ciencia: o mapa do conhecimento (America 500 anos. Raezes e trajeterias, vol. 2)*, Expressao e Cultura, Rio de Janeiro/ EDUSP, Sao Paulo, 1995–6, pp. 837–80. (About the Comparative Studies on the History of Diffusion and Integration of Sciences).

Poggendorff, Johann Christian, 1878, *Geschichte der Physik*, Berlin, 1878, French trans. by E. Bibart et G. de la Quesnerie, *Histoire de la physique*, Dunod, Paris, 1883; reissued, J. Gabay, Paris, 1993.

Rashed, Roshdi, 1983, *Entre arithmétique et algébre. Recherches sur l'histoire des mathématiques arabes*, Belle Lettres, Paris.

——— 1986, *Introduction et édition critique de Sharaf-al-Din-al-Tusi, Oeuvres mathématiques. Algébre et géométrie au XIIé siécle*, 2 vols, Belle Lettres, Paris.

——— 1992, *Optique et mathematiques. Recherches sur l'histoires de la pensee scientifique en arabe*, Aldershot, UK.

——— 1993, *Geometrie et dioprique au Xe siecle. Ibn Sahl, Al-Quhi et Ibn-al-Haytham*, Belle Lettres, Paris.

Ricoeur, Paul, 1950, *Introduction à Ideen I de E. Husserl par le traducteur*, in French trans. of Husserl (1913), 1993 edn., pp. xi–xxxix.

——— 1965, *Le conflit des interprétations*, 2 vols, Seuil, Paris.

Thomson, George, 1955, *Studies in Ancient Greek Society*, vol. 2: *The first philosophers*, London, 1955.

Vernant, Jean-Pierre, 1965, *Mythe et pensee chez les Grecs*, Maspero, Paris, new edn rev. and updated, La Decouverte, Paris, 1985.

Volkov, Alexis, 1994, ed, *Sous les nombres, le monde. Materiaux pour l'histoire culturelle du nombre en Chine ancienne*, in *Extreme-Orient, Extreme-Occident, Cahiers de recherches comparatives* (Presses de l'Université de Vincennes), no. 16, 1994.

Wartofsky, Marx, 1994, Realisme et relativite ontologiques dans la physique: trois étapes dans la constitution historique de l'objet scientifique, Seminar at Equipe REHSEIS, CNRS, Paris, 10 May 1994, (unpublished).

Weber, Max, 1920, *Religionsoziologie*, 1920. French trans. of vol. 1, *L'éthique protestante et l'esprit du capitalisme*, Plon, Paris, 1964.

——— 1922a, *Gesammelte Aufsätze zur Wissenschaftlehre*, 1922, partial French trans., *Essais sur la theorie de la science*, Plon, Paris, 1965.

——— 1922b, *Wirtschaft und Gessellschaft*, Mohr, Tubingen, 1922, French trans. of the 1st book, *Economie et société*, Plon, Paris, 1982.

13

K. Subramaniam

Inscriptions, Facts and Blackboxes: Is Modern Science Radically Different?*

The Unity of Science

In the history and philosophy of science in the second half of the twentieth century Thomas Kuhn's work is among the most influential. Its inclusion in the logical positivists' *Encyclopaedia of Unified Science* notwithstanding, an irony much commented upon, Kuhn's work, and the work that he inspired, led to the picture of science as a disunified enterprise. Kuhn's account of the discontinuities in science was essentially diachronic. Discontinuity for him, was integral to the growth of science. The studies that followed Kuhn's work intensified exploration of the theme of the disunity of science. Some of this work extended it to synchronic aspects, pointing out discontinuities within standard disciplines like physics.[1] As a result, the supposed unity of science, to which the positivist movement was passionately committed is in disarray. What remains is a picture

* I thank G. Nagarjuna, Leena Abraham, Porus Lakdawala and B.M. Udgaonkar for raising questions that I have tried to answer in the essay. This essay has also benefited from the criticism and comments made by the participants at the Needham seminar. I thank Dhruv Raina and an anonymous referee for their criticism and comments which have led to vast improvements in the essay.

[1] See, for example, N. Cartwright. (1983), *How the Laws of Physics Lie*, Oxford, Oxford University Press.

of science as a rather loose federation of 'island empires'.[2]

At first sight the Kuhnian fallout seems to augur well for studies of pre-modern science. The focus on incommensurability and the attention to the way in which meanings were articulated within theories, one might expect, increases tolerance to the starkly different concepts or language used in non-modern science. As a price for this, however, the practice, common among some historians of science, of separating the wheat and the chaff of historical records has to be abandoned. The quick decisive cuts marking what is good and what is worthless in the list of scientific achievements of cultures can no longer be made.[3] Two other kinds of presentism informing the historiography of Indian science are criticized by Basu (1995) in the context of Indian alchemy. One of them is the adoption of the standpoint of modern chemistry in the description and elucidation of the varieties of Indian philosophical atomism. The other is the description of substances mentioned in Indian alchemical texts in modern chemical terms. It would seem that appropriate doses of Kuhnian historiography of science are an antidote to such historical presentism. Basu exemplifies the new perspective of understanding past science in its own terms when he concludes that 'a serious study of Indian alchemy should provide the conceptual apparatus underlying the activity, both in terms of its practices, and knowledge claims' (Basu 1995, p. 21).

The movement away from presentism amounts to an articulation of the differences between modern science and what we take to be Indian science. There is no absolute radical standpoint, for, the definition of the history of Indian science, and identification of the constitutive elements of science in ancient Indian culture, are informed by ideas about a present enterprise called science. Hence, in articulating the unity of a conceptual framework underlying Indian alchemy, the unifying features of modern chemistry and the characterization of

[2] The metaphor is from P. Galison (1995), 'Context and Constraints', in J. Z. Buchwald (ed.), *Scientific Practice: Theories and Stories of Doing Physics,* Chicago, Chicago University Press, pp. 13–41.Galison however points out the similarity of Kuhn's ideas to Carnap's idea of linguistic frameworks, and mentions Carnap's favourable reception of Kuhn's work.

[3] See also Bhargava and Chakrabarti (1995). An intense concern with measuring the achievements of ancient Indian science against modern scientific knowledge is common among practising scientists, some of whom have contributed significantly to the historiography of Indian science.

these in terms of concepts, theories or metaphysics need to be considered. Will an outline of a unified conceptual framework capture the essential features of Indian alchemy? It is problematic enough to characterize the unity of modern science, or even of one of its fields or branches, despite the emphasis laid in modern science on conceptual coherence and theory selection based on experimental evidence. It is here that post-Kuhnian work in the history and philosophy of science becomes significant since it targets attempts to show the unity of modern science in the logical structure of scientific theory, or in the methodology of science or in the content of scientific knowledge.

Post-Kuhnian views about the disunity of science do not pass unchallenged. Recent writers like Hacking (1992) and Galison (1995) locate the unity of modern science in scientific practice rather than in scientific theory. This, however, is a radically different understanding of the unity of science, which often draws technology together with science.[4] In what follows, I shall be chiefly concerned with such characterizations of the unifying and distinguishing features of modern science in terms of categories based on the study of scientific practice. From such an approach to modern science as the backdrop, what features of pre-modern Indian science become significant. What are the radically different aspects of Indian culture that most resemble modern laboratory science? In one of the sections below I shall attempt as a preliminary step, an evaluation of what we know about Indian alchemy from the standpoint of these considerations.

The Laboratory Practice of Science

Ian Hacking argues that much of modern science has been cumulative since the seventeenth century. Hacking is aware that this contradicts 'recent wisdom in the philosophy of science'. His perspective emphasizes the laboratory practice of science. The analysis of laboratory practice reveals that modern science possesses a robust stability.[5] Hacking clarifies that he is not praising stability of science as a virtue, but

[4] The point is made by Galison (1995): 'Constraints on skeptical challenges about, say, the thermodynamics of liquid hydrogen, issued from processes at once "purely scientific and manifestly applied"; technological reliability was part and parcel of physical demonstration. Constraints will not yield to a post hoc division into the scientific, the technological and the social' (p. 19).

[5] Hacking (1992), p. 38.

merely analysing and describing a characteristic which undeniably belongs to modern science. Hacking shows how even revolutionary changes in theory do not disrupt the steady growth of laboratory science. The problems of incommensurability and meaning change that afflict attempts to establish the theoretical unity of science are circumvented by the shift of focus to the arena of experiment. The implications are certainly reassuring since they strengthen the widely shared intuition of scientists that the many recent, radical changes of theory in the last two or three centuries do not disturb the essentially cumulative and progressive character of modern science.

Hacking's stress on the continuity of science from the seventeenth century implies a radical break at that time.[6] What does this stress on continuity imply for pre-modern science? From the standpoint of practice, can connections not be traced across the supposed radical break in science in the seventeenth century? The widely held view, also shared by Needham, that the mathematization of experimental science is the distinguishing feature of modern science is essentially a conceptual point. This view does not privilege scientific practice over scientific theory.

This is a problem inherent in the use of notions such as stability and continuity. We are so accustomed to thinking of science in terms of theory and conceptual structures, that descending to the level of practice seems not only to obscure the break that occurred with modern science, but also to obliterate the distinction between science and other aspects of culture. The move threatens to reduce science to the stream of historical accidents and to discount its privileged status as knowledge. Further, stability is an ambiguous word and might refer to stability of institutions practising science, or the stability of theories and practices, or that of equipment and apparatus. Finally, stressing the stability of modern science without special reference to its theories, contradicts the well-entrenched view that modern science is characterized by rapid change in contrast to the stability of pre-modern traditions of change.

Hacking, analysing the practices of laboratory science, reveals the peculiarly stable character of modern experimental science and, as in many recent studies of science, he cuts across the traditional division between the internal and external aspects of science.[7] The stabilizing

[6] Elsewhere, Hacking says that Paracelsus thought so radically differently from us that we fail to understand him (Hacking 1983, p. 70).

[7] For a discussion about various styles of science and their stabilization at different historical times, see Hacking (1992a).

features that Hacking describes, produce evolution and mutual adaptation of various elements that constitute laboratory practice. Hacking analyses the elements of laboratory practice under three heads: theories, things and marks (or inscriptions). His list of fifteen elements includes background knowledge, topical hypotheses, laboratory apparatus, data processing and interpretation. Stability arises in science 'when theories and laboratory equipment evolve in such a way that they match each other and are mutually self-vindicating'. In a mature science, each of these fifteen elements are 'in such mutual adjustment that there is ... self-vindication' (Hacking 1992, pp. 56, 60).

Hacking's point stresses the plasticity of the resources in the conduct of laboratory science. This is typically true of contemporary experimental science. The point about the plasticity of resources is important, however, even in characterizing the break in experimental science around the seventeenth century. The plasticity and the variety of resources of twentieth-century science are incomparably greater than those of experimental science three centuries ago. Nevertheless, as I shall attempt to show, the techniques of inscription which entered laboratory science laid the foundation for the great expansion in variety and plasticity of the resources of experimental science.

Also of relevance in understanding the break that occurred with the emergence of modern experimental science is the notion of intersection or intercalation of different elements of science. Although these notions have been used to describe late modern science, they are necessary for an adequate grasp of the contrast between the relatively long traditions of pre-modern science and modern science. They also throw into relief the characteristic features that were already emerging with the beginning of modern experimental science. Creating intersections between otherwise unconnected resources is one of the important stabilizing practices in the laboratory. This is well described in Pickering's discussion of how the *goals* of laboratory practice are formed and become stabilized (Pickering 1990). Pickering is among those who disagree with the traditional image of science as having a unitary, monolithic culture. 'Scientific communities typically sustain a whole spectrum of conceptual networks, several at any given level of abstraction, a patchy material culture, too, of artefacts and experimental techniques, a range of work-styles, and so on' (p. 217). The stabilization of goals is a process where imaginative intersections are constructed between these cultural elements. The degree of stabilization increases as the intersections pile up.[8]

Intersection between various cultural elements is one of the central points of interest in historical study. However, the word 'intersection' as used by Pickering means really a motivated intersection, one that is sought after by the social agents themselves. In modern science, these intersections are matched by a corresponding awareness of the resource potential of laboratory products for producing intersections. Not only are the intersections motivated, but the products exported from the laboratory are designed for such intersections. This connects the notion of intersection with Hacking's idea that the various elements of laboratory practice are mutually adapted and fitted to each other. I shall elaborate on this theme in the discussion of blackboxes. The discussion below emphasizes stabilizing practices and intersections to understand the radical difference between pre-modern and modern laboratory science, rather than look for coherent conceptual structures or for the unity in metaphysical beliefs.

Realism and the Radical Break in Laboratory Science

It is widely acknowledged that there was a radical break in laboratory science around the sixteenth-seventeenth century. Crombie views this break as the coming together of theoretical science and the tradition of laboratory experiment (Crombie 1994). In Greek science, experiment referred mainly to medicine, the mathematical *scientia media* (astronomy, music, optics and mechanics), and the arts. The lack of experiments on the science of motion is conspicuous. For Aristotle, this was fundamentally a theoretical inquiry, and must hence develop only by reasoning from first principles. The separation of theoretical science and practical arts was even more pronounced in the early medieval universities. Natural philosophy was pursued quite apart from experiment and aimed at the true demonstration of its principles. 'Experimental arguments and practices were introduced into the university context through the cultural ambience not of natural philosophy but of the sciences closely related to the mathematical arts and especially of optics, acoustics, mechanics and astronomy,

[8] See also Galison (1995) on the importance of the interconnectedness or intercalation of scientific subcultures and its negative implications for the notion of the disunity of science.

through natural magic and through medicine' (Crombie 1994, pp. 314, 317). A reflective tradition that discussed the epistemology of experimental science, its role and place in natural philosophy began in Europe with Grosseteste. The experimental method was elaborated in the writings of Grosseteste, William of Ockham, Galileo, Bacon and Descartes into a 'coherent style of scientific style of scientific argument with decisive points of appeal to observation and experiment in the analysis of a subject matter, the procedure of systematic controls, systematic measurement, and the incorporation into the experimental method of the concept of algebraic functions' (ibid., p. 315).

Crombie's authoritative account describes the growing importance of experiment in natural philosophy. When one turns to an existing tradition of practical laboratory experimentation such as alchemy, here too one finds laboratory practice registering radical changes after the seventeenth century. Indeed, there are some historians who see chemistry as being more continuous with alchemy, especially with regard to the actual laboratory processes and techniques of research employed (Partington 1970). Among those who emphasize change rather than continuity, Shapere (1984) thinks that what changed were the goals of laboratory practice. While the alchemists were searching for a way to make gold, later chemists had an altogether different goal.

Shapere calls the shift from alchemy to chemistry a profound change: an 'alteration of goals and language' of science. He characterizes this change as a movement away from perfectionism to compositionism. The earlier alchemists believed that substances grew to perfection in the earth, from crumbly earth to base metals to gold. The goal of laboratory practice of this period was to mimic the natural growth of metals to perfection within the laboratory. Modern science, on the other hand, approaches matter with a view that material substances can be understood in terms of their constituents. The goal of laboratory practice is now radically transformed. It is to understand substances in terms of their constituents.

Shapere finds the change so profound that he feels uncomfortable in saying that the alchemical tradition was 'even engaged in "investigation of matter", or in "trying to understand material substances"' (Shapere 1984, p. 330). Shapere's account neatly opposes the 'practical' goals of the alchemist to the 'realist' goals of modern chemistry. The idea that the alchemists sought after gold, the chemists after truth (the nature of matter), that the alchemists 'handled' material substances in the laboratory, while modern chemists 'investigated' their nature is inviting, but it simplifies too much. A careful consideration of even modern laboratory practice would show that the goals of laboratory practice are constructed in a complex manner and run in different directions (Pickering 1990). Even in the alchemical tradition there was a long line of dissenters. Avicenna (AD 979–1037), a leading physician known to medieval Europe and whose writings were influential, seriously doubted if metals could be transmuted. It is extremely unlikely that the goal of his laboratory practice was to grow metals to perfection.

In Indian alchemy, the making of gold was relatively less important. The Chinese alchemical tradition, perhaps the oldest among all ancient alchemical traditions, explicitly devalued the pursuit of transmutation (Needham 5–3). It is also misleading to say that the pursuit of immortality by the Chinese alchemists was a practical and not a theoretical pursuit. This tradition had as one of its goals the construction of chemical models of cosmic processes which would serve as objects of contemplation (Needham 5–4, pp. 211ff). It is unjustified to deny that a realist intention lies behind such an undertaking.

We find a variety of positions on the possibility of transmutation in early modern Europe. On the one hand, several natural philosophers seriously worked on transmutation right into the eighteenth century. On the other hand, we find writers in early modern Europe sceptical about such a possibility. Biringuccio's *Pyrotechnica* and other books published in the first half of the sixteenth century reflected the strong alliance of laboratory chemistry with metallurgy. At this time Paracelsus campaigned for combining chemistry with medicine. Practical applications, then as now, provided a context for formulating experimental goals. Such goals, one would think, existed alongside the goal of understanding the nature of substances, for most texts of these periods discuss the constitution of substances. If we are to understand the nature of the radical break in alchemy, or in laboratory practice in general, we need more than the simple scheme of goals or intentions that Shapere suggests.

Nature of the Radical Change

There is indeed a great gulf in the style of discourse which separates the Paracelsian texts from the writings of the experimental natural philosophers of the sixteenth century. The change has been described as one from a 'discourse of patterning' to an analytico-referential discourse (Reiss 1982). For Paracelsus, the signs of nature needed to be 'read' comprehensively. '[Nature] made liverwort and kidneywort with leaves in the shape of the parts she can cure. Do not the leaves of the thistle prickle like needles? Thanks to this sign the art of magic discovered that there is no better herb against internal prickling' (cited in Olson 1994, p. 162). Paracelsus would never give a prescription to his patients without consulting the locations of the stars, because the locations of the stars could give as much information about the nature of an illness as blood pressure and other such facts. Signs, for Paracelsus, were to be interpreted by resemblance, reasoning was by analogy, and a universal vitalism ordained that there was a sympathy among all things (Reiss 1982, p. 49).

For the founding fathers of the Royal Society, such language was unacceptable for it was ambiguous and alluded to occult or mystical meaning. It was unsuitable for serious discourse, which needed to use terms that had a definite reference and followed an analytical and prosaic style. Boyle, for instance, followed a strict division of passages that reported observations and passages that presented speculations, hypotheses and theories. The latter passages were not only to be carefully separated, but also to be marked as hypothetical with such modals as 'probably', 'might be' and 'perhaps'.[9] This approach to representing what was observed in a 'literal' sense found an echo in a variety of intellectual fields. Locke, thought that words are properly used only if they are based on 'obvious appearance' and 'sensible qualities' and anyone who thought otherwise would 'never become a chemist but remain an alchemist' (cited in Olson 1994, p. 175). In the great Dutch tradition of realist painting too we find a vigilant separation of what is plainly observed from what is added by the mind in visual representation.

These changes in discursive and representational style, which had a profound, formative impact on modern science, occur at about the same period and can also be traced to radical changes in the Christian and Jewish traditions of scriptural interpretation. This has led some writers like David Olson to identify the fundamental change that

[9] Olson (1994), p. 174. See also, Sargent (1995, p. 184) on the contrast between Boyle's style and that of Newton who was a younger contemporary.

underlies all of these as a shift from a pre-literate, largely oral society to a literate society (Olson 1996, pp. 6–9). Bacon, who stands as an influential figure at the beginning of these changes, did characterize it in such words when he called for a new discourse which would flow from 'literate experience'. The strong form of the 'literacy hypothesis' linking scientific thinking with the emergence of a literate society must of course be qualified. Any simple causal relation between the spread of literacy in Europe and the ascendance of new forms of discourse is untenable, for, literate subcultures have existed earlier in Europe and elsewhere. In modern Europe, however, the material practices accompanying literacy, the emergence of the new technology of inscription did have an impact on the changed practice of laboratory science.

Printing and the Archival Tradition of Knowledge

The changes in the techniques of inscription accompanying the spread of a literate culture in Europe had a profound significance for science. The style of discourse now informing laboratory practice marked a break not only with the more esoteric texts of European or Arabic alchemy, but also with the 'rational' texts of proto-chemists such as Rhazes, texts which chiefly present procedures or recipes. One of the important reasons for the emergence of a new style of discourse is the change in the approach to reading texts, found in scriptural and general hermeneutics, which raised questions such as 'Why does the author say this?' 'Who is the author addressing?' 'What does the author want the listener or reader to do or to think?' The central achievement in reading texts this way is a new consciousness of what a text *could have meant or could mean* to a putative reader. The strategies of interpretation used in reading reflexively determined the devices that are deployed in writing (Olson 1996, p. 8).

In addition to these factors, there were new resources that became available, specifically in the context of laboratory practice. These resources stem from the invention of movable type, which had a singular impact on society and on science. Printing created institutions that made it possible to view knowledge in an altogether different way. Knowledge could now be separated from those who possessed knowledge and incorporated in the form of inscriptions within a cumulative archival tradition. The division between knowledge and skill became very sharp. Knowledge could be separated from the knower and transported, while skill remained intimately bound to the corporeal body of the knower. The structure of knowledge too could now be altered. Little facts established by consensus among a community of experts could occupy a privileged and secure place. Bacon, among the first to acknowledge the impact of printing, was also a proponent of the new method, in which knowledge could be organized by exploiting the possibilities created by this invention. Hooke, echoing Bacon, urged that 'henceforward the Business of Invention will not be so much the Effect of acute Wit, as of a serious and industrious Persecution' (Partington 1970, p. xiii).

That printing had profound implications for the pursuit of knowledge has been known for long. But it is also true that a study of these implications has been neglected for almost as long. Eisenstein, in her seminal work, explores the extent of this neglect and discovers that it is surprisingly widespread (Eisenstein 1979). The most important consequence of the invention of printing for science was that it became possible to institute an archival tradition that could accurately record inscriptions and at the same time make it available to a widespread audience. 'The advantages of issuing identical images bearing identical labels to scattered observers who could feed back information to publishers enabled astronomers, geographers, botanists and zoologists to expand data pools far beyond all previous limits' (Eisenstein 1979, p. 687). Indeed, the archival tradition gave birth to a new conception of a published fact, which eventually became an important goal of laboratory science.

Inscription, Multiplying Witnesses and Replication

Robert Boyle's role in shaping the new experimental style of science has been admirably described by Shapin and Schaffer (1985). For Boyle, *matters of fact* established through experiment are the proper foundation for knowledge of nature. How do matters of fact obtain their justification? Boyle believed that matters of fact were essentially consensual. They were established by the aggregation of opinions (Shapin and Schaffer 1985, p. 25). This move blurred the rigid distinction between opinion and knowledge, dominant in natural philosophy before the seventeenth century. One could build an 'aggregation of opinions' through a process of multiplying witnesses.

The acceptability of matters of fact, in Boyle's experimental philosophy, depended crucially on multiplying witnesses who would attest to their veracity.

Witnesses could be multiplied in different ways. The experiment could actually be performed in front of a large 'informed' audience. Alternatively, it could be replicated outside the laboratory of its origin. More importantly, witnesses could be greatly multiplied through a process of virtual witnessing. Boyle employed all these methods to multiply witnesses. A large number of witnesses could not be gathered at a single place to view an experiment and, hence, direct witnessing was a limited way of multiplying witnesses. Replication and virtual witnessing depended essentially on the ways in which reports of experiments could be widely disseminated. Boyle chose to publish his experimental findings in the form of letters, in which experimental protocols could be spelt out in a clear fashion so that a reader might 'without mistake, and with as little trouble as possible, be able to repeat such unusual experiments' (ibid. p. 59).[10]

In comparison to replication, the process of virtual witnessing was much more effective in multiplying witnesses. This involved a picturing, in the reader's mind, of the experiment in its laboratory setting. Boyle's manner of inscription of laboratory findings was calculated to produce virtual witnesses. To this effect, he exploited several linguistic devices—the genre of his writings, the speech-acts that marked his propositions, the narrative style he adopted, all aimed at evoking a picture in the mind of the reader of what happened. It was also essential for this process to clearly separate what was actually observed from what was added as inference or speculation. Boyle also used elaborate engravings of the air-pump and its parts to convey vivid images of experimental events. The drawings of the air-pump and its parts were not schematized line drawings but detailed naturalistic pictures with shading and cut-away sections illustrating parts of the device. The more vivid the image that could be conjured up in the mind of the reader, the more complete would be the act of virtual witnessing. The acceptance of a reported matter of fact depended on the details and the feeling of verisimilitude that was communicated.

The methods adopted by Boyle for founding an experimental style of doing science show how important powerful techniques of

[10] Boyle's *New Experiments* was written as a letter to his nephew. The book, in its Latin translations, inspired several attempts to replicate the air-pump experiments.

inscription were. These techniques were made possible with the institutionalization of printing. Consider, for instance, how the replication of experimental results could take place. Replication of some kind is essential to produce an eventual stable inscription of a fact. Virtual witnessing alone cannot produce such stability. However, Boyle's complaint of the failure of his texts to allow successful replication 'despite his care in communicating details of the engine and his procedures', should be noted (Shapin and Schaffer 1985, p. 59). In similar vein, he expressed dissatisfaction with the engravings of pictures, since the designs were being continuously altered and even because some of the engravings were misplaced (ibid., p. 262). It is also true that all the initial attempts to replicate the air-pump without physically witnessing Boyle's experiments in London were unsuccessful.

It is clear that in most cases of scientific experiment, access to texts alone is not sufficient to produce successful replication. However, one must recognize that it is necessary to have such textual inscriptions, not only to directly aid the engineering of equipment, but also to focus attention on the conditions of experiment through its adequate representation. Such representation is achieved not only by detailed textual description but also by pictorial representation using well-developed and stable conventions. This is necessary to allow the experimental apparatus to also become a plastic resource in the experimental programme. It is difficult to imagine these developments taking place without the aid of printing.

Printing provided the means of making accurate copies of inscriptions widely available. It became possible to make copies of not only written texts, but also of diagrams and pictures. This had an important consequence for replication of laboratory results. A key notion that has been discussed by several recent studies of experimental science is that of the experimental regress. There are two aspects to reproduction or replication of experiments—replication of the conditions of experiment and replication of the results of experiment. When an experimenter fails to replicate an experimental result, the difficulty is symmetric with respect to the conditions and the result of the experiment. We do not know whether the experiment failed to work or whether it worked and disproved the result whose replication was attempted. As Collins and Pinch point out, we know that an experiment is working only when the outcome is correct. But when is an outcome correct? Only when the experiment is working fine. The experimenter's regress affects all experimental research at the cutting edge of science (Collins and Pinch 1993, pp. 97f). The important point to note here is that printing creates the conditions where it becomes possible to draw a line between the two kinds of replication.

The ability to make copies of texts, diagrams and pictures soon led to the standardization of equipment. The widespread use of standardized instruments made it possible to divide the burden of replication. One set of texts could carry the burden of recording and reproducing experimental conditions. Another set of texts could record the results of experiment and initiate the process whereby they could turn into facts. Latour (1986) has recorded a variety of statement types that occur in scientific research papers. The bare statement of fact is at one end of the scale. The same statement could at some stage of its life cycle be made as a suggestion or as a speculation, or as a tentative finding, and so on. Cumulativity requires that a sufficient number of statements move to the category of fact and remain there. When it does so, all qualifications to the statement and references to the circumstances of production are eliminated. It is only in the early stages of a statement's life cycle, that the material and experimental conditions which produced it are recorded, and very soon this appendage drops out. It is an important point that the archival tradition of science makes a fundamental distinction between a fact, and the conditions of its production.

This point allows us to see the function of texts inscribing laboratory practice in a better light. When a laboratory result becomes inscribed in a text, it, so to speak, 'escapes' the confines of the laboratory. In doing so it sets up a mechanism which is aimed at eventually perpetuating the life of the laboratory result. Only when the result escapes the laboratory of origin does it become eligible for the status of knowledge and enter into circulation. It circulates by being mentioned, described and by attempts to replicate it. This is true for both the modern scientific tradition and the non-modern laboratory sciences. But the replication of the laboratory results in these two traditions is conducted under entirely different conditions. In modern science the outcome of the interaction between inscription and replication is an inscribed fact. Replication can aim at verifying or correcting a fact. This calls for a clear separation of the conditions of experiment and the experimental result, for putting an end to the experimenter's regress. The archival tradition of modern science has resources sufficient to bring about this separation. In the non-modern laboratory traditions, I suggest, however, that the distinction between replicating an experimental result and the conditions of experiment is blurred.

The texts appended to the non-modern traditions of laboratory science had to discharge a double function on meagre resources. It was necessary, in these traditions too, to ensure that experimenters had their apparatus working well and that they obtained correct results. There certainly would have existed training institutions that passed on experimental know-how and skill from generation to generation. But such educational institutions would also have had to depend on texts and inscriptions (or at least oral recording mechanisms) that they generated. Such texts would circulate in the form of manuscript copies (and perhaps, but not probably, in versified oral reproductions) and essentially be compilations of the experiences of several people. It is difficult to claim that such texts actually sought to record 'facts'. It is equally absurd to say that they merely recorded events which took place in the laboratory. The statements recorded would lie somewhere between these two categories.

We do find statements by Indian alchemists such as the following: 'That which I have heard of learned men and have read in the sastras but have not been able to verify by experiment I have discarded ... those operations which I have according to the directions of my sage teachers, been able to perform with my own hands—those alone I am committing to writing.'[11] This clearly shows that the intentions to record facts were not lacking. But it was hardly possible for the prevailing institutions of inscription to record something of the nature of a consensual fact. Most statements that occur in Indian alchemical texts have the tone of directions for procedure. There is no evidence that a clear distinction was made between the procedures and conditions of experiment and the result as a 'fact'. Thus, texts belonging to manuscript societies did not achieve a clear and distinct idea of a 'fact' as an elementary bit of knowledge, accepted by consensus, upon which further knowledge might be built.

Chinese science is of course an entirely different case in point. Paper was invented in China before the second century AD and printing was underway at least from the eighth or ninth century AD. The printing of canonical works by Buddhists and Taoists was carried on in earnest in the eleventh century AD. Printing was so widely prevalent that, in

[11] Ramachandra in the *Rasendra-chintamani*, quoted in Ray (n.d.).

Needham's view, for the study of Chinese history of science, books are a far more important source than manuscripts. One of the largest encyclopaedias, with 10,000 chapters, the *Thu Shu Chi Chheng*, was compiled and printed, beginning AD 1726. Printing in China, given the large number of characters in the Chinese script, was largely block printing, in contrast to the European movable type.

The several Chinese encyclopaedias that were compiled and printed certainly constitute an archival tradition. One does not find, however, in this tradition, a privileging of the 'published fact', the pivotal category in inscribing the transactions of the laboratory in early modern Europe. Just as China did not, in all probability, have an Aristotle, it did not, it appears, have a Bacon or a Boyle. The Needhamian question of why this did not happen is too complex for us to address in any detail here. One can perhaps discern factors in the European tradition that contributed to developing the style that came to be established in modern laboratory science. One such factor could be the close attention paid to language with the development of hermeneutics, and the separation of the representative functions of pictures and of the text in printed books. The representative function of pictures was developed by the tradition of rational painting into a powerful style, with perspective and other geometric approaches to drawing. It is this style that played an important part in the representation of artefacts and instruments by means of inscriptions. We do find in Chinese alchemical texts good illustrations of apparatus (one of the best collections of illustrations is the *Chin Hua Chhung Pi Tan Ching* of the Sung period—AD1225) (Needham 5–4). The drawings printed, using blocks or woodcuts, in early modern Europe have however representational capabilities of a much higher order.[12]

[12] The role of social factors, such as the type of social organization, is certainly important in attempting to answer Needhamian questions about why a certain development did not take place in some cultures. Needham himself stressed the importance of such factors throughout his career. In the context of printing of scientific texts, the centralized enterprise of printing in China as opposed to private printing houses in Europe, may be an important factor in explaining why experimental science based on the inscription of fact did not develop in China to a level comparable to Europe.

Laboratory Practice in Pre-modern Alchemy

The institution in pre-modern science which in a historical sense is a precursor of the modern laboratory is the alchemical laboratory.[13] The practice of pre-modern alchemy was fairly stable and widespread. Its several traditions maintained a continuity through many generations. However, the manner of stability and the factors which contributed to it are very different from those found in modern science. One factor that was important for survival and continuity was esotericism, itself a correlate of the modern process of professionalization. Esotericism was a radically different strategy from that adopted in modern science, which values widespread access to scientific knowledge. Esotericism enhanced the possibility of survival by controlling entry into the tradition, by ensuring quality of practitioners and thereby the quality of transmission, and by defining a form of life. The function of texts in such traditions was largely to aid transmission by serving as a mnemonic. In India, transmission and retention of texts was largely through oral practices. In the traditional medical system of Ayurveda, for example, the central texts were to be regularly recited and memorized in the course of training (Wood 1985). *Reading* texts was a process that followed internalization of texts in memory, while in pre-modern Europe reading developed into an autonomous process not related directly to memorization. It is most likely that even in laboratory pursuits, such as alchemy, texts were treated in similar fashion.

Another factor stabilizing a laboratory discipline was the links that it developed with institutions already established in a culture. This could also have been intentional and practitioners could align themselves to an existing tradition. The self-ascription of the Indian alchemical tradition, as reflected in the texts from near its high period in the thirteenth century, characterizes it as *rasasastra* concerned with *rasayana*. Rasayana is a term borrowed from a much older tradition of medicine—Ayurveda. It is mentioned as a branch of medicine concerned with rejuvenation in the *Charaka Samhita* (*CS*),[14] a primary text of the Ayurvedic tradition, which pre-dates the thirteenth century

[13] See the interesting etymology of the word 'laboratory' in Shapin and Schaffer (1985, p. 57n). The term was used initially to designate the secret alchemical laboratory.

[14] For a useful scientific synopsis of the *CS*, see Ray (1980).

texts by over a thousand years. The choice of name allows the subsumption of the rasasastra tradition within the broad Ayurvedic tradition, which, presumably, was still robust. Even today, Ayurveda is practised in parts of the country. The rasasastra texts are not a formal part of the Ayurvedic tradition in its received versions. Nevertheless, we do find a discussion of medicines of mineral origin in the *CS*, and increasingly greater space was devoted to such medicines in the later Ayurveda texts (Majumdar 1971, p. 253). Thus, the two-way ascription of disciplinary affinity justifies prima facie our reading that rasasastra belongs to the larger tradition of Ayurveda. This aspect of Indian alchemy resembles the concern with therapeutic effects or macrobiotics central to the Islamic and Chinese traditions of alchemy (Needham 5–3, p. xxxiii).

The rasasastra tradition also drew from the Ayurvedic tradition elements of theory and laboratory processes. A number of laboratory processes mentioned in the *Rasa Ratna Samuccaya* (*RRS*), an important thirteenth-century Indian alchemical text, are also found in the *CS*.[15] A list of processes compiled from the *CS* contains the names for 44 processes (Ray 1980, pp. 115f). A comparison of this list with the list of processes found in Chinese or European alchemical texts shows that several processes are common to all these traditions (Needham 5–4, pp. 4–9), implying that these traditions all drew upon a common stock of processes. In the *RRS*, we also notice that pharmacological properties dominate over physical properties, and are listed for nearly every compound mentioned in the text. This suggests that the action of substances upon human bodies formed part of the essential characterization of substances. Such a view would be easily reconciled with the belief familiar in the ancient world that the human body and its transformations are a microcosmic reflection of the macrocosm. It is also explained partly by the possibility that the human body was used as the primary instrument in the *rasasala* for testing the action of new or unfamiliar substances.

[15] In the *RRS*, the processes described appear to follow a conceptual scheme of the binary opposites of potency and toxicity. They aim at preserving the potency of 'metals' while destroying their toxicity. See Joshi's introduction to the *RRS* (pp. 3–4) and also Basu (1995). The *Rasarnavakalpa (RNK)*, another thirteenth-century alchemical text, presumably pre-dating the *RRS*, describes a series of *kalpas*, in which extracts from specific known medicinal plants are applied to mineral substances to derive new medicines believed to have great power.

The place and function of elements of theory or of metaphysical views in pre-modern laboratory science is quite complex. In Greek alchemy, for instance, theory plays a less important role in comparison with Arabic alchemy. Surprisingly, the atomic theory mentioned and discussed by a great number of Greek thinkers, is completely absent in the whole of Hellenistic proto-chemistry (Needham, 5–4, p. 374). The lack of interest in atomism is also found in Chinese alchemy. Similarly, despite the importance of atomism in the Indian *darsanas*, there appears to be no discussion of it in Indian alchemical texts.

Nathan Sivin's account of the role of theory in Chinese alchemy provides a detailed analysis of its several functions. One of them was to provide modes of control and variation of conditions in the laboratory. Time and duration of processes was an important dimension to be varied and controlled. The techniques of gradually increasing and decreasing the intensity of the fire by use of precisely weighed increments of fuel (fire-phasing), or of varying the distance of the vessel from the fire (heat-phasing), reflect the broad preoccupation with the duration of processes.[16] In the *RRS* (Ch. 11) too, varying the quantity of heat is an important technique, with the use of *putas* of different sizes. Another function of theory was to allow cosmic correspondences to be embodied in apparatus through a variety of means—through spatial orientations, analogical shapes and numerologically defined dimensions of furnaces and reaction vessels.

The use of numbers in laboratory practice stands in interesting contrast to their use in modern quantitative science. Sivin discusses the mention of the different weights of mercury extracted from various kinds of crystalline cinnabar. The first weight mentioned (13 to 14 oz of mercury from 16 oz of the 'best native mercury') is empirically accurate (modern calculation—13.8 oz). The rest of the series of weights are wrong from an empirical point of view, but bear simple ratios to the first weight. Thus the numbers merely seem to follow an anticipated pattern. Contrasting this use of numbers with modern science, Sivin remarks that 'modern chemical theories are essentially mathematical while alchemical theories were numerological'.[17]

[16] Sivin in Needham, (5–4, pp. 266ff).

[17] There are similar instances of such use of quantitative technique from other cultures. One example is found in the tables presented by Ptolemy as results of his experiments on refraction. Contrary to what one would expect, Ptolemy's results show constant second differences of angles of refraction for different incident angles

It is also important to emphasize the exchange relations that laboratories entered into. The institutionalization of modern science in the seventeenth century led to conceiving the product of the laboratory as essentially an addition or correction to the archival store of knowledge. What did the pre-modern laboratory, the rasasala, produce for exchange and wider circulation? We find in the description of the rasasala in the *RRS* directions for storing medicines: 'the prepared medicine should be stored in the N-E quarter of the rasasala' (*RRS*, p. 293). The list of persons to be employed in the rasasala includes physicians 'dedicated to the mercurial process (*rasasangnichantaka vaidya*)', scholars with a sound knowledge of drugs and even persons knowing several languages (*RRS*, pp. 291, 295). However, the rasasala is clearly not a hospital or a nursing home. Hence it would appear that the rasasala is a centre for production of medicines and for preparing new substances which are potentially useful in therapy.[18]

The sorts of practices discussed above did produce and perhaps also sought to produce links or intersections among various elements that constituted laboratory activity. The manner of these intersections stands in deep contrast to the laboratory practice of modern science beginning from the seventeenth century. In pre-modern science, intersections largely left the intersecting elements unchanged. Theoretical and practical resources were borrowed from well-established traditions to provide a framework, a grammar, for the operation of the laboratory. The output of the laboratory was not an inscription. It, rather, aimed at incorporating knowledge of processes and skills in a continuous tradition, renewed by oral modes of communication. In modern science, the laboratory aimed at inscription and the technology of inscription was sufficiently powerful to allow independent representation and inscription of different elements—facts, equipment and apparatus, data and relations between data elements. The framework for inscription was provided by a growing common archival tradition to which contributions from disparate fields by different groups of people could be made. It is the style of inscribing the output of a laboratory for widespread dissemination that allowed a powerful technique to develop. In the

---

and also for different media, a result that he could not possibly have obtained (Crombie 1994, pp. 165ff).

[18] Of a total of 30 chapters in the *RRS*, the last 19 chapters (not translated) are said to deal with therapeutic effects (*RRS*, translator's introduction).

first place this was a technique to produce replicable instruments, but it laid a wider framework for laboratory science for which one might use the term 'blackboxing'.

On Blackboxes

The resources of the archival tradition of modern science are not restricted to inscribed texts, diagrams and pictures. Facts that are inscribed also become encoded into devices that serve as instruments in further laboratory interventions upon 'nature'. At one end of the scale of such instruments is the map, which is simultaneously record and research instrument. At the other end are the sophisticated instruments of microscopy or spectral analysis. At this end of the scale, instruments function largely as blackboxes. The person operating the instrument hardly shares any of the knowledge which went into building the instrument. But the instrument encodes pieces of knowledge that once circulated between laboratories as publishable facts and circulated in the networks of replication. Once the fact is inscribed, it is encoded into an instrument and sent back to the world of laboratories.

It is the blackboxing feature, a sort of hardware encoding of a once-published research finding, that allows the instrument to be used in widely different fields of research. Blackboxes travel far and wide. Analysing the difference between the inscriptions made by an African tribe and by laboratory science, Latour (1990) comments,

> The grey boxes of the *kindani* could not reach further than the very society it mapped out. The blackboxes of the experimenters allow action at a distance. It is because the controversies are locally settled by matters-of-fact, that the rest of society may be moved out of its usual ways. (p. 76).

What does blackboxing do to a device? It produces a sharp cleavage between the circumstances of production of the device and the circumstances of its use. Instruments in non-modern traditions are relatively open. The circumstances of their production are at least indirectly familiar to the user. Skills for using a device draw upon the skills of the one who produces or designs. Similarly, the user has greater flexibility in correcting or improving the instrument. This may be a matter of degree, but its impact on what counts as a matter of fact in a research setting cannot be minimized. If the instrument is blackboxed, the inscriptions it produces are facts. If the instrument is open to modification, then facts become relative. This is another way of arguing that standardization paves the way for the emergence of matters of fact which are consensually agreed upon.

However, there is another side to blackboxes—they shape the reality they intervene in. Since blackboxes represent the limit of modification in the hands of the user, modification and flexibility are distributed among other elements of the laboratory equipment. A similar effect ensues when the blackboxes begin to move out of the laboratory. They exercise their own power of transforming reality. Latour analyses the example of Pasteur's rapid introduction of the anthrax vaccine and the great success that it rapidly achieved. Latour writes, 'if instead of gaping at this miracle we look at how a network is extended, sure enough we find a fascinating negotiation between Pasteur and the farmers' representatives on how to *transform the farm into a laboratory*'.[19] Hacking comments by pointing out that 'we remake little bits of our environment so that they reproduce phenomena first generated in a pure state in the laboratory ... the application of laboratory science to a part of the world remade is not problematic, not miraculous, but rather a matter of hard work' (ibid.). This is the source of the power of blackboxes which derives from the inscriptions of matters of fact. Latour's metaphor is the lever—a little weight moves a great one, a little matter of fact can have great consequences.

The power derives not merely because the knowledge inscribed in them is true, as Bacon had hoped. Even more so, it is the peculiar form of closure and the readiness to be widely applied that is crucial. Heidegger analysed this in his essay on technology (Heidegger 1977). For him, the essence of technology is enframing, which is a mode of experience that reveals every object as a resource or a standing reserve. Heidegger cites the example of the river Rhine, which modern technology has dammed up into water-works and hydroelectric plants. In this way the Rhine becomes a resource that may be dispatched afar. In a traditional form of technology as exemplified by the water-wheel, or a windmill, the technological device is set into nature and is still dependent upon it in a visible manner. There is indeed a striking parallel between Heidegger's notion of a standing reserve or resource and Latour's notion of a blackbox.[20]

[19] Quoted in (Hacking 1992, p. 59), emphasis Latour's.

[20] Elsewhere Latour is critical of Heidegger's account of technology, which he

It was also Heidegger who pointed out the similarity between the experience of science and of technology, the experience which leads to the discovery and inscription of facts on the one hand and the experience of objects as resources or blackboxes on the other. It is striking that both of these form essential features of modern science. What is important is to consider these two aspects together when characterizing modern science, rather than to split them along the science–technology divide, as is the normal practice.

References

Basu, P.K., 1995, 'Spreading of the (Al)Chemical Myth (Word): Scientific Explanation, Historiography of Indian Chemistry, and History of Philosophy', in D.P. Chattopadhyaya and R. Kumar (eds), *Some Aspects of India's Philosophical and Scientific Heritage,* New Delhi.

Bhargava, P.M. and C. Chakrabarti, 1995, 'The Scientific Temper and the Scientific Method in Science in India through History, with Special Reference to Biology', in D.P. Chattopadhyaya and R. Kumar (eds), *Mathematics, Astronomy and Biology in Indian Tradition,* New Delhi.

Cartwright, N., 1983, *How the Laws of Physics Lie,* Oxford.

Collins, H., and T. Pinch, 1993, *The Golem: What Everyone Should Know about Science,* Cambridge.

Crombie, A.C., 1994, *Styles of Scientific Thinking in the European Tradition,* 1, London.

Eisenstein, E.L., 1979, *The Printing Press as an Agent of Change,* Cambridge.

Galison, P.L., 1995, *Context and Constraints, Scientific Practice: Theories and Stories of Doing Physics,* J.Z. Buchwald (ed.), Chicago.

Hacking, I., 1983, *Representing and Intervening,* Cambridge.

——— 1988, 'On the Stability of the Laboratory Sciences', *Journal of Philosophy,* 85, 507–14.

——— 1992, 'The Self-vindication of the Laboratory Sciences', in A. Pickering (ed.), *Science as Practice and Culture,* Chicago.

——— 1992, '"Style" for historians and philosophers', *Studies in the History and Philosophy of Science,* 23, 1–20.

interprets as a form of technological determinism (Latour, p. 30). Latour allows for his blackboxes to be reversed when, for example, they cease to function in the expected manner. The analysis is again reminiscent of Heidegger's description of the ready-to-hand object which is absent or is malfunctioning.

Heidegger, M., 1977, 'The Question Concerning Technology', in *The Question Concerning Technology and Other Essays*, New York.

Latour, B., 1990, 'The Force and Reason of Experiment, in Experimental Inquiries', in H.E. Le Grand, *Experimental Inquiries*, Dordrecht.

——— 1994, 'On Technical Mediation—Philosophy, Sociology, Geneology', *Common Knowledge*, 3 (Fall), 29–64.

Latour, Bruno and Steve Woolgar, 1986, *Laboratory Life: The Construction of Scientific Facts,* Princeton.

Majumdar, R.C., 1971, 'Medicine', in D.M. Bose, S.N. Sen and B.V. Subbarayappa (eds), *A Concise History of Science in India*, New Delhi.

Needham, Joseph, 1980, *Science and Civilization in China*, Cambridge. (Volume and part numbers are indicated in the text as follows: volume 5, part 4 as [Needham 5–4].)

Olson, D., 1994, *The World on Paper*, Cambridge.

——— 1996, Cognitive Consequences of Literacy, *Vivek*, a quarterly in artificial intelligence, 9.2, 3–17, Lecture notes written by K. Subramaniam.

Partington, J.R., 1970, *Chemistry as Rationalized Alchemy, A History of Chemistry*, vol. 1, London.

Pickering, A., 1990, 'Openness and Closure: On the Goals of Scientific Practice', H.E. Le Grand (ed.), *Experimental Inquiries*, Dordrecht, 215–39.

Ray, P.C., 'Chemistry in Ancient India', in Debiprasad Chattopadhyaya (ed.), *Studies in the History of Science in India,* vol. 1, Editorial Enterprises, New Delhi.

Ray, P. and H.N. Gupta, 1980, *Charaka Samhita: A Scientific Synopsis*, New Delhi.

Reiss, T.J., 1982, *The Discourse of Modernism*, Ithaca.

Sargent, R.M., 1995, *The Diffident Naturalist: Robert Boyle and the Philosophy of Experiment*, Chicago.

Shapere, D., 1984, 'Alteration of Goals and Language in the Development of Science', *Reason and the Search for Knowledge*, Dordrecht, 325–41.

Shapin, S. and S. Schaffer, 1985, *Leviathan and the Air-pump: Hobbes, Boyle and the Experimental Life,* Princeton.

Wood, A.E., 1985, *Knowledge Before and After Printing: The Indian Tradition in Changing Kerala,* New Delhi.

Index